Flow Through Heterogeneous Geologic Media

This textbook integrates classic principles of flow through porous media with recently developed stochastic analyses to provide new insight on subsurface hydrology. Importantly, each of the authors has extensive experience in both academia and the world of applied groundwater hydrology. The book not only presents theories but also emphasizes their underlying assumptions, their limitations, and the potential pitfalls that may occur as a result of blind application of the theories as "cookie-cutter" solutions. The book has been developed for advanced-level courses on groundwater fluid flow, hydraulics, and hydrogeology, in either civil/environmental engineering or geoscience departments. It is also a valuable reference text for researchers and professionals in civil/environmental engineering, geology, soil science, environmental science, and petroleum and mining engineering.

TIAN-CHYI "JIM" YEH is a professor in the Department of Hydrology and Water Resources at the University of Arizona; a joint professor in the Department of Resources Engineering at the National Cheng Kung University, Taiwan, Republic of China; and an adjunct professor in the Department of Earth and Environmental Sciences at the University of Waterloo. He also holds an Overseas-Renowned Professorship from the Chinese Department of Education at Jilin University, China. Dr. Yeh has more than 30 years of experience in stochastic/numerical analysis and laboratory/field investigations, as applied to heterogeneity effects on flow and solute transport in the saturated and unsaturated geologic media. He pioneered stochastic analysis of the effects of spatial variability on flow in unsaturated geologic media; he discovered and developed the theory of moisture-dependent anisotropy for unsaturated hydraulic conductivity, which has significantly advanced our understanding, characterization, and prediction of unsaturated and saturated zone processes. Dr. Yeh also was the first to explore the conditional simulations and inverse modeling of flow and transport processes in variably saturated geologic media. He developed a stochastic successive linear estimation technique as an innovation that overcomes difficulties of the traditional inverse modeling technique. This innovation has led to his pioneering development of a new generation of aquifer characterization method (hydraulic tomography) and stochastic fusion methods to image the subsurface heterogeneity. Success of the stochastic fusion method further brings about his vision of exploiting naturally occurring stimuli (storm, earthquake, river stage, lightning, etc.) as energy sources for basin-scale subsurface tomographic surveys. He believes this new concept is the future of hydrologic sciences and other disciplines of environmental sciences and engineering.

RAZIUDDIN KHALEEL is a senior hydrogeologist at INTERA, Inc., and an adjunct professor in the Department of Civil and Environmental Engineering at Washington State University, Richland. He has 48 years of experience in the areas of numerical modeling and groundwater, vadose zone, and surface water hydrology. Dr. Khaleel has taught numerous graduate and undergraduate courses in groundwater, fluid mechanics, hydrology, groundwater hydraulics, numerical modeling, and vadose zone hydrology at Washington State University; the New Mexico Institute of Mining and Technology, Socorro; and the Indian Institute of Technology, Roorkee. He has published more than 50 refereed journal articles, several book chapters, and numerous papers in symposium proceedings. Dr. Khaleel also served as UNESCO and UNDP Consultant to the government of India; the School of Hydrology, Indian Institute of Technology; and a number of companies dealing with nuclear waste repository programs in Taiwan and Japan.

KENNETH C. CARROLL is an assistant professor in the Department of Plant and Environmental Sciences and the Water Science and Management Graduate Program at New Mexico State University. He teaches multidisciplinary classes, including geohydrology, to undergraduate and graduate students in both science and engineering. His research has focused on hydrogeology, contaminant transport, and groundwater remediation. Dr. Carroll has published more than 30 refereed journal articles and symposium proceedings papers. His teaching and research interests cover a broad range of areas that pertain to the coupling of hydrobiogeochemical processes that mediate transport and exchange of water and chemicals in the hydrosphere, lithosphere, and biosphere.

FLOW THROUGH HETEROGENEOUS GEOLOGIC MEDIA

Tian-Chyi "Jim" Yeh

University of Arizona, Tucson, Arizona

Raziuddin Khaleel

INTERA, Inc., Richland, Washington

Kenneth C. Carroll

New Mexico State University, Las Cruces, New Mexico

CAMBRIDGE
UNIVERSITY PRESS

CAMBRIDGE
UNIVERSITY PRESS

32 Avenue of the Americas, New York, NY 10013-2473, USA

Cambridge University Press is part of the University of Cambridge.

It furthers the University's mission by disseminating knowledge in the pursuit of education, learning, and research at the highest international levels of excellence.

www.cambridge.org
Information on this title: www.cambridge.org/9781107076136

First published 2015

Printed in the United States of America

A catalog record for this publication is available from the British Library.

Library of Congress Cataloging in Publication Data
Yeh, T. C. (T.-C. Jim)
Flow through heterogeneous geological media / T.-C. Yeh, University of Arizona, Tucson, Raziuddin Khaleel, Fluor Government Group, Richland, WA, Kenneth C. Carroll, Pacific Northwest National Laboratory, Richland, WA.
pages cm
Includes bibliographical references and index.
ISBN 978-1-107-07613-6 (Hardback : alk. paper) 1. Groundwater flow–Mathematical models. 2. Porous materials–Permeability–Mathematical models. 3. Transport theory–Mathematical models. I. Khaleel, Raz (Raziuddin) II. Carroll, Kenneth C. III. Title.
GB1197.7.Y46 2015
551.49–dc23 2014050307

ISBN 978-1-107-07613-6 Hardback

Contents

Preface and Acknowledgments

Unlike traditional groundwater hydrology textbooks, this book integrates classical principles of flow through porous media with recently developed stochastic analysis to provide new insight into subsurface hydrology. It is not a collection of "cookbook" recipes for solving groundwater flow problems as are many currently available textbooks. On the contrary, it examines classical principles of groundwater flow in geologic media in a stochastic framework; it explains their limitations to real-world problems and provides improved solutions and a better understanding of the underlying principles.

We believe that there is a lack of understanding of scale issues related to observations, theories, and processes, as well as uncertainty in hydrogeologic science. Likewise, there is a fundamental knowledge gap between the stochastic groundwater theory and its applications to applied, real-world problems. As a result, this book is designed to illustrate to readers what different laws and conceptual models tell us in relation to our observation scales, why we should care about stochastic theories, and why we need to consider using stochastic subsurface hydrology principles for real-world problems.

Specifically, this book starts in Chapter 1, with a discussion the fundamental fluid mechanic concepts, built upon the control volume (CV) at a scale that is larger than many molecules, but still smaller than a pore in porous media. By expanding the CV to scales larger than many pores, we then discuss in Chapters 2 and 3 Darcy's law under variably saturated conditions, and we define hydraulic properties for flow through porous media at laboratory scales, using the fluid mechanics principles. New concepts of spatial REV (representative elementary volume) and ensemble REV are introduced. Because of the average nature of the CV, spatial REV, and ensemble REV concepts, the importance of scale consistency among our interests, observations, hydraulic properties, and Darcy's law is then emphasized.

In order to quantify multiscale spatial variability of the laboratory-scale hydraulic properties in a field, stochastic processes or random field concepts are presented in

Chapter 4. These concepts facilitate a quantitative method to describe spatially varying hydraulic properties of a field in terms of their spatial statistics (i.e., the most likely value, their standard deviation, as well as their spatial fabrics – sizes of layers or stratifications). In Chapter 5, different conceptual models are discussed for characterizing and modeling a heterogeneous porous medium, including equivalent homogeneous, geologic, and highly parameterized heterogeneous conceptual models, satisfying different scales of our interests. The governing flow equations based on the heterogeneous conceptual model are also presented thereafter. Afterward, the conventional effective properties (e.g., hydraulic conductivity anisotropy and tensor) and mathematical models for the equivalent homogeneous conceptual model are presented in Chapter 6. At the end of Chapter 6, we discuss the prediction resolution of different conceptual models in relation to CV, REV, and our observation scale.

Chapters 7 and 8 examine traditional analyses of aquifer tests for confined, leaky, and unconfined aquifers, and boundary effects, which are built upon the equivalent homogeneous conceptual models. While more realistic numerical models for three-dimensional variably saturated flow through heterogeneous geologic media may replace these analyses, the two chapters aim to illustrate the ingenuity of past scientists in overcoming limitations of their tools (analytical solutions) to estimate aquifer properties and to understand their influence on large-scale flow and transport behaviors of an aquifer in a very conceptual sense. These chapters also expose scale issues and nonrepresentative characteristics of the estimated properties associated with applications of the conventional aquifer analysis methods to real-world heterogeneous aquifers.

Lastly, Chapter 9 introduces stochastic modeling approaches including the unconditional approach and the conditional stochastic approach to derive the best unbiased estimates of aquifer properties, which can yield unbiased predictions of flow and head fields at different resolutions and levels of uncertainty. At the end, a new approach to conduct and interpret pumping tests (joint interpretation of sequential pumping tests, or *hydraulic tomography*) is discussed. This new approach allows us to maximize the utility of existing well facilities so as to reduce uncertainty associated with our estimates of properties and predictions of flows.

The first author, Tian-Chyi Jim Yeh, would like to take this opportunity to express his deepest appreciation for sacrifices made by his beloved wife, Mei-Lin Yeh, for the past four decades, since the author started pursuing his science career in the United States of America. Undoubtedly, without Mei-Lin's endurance, support, and encouragement, the thoughts and philosophy of this book would not have evolved and materialized. In every sense, she is the real inspiration behind the book. Supports for Yeh from the Chinese Department of Education as an Overseas-Renowned Professor at Jilin University, China, as well as the NSF EAR grant 1014594 and the Strategic Environmental Research and Development Program (SERDP) under grant ER-1365 and Environmental Security and Technology Certification Program (ESTCP) under grant ER201212, USA, are also appreciated.

We would also like to acknowledge Mr. Ravindra Dwivedi's meticulous editing and proofreading of the manuscript. The proofreading of Professors Ye Zhang, Frank Tsai, and Junfeng Zhu is also acknowledged.

1

Fluid statics and dynamics

1.1 Introduction

Our experience suggests that most students from geosciences, natural sciences, and other backgrounds are not required to take a fluid mechanics class, and most engineering students who took a fluid mechanics class earlier may still need to reinforce their physical understanding of the fluid mechanics concepts. Thereby, this chapter is intended to provide readers with some basics of fluid mechanics (particularly, in terms of physics) that are essential to understand and appreciate flow through geologic media.

Fluid Mechanics is the study of the forces on fluids. These fluids can be either a gas or a liquid. Fluid Mechanics includes both **fluid statics** (the study of fluids at rest) and **fluid dynamics** (the study of fluids in motion). Notice that the fluid mechanics serves as the fundamental principles in a number of disciplines in science and engineering. For instance, atmospheric science is built upon fluid mechanics, as is cardiology – the study of the blood flow through our veins and arteries. The study of the infiltration of water and its subsequent movement in unsaturated subsurface media (vadose zone hydrology) also relies on fluid mechanics, as do studies of the movement of groundwater in geologic media (groundwater hydrology). Other disciplines include surface hydrology (the study of movement of water on land surfaces, in canals and rivers, and in oceans, lakes, estuaries, etc.), and petroleum engineering that studies the movement of oil and gas in geologic reservoirs. Examples also include hydraulics which studies fluid motions in pipes or man-made conduits (conduit flow systems).

This chapter first introduces readers to the basic continuum assumption (i.e., volume averaging concept). Then, definitions of fluid properties and their units follow. Forces of static fluids and the relationship between forces and energies are discussed subsequently. The chapter then examines fluid dynamics, in which the fixed and moving coordinate systems are introduced, the time derivatives associated with these two coordinate systems are brought forth, and fluid acceleration is investigated. Using force balance and acceleration concepts, we derive the Bernoulli equation for flow through pipe; we elucidate its physical meaning in term of conservation of energy, which leads to the concept of total, pressure, velocity, and potential heads. Head loss due to viscous forces ignored by the Bernoulli equation then explains the limitation of the equation for flow through real-world systems. Subsequently, the influence of the relative strength of the viscous force to the inertia force of the fluid (i.e., Reynolds Number) is discussed on flow behavior (i.e., laminar or turbulent flow). At the end, this chapter sets the stage for the discussion of flow through geologic media in next few chapters, which is generally considered as laminar flow.

1.1.1 Fluids and Solids

Discussion of the states of a matter is a way to explain the definition of fluids and solids. The states of a matter are also known as phases of the matter or states of an aggregation, which are: solid, liquid, and gas. Under a solid state, the molecules of matters are limited to only vibration about a fixed position. This restriction gives the material both a definite volume and a definite shape. Any material under this condition is called solid phase. As some energy is added to a solid, its molecules begin to vibrate more rapidly until they break out of their fixed positions. The solid then becomes a liquid. The molecules of a liquid are free to move throughout the liquid, but are held from escaping the liquid surface by the intermolecular forces. This gives a liquid a definite volume, but no definite shape. As more energy is added to a liquid, a small number of molecules gain enough energy to break away completely, and escape into the adjacent space. Finally, a threshold temperature is reached at which molecules throughout the liquid are becoming energetic enough to escape. Then, vapor bubbles form and rise to the surface, and the liquid becomes a gas. The molecules of a gas are free to move in every possible way; a gas has neither a definite shape nor a definite volume but expands to fill any container in which it is placed. Both liquids and gases are considered to be fluids.

Strictly speaking, there is no quantitative criterion to clearly distinguish solids from fluids. Nevertheless, a fluid is defined as a substance whose particles can move and change their relative position easily in contrast to a solid. Specifically, a fluid is a substance that deforms continuously without substantial resistance, when a shear stress (explained later) is applied to it, no matter how small the magnitude of the applied shear stress is. A solid can be defined as a material of which the shape or the relative position of the constituent elements changes only by an insignificantly small amount, when forces acting on it change, either in magnitude or direction. Thus, the distinction between solids and fluids is rather relative.

While fluids can mean liquids or/and gas, in the context of understanding the basic processes of subsurface hydrology, we are typically interested in the liquid phase of water when we deal with problems related to water resources. Notice that in certain cases we may have to deal with liquid and/or gas phases of nonaqueous fluids in the case of groundwater pollution problems.

1.1.2 Dimensions and Units

Variables used in fluid mechanics are expressed in terms of basic dimensions (e.g., mass, length, and time), which in turn are quantified by three basic units: English (US), Metric (rest of the world) and SI–International system units (scientific community). Before introducing fluid properties, we will review the dimensions and units that are of interest in the study of subsurface hydrology.

A fundamental variable in fluid mechanics is force; its SI unit is Newton (N), after Sir Isaac Newton (1643–1727). From fundamental physics, we are all familiar with Newton's second law: the relation between the mass m of an object, its acceleration $\mathbf{a}$, and the applied force $\mathbf{F}$ is $\mathbf{F} = m\mathbf{a}$. Acceleration (the rate of change in velocity), $\mathbf{a}$, and force, $\mathbf{F}$, are vectors (as indicated by their bold font symbols); the direction of the force vector is the same as the direction of the acceleration vector. Scalars are quantities that

Table 1.1. *Units of measurement of fundamental dimensions in three systems*

Fundamental Dimension	International System (SI) Units	Metric or Centimeter-gram-second (cgs) Unit	English (Imperial) Unit
Mass	kilogram (kg)	gram (g)	slug
Force	newton (N)	dyne (dyn)	pound (lb)
Length	meter (m)	centimeter (cm)	foot (ft)
Time	second (s)	second (s)	second (s)
Temperature	kelvin (K)	degree Celsius (°C)	degree Fahrenheit (°F)

Table 1.2. *Systéme International (SI) units and prefixes*

Unit	Abbreviation	Prefix	Symbol	Multiplication factor
Joule (N·m)	J	Tera	T	10^{12}
Kelvin	K	Giga	G	10^{9}
Kilogram	kg	Mega	M	10^{6}
Meter	m	Kilo	k	10^{3}
Newton	N	Milli	m	10^{-3}
Pascal (N m^{-2})	Pa	Micro	μ	10^{-6}
Radian	rad	Nano	n	10^{-9}
Second	s	Pico	p	10^{-12}
Watt (J s^{-1})	W	Femto	f	10^{-15}

require only magnitude, whereas vectors are quantities that require not only a magnitude, but also a direction to specify them completely.

Recall that mass is a measure of how much material an object contains, but weight is a measure of the gravitational force exerted on that material in a gravitational field. Thus, mass and weight are proportional to each other, with the acceleration due to gravity being the proportionality constant. The dimension of **F** follows directly from Newton's second law, as shown in the following:

$$\text{Force}: \ \mathbf{F} = m\mathbf{a} \tag{1.1.1}$$

Dimensions of force are denoted by $[MLT^{-2}]$ where M, L, and T denote dimensions of mass, length, and time, respectively. The Newton (N) or kg·ms^{-2} is often used as its units. For instance, a force of 1 Newton indicates the force applied on a substance of 1 Kg mass to accelerate it by 1 m/s^2.

Three systems of units listed in Table 1.1 have been used widely in various branches of science and engineering. The SI units (Table 1.2) are the most popular. Some useful prefixes in the metric system are also given in Table 1.2.

1.1.3 Fluid as a Continuum

Like any scientific study of the real world systems, fluid mechanics makes some basic assumptions about the materials being studied. For example, fluids are composed of molecules that collide with one another. The properties of a fluid in fluid mechanics refer to properties of a point in the fluid. However, if we picked a point at random in a fluid, that point might be on the inside of an atomic particle or alternatively might be in the space between atomic particles or molecules due to the discrete nature of the fluid matter. The fluid "properties" associated with a point as being defined in fluid mechanics therefore would depend upon the location of the measurement point. This can create a discontinuous distribution of properties in space for a substance. This discontinuity violates the assumption of smoothness and continuity embedded in calculus and differential equations. To avoid this problem, we have to adopt the ***continuum assumption*** in the analysis of fluid mechanics. In other words, the fluid is idealized macroscopically as being continuous throughout its entirety. The molecules are pictured as being "smeared" or "averaged" to eliminate spaces between atomic particles and molecules. Although mathematically we define a physical or chemical attribute of the fluid at a point, we should remember that the "point" actually represents a small volume in the real space over which the average fluid properties are defined. The spatial volume is large enough to contain many molecules (e.g., there are 6.022×10^{23} molecules in a mole of gas, also called Avogadro's Number).

Adopting the continuum assumption, fluid mechanics considers fluids to be continuous. That is, fluid properties such as density and viscosity are taken to be well-defined at infinitely small points, and are continuous from one point to another. Thus, the fact that the fluid is made up of discrete molecules is ignored. The same continuum concept also applies to the definition of forces, temperature, pressure, energy, and other variables of interest in fluid mechanics. For example, in considering the action of a set of forces on a fluid, one can either account for the behavior of each and every molecule of fluid in a given flow field (can you imagine considering behavior of 6.022×10^{23} molecules together in a single mole of a substance) or simplify the problem by considering the averaged effect on the molecules in a given volume. The continuum assumption permits fluid mechanics to use the latter approach. The continuum hypothesis is an approximation, however. Therefore, it may lead to results which are not of desired accuracy or behavior. That being said, under the right circumstances or observation scales, the continuum hypothesis produces extremely accurate results.

1.1.4 Fluid Properties

With the introduction of the continuum assumption, definitions of fluid properties (i.e., specific weight, mass density, specific gravity, viscosity) are given as follows.

Specific weight, γ. Specific weight is the gravitational force (Mass of the fluid, M, times gravitational acceleration, g) per unit volume (V) of fluid, or simply the weight (W) per unit volume:

Table 1.3. *Densities of some common fluids*

Fluid	**Density (g/cm^3) at 20°C and 1 atm, unless noted otherwise**
water	0.998
water (0°C)	1.000
water (4°C)	1.000
water (100°C)	0.958
gasoline	0.726
ethyl alcohol	0.791
air	1.204
sea water	1.030
tetrachloroethene	1.622
glycerin	1.260
mercury	13.546

$$\frac{Mg}{V} = \frac{W}{V} \tag{1.1.2}$$

Specific weight is generally represented by the symbol γ (gamma). Water at 20°C has a specific weight of 9.79×10^3 N/m^3. In contrast, the specific weight of air at the same temperature and at atmospheric pressure is 11.9 N/m^3.

Mass density, ρ. The mass per unit volume is *mass density* or simply *density*; it has the units of kilograms per cubic meter and is given the symbol ρ (rho). Because specific weight is weight per unit volume, mass density will be given by the specific weight at the earth's surface divided by g. The mass density of water at 10°C is 1000 kg/m^3. For air at 20°C, the mass density is 1.20 kg/m^3 (see Table 1.3 for densities of some common fluids).

Specific gravity, SG. The ratio of the specific weight of a given fluid to the specific weight of water at a standard reference temperature is defined as *specific gravity*, SG. The standard reference temperature for water is often taken as 4°C, where the specific weight of water at atmospheric pressure is 9810 N/m^3. With this reference, the specific gravity (SG) of mercury with a specific weight of 133 kN/m^3 at 20°C is

$$SG_{Hg} = \frac{133\ \text{kN/m}^3}{9.81\ \text{kN/m}^3} = 13.6 \tag{1.1.3}$$

Because SG is a ratio of specific weights, it is dimensionless and independent of the system of units used.

Dynamic Viscosity (μ) and Kinematic Viscosity (υ). Viscosity is a fluid property, which describes the fluid's resistance to shearing or tangential motion under an applied shear stress (see Section 1.2.1). Of course, different fluids deform at different rates. For example, oil and honey deform slower than water, and therefore we intuitively recognize that both are more viscous than water. But which one is more

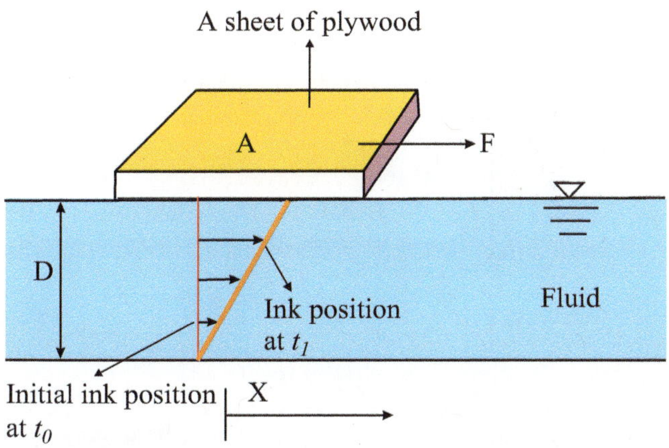

Figure 1.1. An illustrative experimental set-up demonstrating the relation between shear stress and velocity gradient.

viscous (oil or honey)? To differentiate viscosities of different fluids, quantitative descriptors for the viscosity of different fluids are needed.

Consider a sheet of plywood floating on a static water body in a tank (Figure 1.1). First, we use an ink pen to draw a vertical line from the bottom of the plywood to the bottom of the water body, and we then apply a force tangentially to the plywood. We expect to observe the phenomenon that the ink position will be displaced as shown in the figure. Based on the displacement of the ink, we can define the velocity of the movement of water, v, at any vertical location (z), that is,

$$v(z) = \frac{\Delta x(z)}{\Delta t} \tag{1.1.4}$$

where Δx(z) is the horizontal displacement at elevation z over a time interval Δt.

From the experiment, we will find that the maximum velocity exists at the top surface of the water body and the minimum velocity at the bottom. This difference is attributed to high friction between the fluid and the tank walls and the low friction between the fluid and the air. Now, we define the velocity gradient at a point in the fluid as

$$\alpha(z) = \frac{v(z)}{z} \tag{1.1.5}$$

where z is the vertical distance from the bottom of the tank to the elevation where the velocity is defined. If the displacement of the ink increases linearly with z, the velocity gradient then will be constant.

Suppose we repeat this experiment several times with different forces, F, and plot the shear stress (F/A) versus $\alpha(z)$ (or the velocity gradient), we will observe a relationship showing that the value of α increases with the stress (F/A) as illustrated in Figure 1.2. That is,

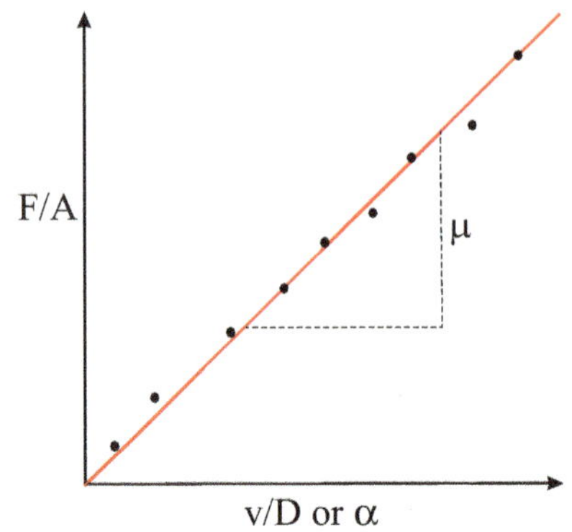

Figure 1.2. An illustration of the relationship between shear stress and the vertical velocity gradient of a Newtonian fluid. D is the depth of the fluid (see Figure 1.1).

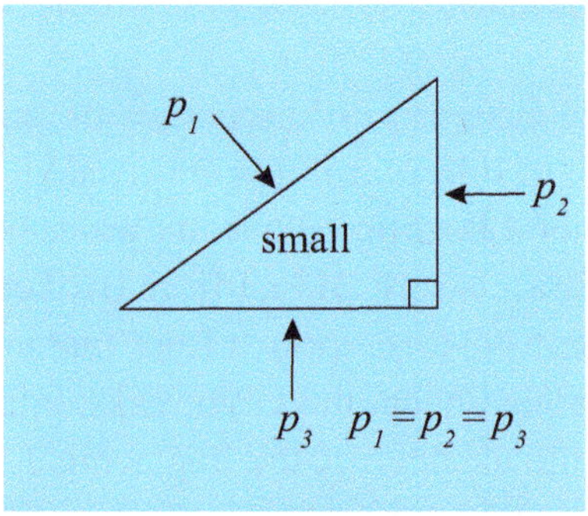

Figure 1.3. Pressure acts equally in all directions in fluids.

$$\frac{F}{A} \propto \alpha(z) \tag{1.1.6}$$

This relationship has been used to classify various fluids into categories such as Newtonian or non-Newtonian fluids. Specifically, a **Newtonian fluid** (e.g., water, air, and so on.) is a fluid whose shear stress is linearly proportional to the velocity gradient in the direction perpendicular to the plane of shear (Figure 1.3). The constant of proportionality is known as viscosity. On the other hand, a **non-Newtonian fluid** is a fluid in which the relationship between the shear stress and the velocity gradient is nonlinear (such as plastics, blood, body wash and hand soap). That is, the viscosity changes with the applied shear force.

In subsurface hydrology, we generally deal with water, which is considered to be a Newtonian fluid. Therefore, *F/A* is a linear function of *v(z)/z* for water, and we can explicitly write the relationship as

$$\frac{F}{A} = \mu \frac{v(z)}{z} \tag{1.1.7}$$

Table 1.4. *Dynamic and kinematic viscosity of water in imperial units*

Temperature t (°F)	**Dynamic Viscosity μ (lb s/ft^2) $\times$ 10^{-5}**	**Kinematic Viscosity v (ft^2/s) $\times$ 10^{-5}**
32	3.732	1.924
40	3.228	1.664
50	2.730	1.407
60	2.344	1.210
70	2.034	1.052
80	1.791	0.926
90	1.580	0.823
100	1.423	0.738
120	1.164	0.607
140	0.974	0.511
160	0.832	0.439
180	0.721	0.383
200	0.634	0.339
212	0.589	0.317

where μ (mu) is the constant of proportionality (i.e., the slope of the linear relationship between *F/A* and α) and is called the **dynamic viscosity** of the fluid. If we repeat the experiment using a different fluid, the slope of the line may change, reflecting its different resistance to the applied shear stress. We can replace the slope $v(z)/z$ by a differential dv/dz, where z is the elevation. Also replacing *F/A* by the symbol τ (tau) for shear stress, we obtain

$$\tau = \mu \frac{dv}{dz} \tag{1.1.8}$$

Equation (1.1.8) is called **Newton's Law of viscosity**.

Now let us examine the dimension of each variable of the preceding equation.

Note M, T, and L stand for the dimension of mass, time, and length, respectively.

τ is the shear stress $= \frac{F}{A} = \frac{m\,a}{L^2} = \frac{[\mathrm{M}][v\mathrm{T}^{-1}]}{[\mathrm{L}^2]} = [\mathrm{M}]\left[\frac{\mathrm{L}}{\mathrm{T}^2}\right]\cdot\left[\frac{1}{\mathrm{L}^2}\right] = \left[\frac{\mathrm{M}}{\mathrm{LT}^2}\right]$

μ = dynamic viscosity $\left[\frac{\mathrm{M}}{\mathrm{LT}}\right]$

$\frac{dv}{dz}$ = velocity gradient $\left[\frac{\mathrm{LT}^{-1}}{\mathrm{L}}\right] = \left[\frac{1}{\mathrm{T}}\right]$

The units for the corresponding variables are

Force: $F = ma$ 1 Newton (N) = 1 $\frac{\mathrm{kg\,m}}{\mathrm{s}^2}$

Pressure or stress: $p = F/A$ 1 Pascal (Pa) = 1 $\frac{\mathrm{N}}{\mathrm{m}^2} = 1\ \frac{\mathrm{kg}}{\mathrm{ms}^2}$

Viscosity: $\mu = \frac{\tau}{dv/dz}$ 1 Centipoise (cP) = $10^{-2}P = 10^{-3}\,\frac{\mathrm{N}}{\mathrm{m}^2}s = 10^{-3}$Pa-s

Note that viscosity varies as a function of temperature, and it generally decreases as temperature increases, or vice versa. Instead of dynamic viscosity, the kinematic viscosity is also used occasionally (Table 1.4).

Table 1.5. *Dynamic and kinematic viscosity of water in SI units*

Temperature T (°C)	Dynamic Viscosity μ (Pa s, N s/m^2) $\times 10^{-3}$	Kinematic Viscosity ν (m^2/s) $\times 10^{-6}$
0	1.787	1.787
5	1.519	1.519
10	1.307	1.307
20	1.002	1.004
30	0.798	0.801
40	0.653	0.658
50	0.547	0.553
60	0.467	0.475
70	0.404	0.413
80	0.355	0.365
90	0.315	0.326
100	0.282	0.29

Table 1.6. *Densities and specific weight of water in imperial units*

Temperature T (°F)	Density ρ (slugs/ft^3)	Specific Weight γ (lb/ft^3)	Specific Weight γ (lb/US gallon)
32	1.940	62.42	8.3436
40	1.940	62.43	8.3451
50	1.940	62.41	8.3430
60	1.938	62.37	8.3378
70	1.936	62.30	8.3290
80	1.934	62.22	8.3176
90	1.931	62.11	8.3077
100	1.927	62	8.2877
120	1.918	61.71	8.2498
140	1.908	61.38	8.2048
160	1.896	61	8.1537
180	1.883	60.58	8.0969
200	1.869	60.12	8.0351
212	1.860	59.83	7.9957

Kinematic Viscosity (υ) = the ratio of dynamic viscosity (μ) to mass density (ρ).

$$\upsilon = \frac{\mu}{\rho} = \left[\frac{L^2}{T}\right] \tag{1.1.9}$$

Tables 1.5, 1.6, and 1.7 list water properties at different temperatures and their units.

1.2 Fluid Statics

Fluid statics is concerned with the balance of forces on a fluid parcel which stabilizes it and brings it to rest. In the following, some fundamental concepts related to fluid statics will be discussed.

Table 1.7. *Density and specific weight of water in SI units*

Temperature T (°C)	**Density ρ (kg/m^3)**	**Specific Weight γ (kN/m^3)**
0	999.8	9.806
4	1000	9.807
10	999.7	9.804
20	998.2	9.789
30	995.7	9.765
40	992.2	9.731
50	988.1	9.690
60	983.2	9.642
70	977.8	9.589
80	971.8	9.530
90	965.3	9.467
100	958.4	9.399

1.2.1 Forces on Fluids

A **force** is any influence that can cause an object with mass to accelerate. Force has both magnitude and direction, thus it is a vector quantity. Generally, forces can be classified as 1) body forces and 2) surface forces. Body forces are defined as the forces that do not require direct contact with the fluid; they "act at a distance" (for examples, gravitational forces, or electric magnetic forces). On the other hand, surface forces represent those types of forces caused by direct contact between substances (fluid particles or fluid and solid). They can be a tangential force (e.g., a shear force), a compressive (a normal inward force) or a tensile force (a normal outward).

Surface Tension (σ) Surface tension plays an important role in flow through geologic media under unsaturated conditions, and it is an effect within the surface layer of a liquid that causes the layer to behave as an elastic sheet and to allow insects (such as the water strider) to walk on water. Surface tension is caused by the attraction between the liquid molecules due to various intermolecular forces. In the bulk of the liquid each molecule is pulled equally in all directions by neighboring liquid molecules, resulting in a net force of zero. At the surface of the liquid, the molecules are pulled inwards by attractions to other molecules deeper inside the liquid. Because there are no liquid molecules outside of the surface to balance these forces, the surface molecules are subject to an inward force of molecular attraction which is balanced by the resistance of the liquid to compression. There may also be a small outward attraction caused by air molecules, but as air is much less dense than the liquid, this force can be considered negligible.

Dimensional analysis and the work-energy theorem show that the units of surface tension ($N \cdot m^{-1}$) are equivalent to joules per square meter ($J \cdot m^{-2}$)–energy per unit area. Surface tension, measured in Newton per meter ($N \cdot m^{-1}$), is represented by the symbol σ. It is defined as the force along a line of unit length perpendicular to the

Table 1.8. *Several common units for fluid pressure*

Pressure Units					
	pascal (Pa)	**bar (bar)**	**atmosphere (atm)**	**torr (Torr)**	**pound-force per square inch (psi)**
1 Pa	$\equiv$ 1 N/m^2	10^{-5}	9.8692×10^{-6}	7.5006×10^{-3}	145.04×10^{-6}
1 bar	100,000	$\equiv 10^6$ dyn/cm^2	0.98692	750.06	14.5037744
1 atm	101,325	1.01325	$\equiv$ 1 atm	760	14.696
1 torr	133.322	1.3332×10^{-3}	1.3158×10^{-3}	$\equiv$ 1 Torr; $\approx$ 1 mmHg	19.337×10^{-3}
1 psi	6.894×10^3	68.948×10^{-3}	68.046×10^{-3}	51.715	$\equiv$ 1 lbf/in^2

Example reading: 1 Pa = 1 N/m^2 = 10^{-5} bar = 9.8692×10^{-6} atm, etc.

Table 1.9. *Conversion of some common pressure units*

Name of unit	**Unit**	**Conversion**
Pascal	Pa	1 Pa = 1 N/m^2
Bar	bar	1 bar = 0.1 MPa
Water column (meter)	mH_2O	1 mH_2O = 9 806.65 Pa
Atmospheric pressure	atm	1 atm = 101 325Pa
Mercury column (meter)	mHg	1 mHg = 1 /0.76 atm

surface, or work done per unit area. If a surface with surface tension σ is increased by a unit area, then the increase in the surface's stored energy is also equal to σ. This means that surface tension can also be considered as surface energy.

Pressure (symbol: p or P) is the force per unit area applied to an object in a direction perpendicular to the surface. At every point in a static fluid parcel there exist certain pressure intensity. Specifically, the pressure intensity, usually called pressure, is defined as follows

$$p = \lim_{\Delta A \to 0} \frac{\Delta F}{\Delta A} = \frac{dF}{dA} \tag{1.2.1}$$

where F is the normal force acting over the surface area A. Pressure intensity is a scalar quantity; that is, it has magnitude only and acts equally in all directions (for example, pressure intensities along all faces of the triangular-shaped fluid element in equilibrium in Figure 1.3 are equal). Some common units for fluid pressure are given in Table 1.8, and pressure unit conversions are presented in Table 1.9

There are two methods to express the pressure: one is based on the **absolute pressure** and the other based on the gauge pressure. **Gauge pressure** is the pressure relative to the local atmospheric or ambient pressure. The relationship between the two pressures is then: gauge pressure = absolute pressure – local atmospheric pressure.

Pressure has the following three characteristics:

1. The pressure of a fluid always acts perpendicular to the surface in contact with the fluid.
2. The values of the pressure acting at any point in a fluid at rest are equal regardless of its direction (Figure 1.3). That is, pressure is a scalar.
3. The pressure applied to a fluid in a closed vessel is transmitted to all parts at the same pressure value as that applied (Pascal's law).

1.2.2 Hydrostatic Equation

As mentioned previously, fluid statics is concerned with the balance of forces which stabilizes a fluid and brings it to rest. In a fluid at rest, the fluid pressure increases with the depth, and this relationship can be described by a hydrostatic equation. We will derive the following equation for two types of fluids: an incompressible fluid and a compressible fluid.

(a) Incompressible fluid

Mathematically, incompressibility of a fluid is expressed by saying that the density of a fluid parcel does not change as it moves. That is, an incompressible fluid has a constant density in time and space. All fluids are compressible to some extent: changes in pressure at a given temperature will result in changes in density. However, in many situations the changes in pressure are sufficiently small that the changes in density are negligible. In this case the flow can be treated as an incompressible flow.

Consider a slice of fluid with a thickness of Δz and surface area A in a tank of static fluid as shown in Figure 1.4.

In Figure 1.4, Δp represents the pressure difference between pressures acting on the top and bottom of the slice. A is the area of the slice. Because the slice of the fluid is at rest, we conclude that there are no net forces in the horizontal direction and in the vertical direction. These conditions imply that downward forces must equal upward forces:

$$(p + \Delta p)A + \rho g(A\Delta z) = pA \tag{1.2.2}$$

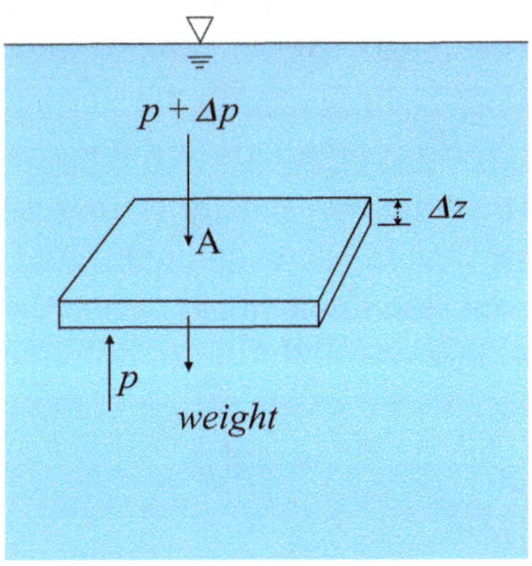

Figure 1.4. Forces acting upon a slice of a fluid under a hydrostatic condition.

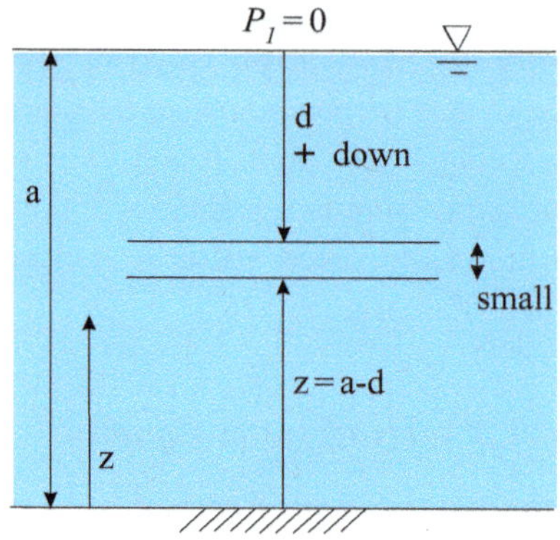

Figure 1.5. Force balance on a slice of a fluid.

Simplification of equation (1.2.2) leads to

$$\frac{\Delta p}{\Delta z} = -\rho g \text{ or } \frac{dp}{dz} = -\rho g \tag{1.2.3}$$

if ρ constant. Equation (1.2.3) specifies the rate of change in pressure with depth. The negative sign indicates that pressure increases with depth (positive downward). If ρ is a constant, we can integrate both sides of the equation (using the setup in Figure 1.5: the z coordinate is positive upward with $z = 0$ at the depth a):

$$\int_{p}^{p_1} dp = -\rho g \int_{a-d}^{a} dz \tag{1.2.4}$$

where $p_1 = 0$ representing the gauge pressure (relative to the atmospheric pressure).

The result of the integration is

$$p = \rho g d = \gamma d \tag{1.2.5}$$

This equation is called the **hydrostatic pressure equation**. Under the hydrostatic conditions, the pressure at a depth in a fluid of a constant density is the product of the unit weight of the fluid and the depth. Thus, the pressure increases linearly with depth.

(b) Compressible Fluid

Unlike an incompressible fluid whose density is invariant with time and space, a compressible fluid has a density which varies with fluid pressure and temperature. However, under an isothermal condition (means constant temperature), density is a function of pressure only so that

$$\frac{dp}{dz} = -\rho(p)g \tag{1.2.6}$$

where density, ρ, is a function of pressure only. To determine the hydrostatic pressure for the conditions in Figure 1.5, we integrate both sides of Equation (1.2.6):

$$\int_{p}^{p_1} \frac{dp}{\rho(p)} = -g \int_{a-d}^{a} dz, \tag{1.2.7}$$

and a functional relation for $\rho(p)$ in the equation (1.2.7) must be specified. That is, an equation of state (under the isothermal condition) must be specified. The equation of state can have multiple forms and one such form is:

$$\rho = \rho_o \exp[\beta(p - p_o)] \tag{1.2.8}$$

where ρ_o is the fluid density at a reference pressure p_o, and β is the fluid compressibility which is defined as:

$$\beta = -\frac{1}{V}\frac{dV}{dp} \tag{1.2.9}$$

where V is the total volume of the fluid.

1.2.3 Various Forms of Energy and Their Relationship to Force

Before we discuss fluid dynamics, it is necessary to relate forces to energy that will be used extensively in the fluid dynamics. **Energy** is a measure used by scientists to describe the capacity of a system (e.g., a volume of fluid) to do work. In turn, **work** is defined as the result of application of a force (gravitational, electromagnetic, etc.) through a physical distance. Due to the existence of a variety of forces in nature, energy and work have many different forms (e.g., gravitational, electric, heat, etc.), most of which can be grouped into two major categories: kinetic energy and potential energy. **Kinetic energy** is energy due to the motion of a volume of a fluid. **Potential energy** is the energy due to the position of a volume of fluid relative to its another position. On the other hand, the **internal energy** of a volume of fluid refers to invisible microscopic energy on atomic and molecular scale. That is, it is the total of the kinetic energy due to the microscopic motion of atoms (translational, rotational, and vibrational) and the potential energy associated with the vibrational and electric energy of atoms within molecules or crystals. It also includes the energy in all the chemical bonds, and the energy of the free, conduction electrons in metals. **Heat** is defined as energy in transit. Generally, heat is a form of energy transfer associated with the different motions of atoms, molecules, and other particles that comprise matter when it is hot and when it is cold, while the hot and cold concepts can only be explained in a relative sense.

Energy can be positive or negative because it can be either work done on an object by a force, or work done by the object against a force. Furthermore, we must emphasize again that the kinetic or the potential energy defined in fluid mechanics is the energy due to the motion or the position of fluids in a volume that contains many molecules: macroscopic energy. On the other hand, internal energy denotes the total energy of a volume of a fluid due to the kinetic and the potential energy of molecules and atoms within the volume: microscopic energy.

1.3 Fluid Dynamics

With the introduction of moving fluid and the associated kinetic energy in the preceding section, the basic concepts underlying the study of dynamics or kinematics of fluid are discussed next. First, a control volume approach is introduced, which will be used extensively in later chapters. A **control volume (CV)** can be thought of as an arbitrary volume in which attributes (e.g., mass, energy, etc.) of fluid are examined and balanced. The control volume approach is used to derive the mass continuity equation, which is the basic equation to all the fluid flow problems. A **control surface** is the surface surrounding that control volume.

Before proceeding to the discussion of fluid dynamics, we will define some terminologies first. In fluid dynamics, **streamlines** are a family of curves that are instantaneously tangent to the velocity vectors of the flow field. They show the direction a volume of fluid at a point in space and time will travel. **Pathlines** are the trajectory that a small particle would make if it followed the flow of the fluid in which it was embedded. If the flow is steady, pathlines are identical to streamlines. A scalar function whose contours define the streamlines is known as the stream function. Fluid flow is described in general by a vector field in three (for steady flows) or four (for non steady flows including time) dimensions. In a steady flow, the path lines and streamlines are identical.

1.3.1 Coordinates: Lagrangian and Eulerian viewpoints

There are two coordinate systems that can be used for describing fluids in motion. One reference framework – the **Lagrangian viewpoint** – has to do with considering an individual fluid particle for all times. The other reference framework – the **Eulerian viewpoint** – focuses on a certain point in space and considers the motion of fluid particles that pass through that point as time goes on. A further explanation of the two reference frameworks is given as follows using an example.

Lagrangian coordinates or viewpoint (moving framework): The coordinates of a moving particle are represented as functions of time and its initial position. This coordinate system is also called a "Moving Framework." To illustrate the concept of a moving framework, imagine that you are on a Pacific cruise (probably not for any graduate student!). You don't have much to do except enjoy the view and write your diary. One day, you saw a spectacular island and you wanted to write about it in your diary. How would you describe the location of the island?

First, let us recognize what the facts are

- You know where you boarded the ship,
- You know how long you have been on the ship, and
- You do not know the latitude and the longitude of the ship.

Based on these facts, you would probably write "*I saw a spectacular island on the sixth day of the cruise starting from San Diego*." That is, you used the initial position of the ship and the time you saw the island to describe the location of the island. In fact, you used the Lagrangian coordinates to describe the location of the island. Mathematically, the Lagrangian coordinates of the ship are as follows:

$$\xi(\xi_0, \eta_0, t) \tag{1.3.1}$$

where ξ_0 and η_0 are the initial position, and t is the time. Notice that the Lagrangian coordinate system does not tell you exactly where you are, unless the ship's travel path and velocity are known. Nonetheless, this coordinate system is useful in the analysis of fluid dynamics as demonstrated later. Notice that this coordinate system is also called "the convective coordinates system."

Eulerian coordinates or viewpoint (fixed framework): We will now examine an alternate way of describing the location of the island during the cruise (see the preceding section). If you were the captain of the ship, you would have written: *I saw an island at latitude 20°N and longitude 150° W*. The latitude gives the location of a place on Earth north or south of the Equator and longitude describes the location of a place on Earth east or west of a north-south line called the Prime Meridian. In essence, the location of the island is referring to a fixed location in space. Thus, the coordinate system has a fixed framework. The coordinates of the ship are then described by (x, y), which are, of course, independent of time. Such a coordinate system is called the Eulerian coordinates. Generally, the Eulerian coordinate system is more convenient to use, but when dealing with material front movements (for example, balance of forces in a control volume that is moving), we prefer the Lagrangian approach, whose usefulness would be apparent in the later chapters.

1.3.2 Time Derivatives

So far we employed the Eulerian and Lagrangian approaches to describe an observation using different frameworks. They can also be applied to describe the rate of change in observations as well. Next, we discuss the rate of change of an observation in time of an event (i.e., the time derivative of an attribute of a process). First, let's define the partial time derivative, the total time derivative, and the substantial derivative.

1) ***Partial time derivative:*** Suppose we stand on a bridge and count the number of fish passing a selected length under the bridge. If we define the number of fish per unit length of the bridge as the fish concentration $c(x, y, t)$, where x and y are the coordinates of the bridge, the rate of change in fish concentration observed at the bridge is then

$$\left.\frac{\partial c(x,y,t)}{\partial t}\right|_{x,y} \tag{1.3.2}$$

where the derivative is evaluated at the location (x, y) of the observation. Here, we use the partial derivative because we are observing how the concentration changes with time at a fixed position in space. We thus hold x and y as constants and take the derivative of c with respect to time. This is the Eulerian approach.

2) ***Total time derivative:*** Suppose now that instead of standing on the bridge, we get in a motorboat and speed around on the river, sometimes going upstream, sometimes across the current, and perhaps sometimes downstream. If we report the change of fish concentration (say, the number of fish per length along a line perpendicular to the boat) with respect to time, the number we report must also reflect the motion of the boat. In other words, the coordinates of observation, x and y are also functions of a single variable, t (i.e., in the fish concentration $c\,(x(t), y(t), t)$ Therefore, the observed change of fish concentration is no longer a partial derivative of c with respect to time but a total derivative.

$$\frac{dc}{dt} = \frac{\partial c}{\partial t} + \frac{\partial c}{\partial x}\frac{dx}{dt} + \frac{\partial c}{\partial y}\frac{dy}{dt} \tag{1.3.3}$$

where $\frac{dx}{dt}$ and $\frac{dy}{dt}$ are the components of the velocity of the boat in the x and y directions.

3) ***Substantial derivative:*** Now, suppose that we get into a canoe and simply float along the stream counting fish. Now the velocity of the observer is just the same as the velocity of the stream, $\boldsymbol{v}$. The rate of change in fish concentration is then:

$$\frac{Dc}{Dt} = \left.\frac{\partial c}{\partial t}\right|_{x,y} + v_x\frac{\partial c}{\partial x} + v_y\frac{\partial c}{\partial y} \tag{1.3.4}$$

where v_x, and v_y are components of the local fluid velocity $\boldsymbol{v}$. Dc/Dt, the derivative of $c(x,y,t)$ with respect to time is called the substantial derivative (or material derivative). This is basically the Lagrangian description of the change in concentration. The substantial derivative is just a special application of the total derivative to fluid mechanics. That is, the coordinates in substantial derivative always moves at the velocity of fluids, whereas in the total derivative, it can be something else, such as the velocity of a motor boat, as in the example presented in the last section.

1.3.3 Acceleration

Instead of the concentration of fish, we can use velocity of a substance (for example, the velocity of fish) as the variable in the time derivative. In this case, we are looking at the change in velocity in the x direction only, which is acceleration. Acceleration is defined as the rate of change in velocity, that is,

$$a = \frac{\Delta v}{\Delta t} = \frac{Dv}{Dt} \tag{1.3.5}$$

Here, we use the substantial derivative to represent the rate of change in velocity. This means that the acceleration is the velocity change perceived by a particle of the moving fluid. In other words, the change in velocity is described in the Lagrangian framework. A practical example for this would be the acceleration of your car that you perceive as you step on the gas pedal on a road that goes downhill. Therefore, equation (1.3.5) can be rewritten as:

$$a = \frac{Dv}{Dt} = \left.\frac{\partial v}{\partial t}\right|_x + v\frac{\partial v}{\partial x} \tag{1.3.6}$$

where the term on the left-hand side of the equation is the acceleration observed in the Lagrangian framework, and the first term and the second term on the right-hand of the equation represents **local acceleration** and **convective acceleration**, respectively, as described in the Eulerian framework. In fluid mechanics, the local acceleration describes the change in fluid velocity with time at a fixed point. If the local acceleration equals zero, the flow is defined as **steady-state flow**. That is, if at any given point the velocity does not vary in magnitude or direction with time, then the flow is steady. The convective acceleration on the other hand describes the fluid velocity change caused by the change in fluid particle location. If the convective acceleration is zero, the flow is called **uniform flow**. In uniform flow the velocity does not change from point to point along any of the streamlines in the flow field. Therefore, it follows that the streamlines depicting such flow must be straight and parallel. If they are not straight, the flow is nonuniform. In nonuniform flow the flow velocity changes from point to point along the streamline. Therefore, the flow pattern consists of streamlines that are curving in space representing a converging or/and diverging flow conditions.

1.3.4 Bernoulli Equation

In 1738, Daniel Bernoulli, a Swiss mathematician, presented the equation that is most frequently used in engineering hydraulics. The equation is referred to as the Bernoulli Equation. The Bernoulli Equation can be considered to be a statement of the conservation of energy principle appropriate for flowing fluids. The equation will be derived next.

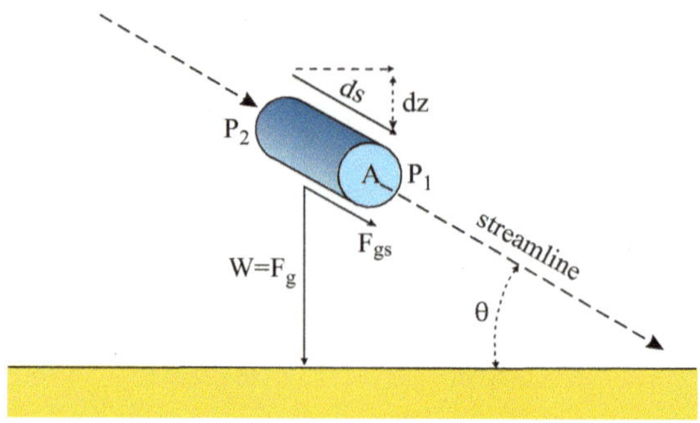

Figure 1.6. Force components in an inclined water hose.

Consider a control volume consisting of a portion of a flow field having a cross-sectional area, A, and the hose length, ds, inclined at an angle θ with the horizontal (see Figure 1.6). Flow is occurring within the hose from left to right with the fluid pressures at locations 1 and 2 being p_1 and p_2, respectively. Consider the forces acting on the fluid in a control volume of the water body within the hose. With a pressure drop between locations 1 and 2 within the hose, the surface force, F_p and the body force, F_g are as follows.

- Surface force: $F_p = (p_1 - p_2)A$
- Body force: $F_g = W = \rho g A\, ds$

With V being the fluid volume, the weight, W, of water in the hose segment between locations 1 and 2 is $W = \rho g V$, so that the body force is: $F_g = W = \rho g A ds$. Note that the contribution of tangential forces (which are also the frictional forces) is assumed to be small compared with the contribution of surface and body forces. Thus, tangential stress due to the viscosity of the water, which works against both the surface and the body forces, is neglected.

With dp/ds being the change in fluid pressure per unit change in length (s), the pressure p_2 is given as

$$p_2 = p_1 + \frac{dp}{ds} ds \tag{1.3.7}$$

The surface force F_p can thus be written as

$$F_p = \left(p_1 - \left(p_1 + \frac{dp}{ds} ds\right)\right) A = -\frac{dp}{ds} dsA \tag{1.3.8}$$

where the negative sign indicates that the force is in the down-gradient direction (from p_1 –high pressure region- to p_2 –low pressure region).

The force due to weight of the fluid along the hose (s) is F_{gs}, and it can be written as

$$F_{gs} = F_g \sin\theta = F_g\left(-\frac{dz}{ds}\right) = \rho g A ds\left(-\frac{dz}{ds}\right) \tag{1.3.9}$$

According to Newton's Second Law, the sum of the forces, F_{gs} and F_p, acting on the fluid must be equal to the product of mass, m and acceleration, a_s, of the fluid within the hose, if the effects of viscosity of the fluid are negligible.

$$F_{gs} + F_P = ma_s \tag{1.3.10}$$

That is,

$$-\rho gAds\left(\frac{dz}{ds}\right) + \left(-\frac{dp}{ds}Ads\right) = (\rho Ads)a_s \tag{1.3.11}$$

Notice that the force balance is carried out for a control volume of the water that is moving with the velocity of water. Therefore, the acceleration is in a moving coordinate system and can be written in a fixed coordinate system as

$$a_s = \frac{\partial v}{\partial t} + v\frac{\partial v}{\partial s} \tag{1.3.12}$$

If we assume the flow is steady (velocity is invariant in time), the acceleration is

$$a_s = v\frac{dv}{ds} \tag{1.3.13}$$

Substituting this into equation (1.3.11), we have

$$-\rho gAds\left(\frac{dz}{ds}\right) + \left(-\frac{dp}{ds}Ads\right) = (\rho Ads)\left(v\frac{dv}{ds}\right) \tag{1.3.14}$$

Then we divide both sides of equation (1.3.14) by ρdsA and rearrange the resultant equation, we obtain

$$v\frac{dv}{ds} + g\frac{dz}{ds} + \frac{dp}{\rho ds} = 0 \tag{1.3.15}$$

According to the rule of differentiation:

$$v\frac{dv}{ds} = \frac{1}{2}\frac{dv^2}{ds} \tag{1.3.16}$$

Combining equations (1.3.15) and (1.3.16) and dividing throughout by g,

$$\frac{d}{ds}\left(\frac{v^2}{2g} + z + \frac{p}{\rho g}\right) = 0 \tag{1.3.17}$$

Integrating the ordinary differential equation (1.3.17), we obtain

$$\frac{v^2}{2g} + z + \frac{p}{\rho g} = C \tag{1.3.18}$$

where C is a constant and ρ is constant in time and space. We thus have derived the **Bernoulli equation,** which relates the pressure, velocity, and elevation between any two points along a streamline in the flow field for flow that is steady, nonviscous, and incompressible.

1.3.5 Heads (Energy/Weight)

In order to elucidate the physical meaning of the Bernoulli equation, we multiply the equation by the gravity acceleration constant, g, to obtain the following expression.

$$\frac{v^2}{2} + zg + \frac{p}{\rho} = Cg \tag{1.3.19}$$

We examine the dimensions of each term in equation (1.3.19).

$$\left[\frac{L^2}{T^2}\right] + [L]\left[\frac{L}{T^2}\right] + \frac{\left[\frac{M}{LT^2}\right]}{\left[\frac{M}{L^3}\right]} \text{ or } \left[\frac{L}{T}\right]^2 + \left[\frac{L}{T}\right]^2 + \left[\frac{L}{T}\right]^2$$

Recall that

$$\text{kinetic energy} = \text{work} = mv^2/2 = [M]\left[\frac{L}{T}\right]^2$$

where m is mass.

$$= \textit{force} \times \textit{distance} = [M]\left[\frac{L}{T^2}\right] \times [L] = [M]\left[\frac{L}{T}\right]^2$$

$$\frac{\textit{energy}}{\textit{mass}} \Rightarrow \left[\frac{L}{T}\right]^2$$

Based on the dimensional analysis, we can conclude that each term in the Bernoulli Equation represents energy per unit mass of fluid. Specifically,

$$\begin{aligned} \frac{v^2}{2} &\Rightarrow \textit{kinetic energy/mass} \\ zg &\Rightarrow \textit{potential energy/mass} \\ \frac{p}{\rho} &\Rightarrow \textit{flow work/mass} \end{aligned}$$

That is, each term represents the energy required to bring a unit mass of fluid to a specific state. For instance, the term zg represents the energy required to bring a unit mass of fluid from elevation z = 0 to z = z.

Now, examine the physical meaning of equation (1.3.18),

$$\frac{v^2}{2g} + z + \frac{p}{\rho g} = \text{constant} = C \text{ in which } \frac{v^2}{2g} = [\mathrm{L}], \quad z = [\mathrm{L}] \quad \frac{p}{\rho g} = [\mathrm{L}] \quad (1.3.20)$$

which is exactly the same as that of equation (1.3.19), with the exception that each energy term is expressed in units of length, which can be determined by simply using a ruler. Each term in equation (1.3.20) represents energy per unit weight of the fluid, which is defined as "head". Therefore,

$$\begin{aligned} \frac{v^2}{2g} &\Rightarrow \text{ velocity head [L]} \\ z &\Rightarrow \text{ potential head } (z) \text{ [L]} \\ \frac{p}{\rho g} &\Rightarrow \text{ pressure head } (h) \text{ [L]} \\ C &\Rightarrow \text{ total head } (H) \text{ [L]} \end{aligned}$$

After all, the Bernoulli equation simply says that the sum of these three energies per unit weight–in the unit of length–at any point in the fluid should be equal to a constant value, which is called **the hydraulic head.** In other words, the Bernoulli equation is an **energy conservation equation.** An important implication of the preceding analysis of the Bernoulli equation is that if the flow is steady and free of resistance (i.e., idealized fluid with zero viscosity), the total head anywhere in the system must be the same. This is analogous to the conservation of energy in fundamental physics: if an object is ejected horizontally on a frictionless plane in a chamber under perfect vacuum, the total energy must be conserved. The object initially accelerates converting kinetic energy to the other two energies. Once it reaches a steady condition, no further acceleration takes place and the object moves at a steady velocity, while total energy is conserved.

According to the Bernoulli equation, for horizontal fluid flow, an increase in the velocity of flow will result in a decrease in the static pressure, as there will be no change in the potential energy. The most practical example of this is in the case of an airfoil. The shape of an airplane wing is such that air flowing over the top of the wing must travel faster than the air flowing under the wing, and so there is less pressure on the top than on the bottom of the air foil, and thus the pressure difference results in air lift to the airplane.

Before we proceed further, let's review all the assumptions used to derive the Bernoulli equation

- No friction, and isothermal condition
- Incompressible, homogeneous fluid: ρ is invariant with p and in time and space
- Steady flow: constant velocity in time (no local acceleration)
- The equation is only valid along the streamline (Why ?)

For compressible fluids, the fluid density is a function of pressure and temperature, $\rho(p,T)$. In the following analysis, we assume that the fluid is under isothermal conditions, that is, $\rho(p)$. The Bernoulli equation then becomes

$$\frac{v^2}{2g} + z + \int_{p_0}^{p} \frac{dp}{\rho g} = \text{Constant} \tag{1.3.21}$$

where ρ must be a function of pressure, otherwise the value $\int_{p_0}^{p} \frac{dp}{\rho g}$ is indeterminate.

In other words, the isothermal condition is a must, so that ρ is strictly a function of pressure.

1.3.6 Water Manometer (Piezometer)

As mentioned before, the energy per unit weight of a fluid has the dimension of length. We therefore can determine the energy state of a fluid at different locations using a ruler. To illustrate this possibility, we introduce a water manometer or piezometer, which can be used to measure the energy state at any location within a fluid.

A manometer is a device that is generally made of a tube full of water (or some other type of fluid such as mercury); one end of the tube is connected to a flow system (e.g., a pipe that is full of water) and the other end is open to atmosphere. As illustrated in Figure 1.7, a water manometer is installed on a water column at point B and the water level in the manometer rises up to point C. If we chose point A as the datum, the distance between point A and point B then defines the elevation head, z; the distance between points B and C represents the pressure head, $h = p/\rho g$; the total head is denoted by the distance between points A and C. Notice that pressure head at any location with the same elevation in the manometer should be the same according to the Bernoulli equation. For example, the pressure heads at locations 1 and 2 are the same since the elevations of the two locations are the same. The pressure head at point 3 of the manometer is equal to the atmospheric pressure head, which is normally set to zero (in a gauge pressure sense). As a result, the pressure heads at locations 1 and 2 must be equal, and they are represented by the distance between point B and point C.

1.3.7 Head Loss

The preceding sections show how the Bernoulli equation was derived and discusses its physical implication: the total energy in a system must be conserved. In addition, we have introduced the manometer that can be used to determine the energy state of a fluid. We therefore are ready to test the validity of the equation. We will conduct an experiment (Figure 1.8), which consists of a long horizontal pipe of a uniform radius, a water reservoir on one of its end, and an opening on the other end.

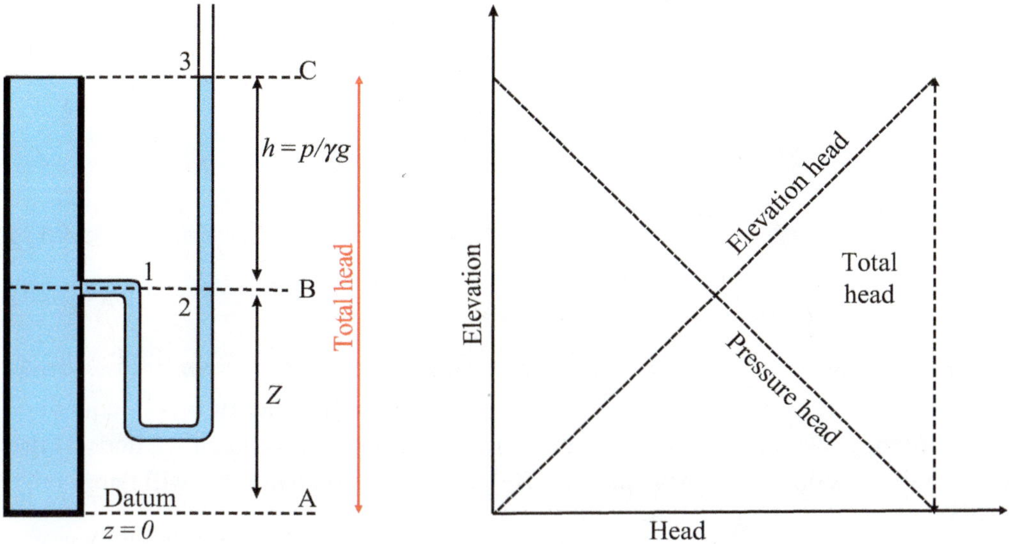

Figure 1.7. A schematic illustration of a water manometer, and pressure, elevation, and total head and their distribution along the column.

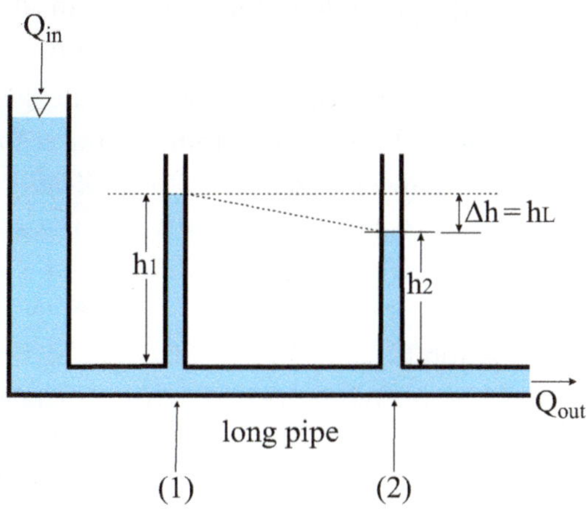

Figure 1.8. Head distribution along a pipe under a steady flow.

Along the pipe, we will install two manometers at locations 1 and 2 to determine the energy of flowing water in the pipe. The water reservoir is kept at a constant water level so that the flow in the pipe remains steady. Under these conditions, we know that Q_{in} must equal Q_{out}, implying steady state flow; the elevation heads at points 1 and 2 are the same (z_1=z_2), and the cross-sectional area of the pipe at the two points are identical (A_1=A_2). Therefore, we can conclude that the velocity at the two points must be the same: $v_1 = v_2$, implying uniform flow. According to the Bernoulli equation, we then should have a relationship of the following form:

$$\frac{p_1}{\rho g} + z_1 + \frac{v_1^2}{2g} = \frac{p_2}{\rho g} + z_2 + \frac{v_2^2}{2g} = \text{constant} \tag{1.3.22}$$

Or, a relationship showing that the pressure heads at the two points must be equal:

$$\frac{p_1}{\rho g} = \frac{p_2}{\rho g} \quad \Rightarrow \quad h_1 = h_2 \tag{1.3.23}$$

But the experiment shows $h_1 > h_2$, implying some loss of energy, as water flows from location (1) to location (2). This result obviously contradicts the Bernoulli equation. In other words, we must have forgotten some important factors when we derived the Bernoulli equation. In order to make the Bernoulli equation work, we will temporally add a term, called head loss, h_L, so that

$$\frac{p_1}{\rho g} + z_1 + \frac{v_1^2}{2g} = \frac{p_2}{\rho g} + z_2 + \frac{v_2^2}{2g} + h_L = H \tag{1.3.24}$$

This head loss is attributed to the fact that we assumed the fluid is nonviscous (no friction) when we derived the Bernoulli equation. If we examine the velocity field in a frictionless and a frictional system (Figure 1.9), we find that velocity profile must be uniform in the former system. In the latter, the profile is parabolic, with the maximum velocity at the center of the pipe and minimum at the walls of the pipe. The radius of the pipe used in the experiment is small and water is somewhat viscous. During the movement of water from point 1 to point 2, there must be some frictional forces (due to shear stress) between the fluid and the wall of the pipe. The fluid has to overcome these forces and thus energy is dissipated in the system during flow from location (1) to location (2). The energy dissipating in the system is called the **head loss h_L**. Results of various laboratory experiments have concluded that the head loss can be expressed as

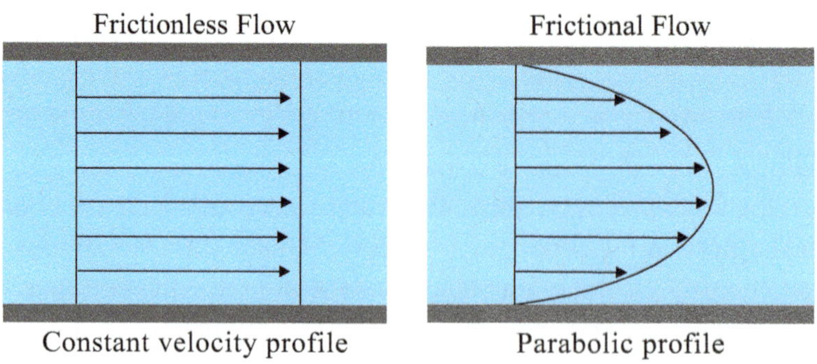

Figure 1.9. Illustrations of velocity profiles in an ideal frictionless flow and a frictional flow in a pipe.

$$h_L = f \frac{L}{d} \frac{v^2}{2g} \tag{1.3.25}$$

where L is the length of the pipe, d is the diameter of the pipe, v is the velocity of the moving fluid in the pipe, and f denotes the friction factor.

In addition, from the velocity profile and the viscosity experiment, we can conclude the head loss is caused by the shear stress between wall and fluid, thus:

$$h_L \propto \tau = \mu \frac{dv}{dy} \tag{1.3.26}$$

Therefore, the friction factor (f) must be related to the viscosity of the fluid. We subsequently can conclude that the head loss (h_L) is a function of 1) viscosity of the fluid, 2) roughness of the wall, 3) fluid velocity, and 4) diameter and the length of the pipe. In fact, this shear stress, which is responsible for the head loss in the system, represents the "viscous force" in the system. As a matter of fact, we did neglect this viscous force when we derived the Bernoulli equation:

$$\underbrace{F_P}_{\text{Pressure Force}} + \underbrace{F_{gs}}_{\text{Gravity Force}} + \underbrace{F_v}_{\underline{\text{Viscous Force}}} = \underbrace{ma}_{\text{Inertial Force}} \tag{1.3.27}$$

Note that F_v is small along the center-line of the water hose. If this expression does not impress you, well, how about this:

$$-\nabla p + \rho g + \mu \nabla^2 v = \rho \frac{Dv}{Dt} \tag{1.3.28}$$

where p is the pressure, g is gravitational acceleration, μ is the viscosity, v is the velocity, ρ is the density of the fluid. Equation (1.3.28) is called the Navier-Stokes equation. Don't be overwhelmed by the complicated mathematical expression in equation (1.3.28)! The Navier-Stokes equation simply describes the force balance as does Equation (1.3.27).

The preceding discussion points to a fact that if a fluid flows through a frictional pipe, some of its total energy must be used to overcome the frictional forces, and this energy in turn is converted into the internal energy or heat. This results in the loss of the total head. Therefore, it is logical to draw the conclusion that because of friction, a fluid must flow from a high total head region to a low total head region, which is in fact the fundamental basis of hydrogeology.

1.3.8 Inertial and Viscous Forces and Flow Behavior

The Navier-Stokes equation discussed in the previous section is merely an equation that describes the balance of different inertial and viscous forces. The balance between these forces certainly is reflected in the flow behavior. Osborne Reynolds

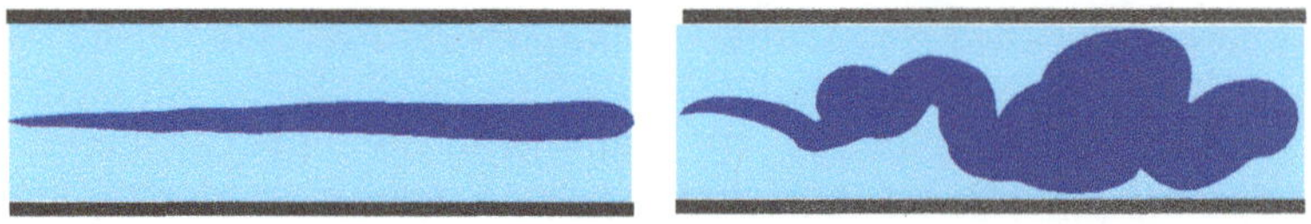

Figure 1.10. Migration of a dye tracer in a laminar flow and turbulent flow in a pipe.

(1843–1912) empirically defined a dimensionless number *Re* to classify flow behaviors into two categories 1) laminar flow and 2) turbulent flow. The *Re* is called the Reynolds Number and is defined as

$$Re = \frac{v\ell\rho}{\mu} \tag{1.3.29}$$

where v is a characteristic velocity; ℓ denotes a characteristic length; ρ and μ are density and dynamic viscosity of the fluid, respectively. Notice that this equation is an empirical equation, meaning it is a result of many experiments.

Visually, laminar and turbulent flows can be demonstrated from dye tracer experiments. Figure 1.10 illustrates a dye tracer movement in a laminar flow condition (figure on the left) and a turbulent flow condition (figure on the right). If *Re* is small, say less than 50, we have laminar flow and the tracer tends to flow in a straight line fashion (i.e., the flow is characterized by smooth, uniform fluid motion). On the other hand, if *Re* is large, say greater than 5000, we then have turbulent flow and the tracer migration tends to be chaotic (i.e., the flow exhibits random eddies, vortices and other flow fluctuations). However, there is no universal cutoff value for *Re* to define laminar and turbulent flows in all flow systems.

1.3.9 Physical Meaning of the Reynolds Number (Re)

While the formula for the Reynolds number provides a measure to classify fluid flow behavior, the physics behind the formula can be revealed by examining the dimensions of inertial and viscous forces.

$$\text{Inertial force} \propto ma = \left[\rho \mathrm{L}^3\right]\left[\frac{\mathrm{L}}{\mathrm{T}^2}\right] = \left[\frac{\rho \mathrm{L}^4}{\mathrm{T}^2}\right] \tag{1.3.30}$$

$$\text{Viscous force} \propto \tau A = \mu\left[\frac{v}{\mathrm{L}}\right]\left[\mathrm{L}^2\right] = \mu v \mathrm{L} \tag{1.3.31}$$

The ratio of the inertial force to the viscous force thus takes the following form

$$\begin{aligned}\frac{\text{Inertial force}}{\text{Viscous force}} &\propto \frac{\rho[\mathrm{L}^3]v/[\mathrm{T}^2]}{\mu v[\mathrm{L}]} = \frac{\rho \mathrm{L}\left[\frac{\mathrm{L}}{\mathrm{T}}\right]}{\mu} \\ &\propto \frac{\rho[\mathrm{L}]v}{\mu} = Re\end{aligned} \tag{1.3.32}$$

This dimensional analysis shows that the Reynolds number is directly related to the ratio of inertial to viscous forces within a fluid. Therefore, when

$$\text{Viscous force} >> \text{inertial force} \Rightarrow \text{Small } Re \Rightarrow \text{Laminar flow.}$$

In other words, the acceleration of fluid particles due to inertial force tends to be damped out by the viscous force and the flow becomes a laminar flow. When

$$\text{Viscous force} << \text{inertial force} \Rightarrow \text{Large } Re \Rightarrow \text{Turbulent flow.}$$

The viscous force cannot restore the deviating path of the fluid particle due to acceleration and the flow behavior as such becomes chaotic.

More precisely, a large Reynolds number indicates that viscous forces are not important to the flow. With a low level of viscosity, small-scale fluid motion perturbations are undamped – there is not enough viscosity to dissipate their chaotic motions. In contrast, a low Reynolds number indicates that viscosity is important to the flow dynamics, which implies that the small-scale perturbations are damped out, and only the large-scale fluid motion remains.

While the interpretation of the physical meaning of the Reynolds number is rather rudimentary, it will eventually help us understand the reason why Darcy's Law works in groundwater systems.

Quiz:

1. According to the Bernoulli equation (considering head loss), can you explain why water can flow uphill now?
2. If the pipe in Figure 1.8 inclines at angle with the horizontal while the flow rate is the same, will the head loss be different?

1.4 Summary

- Fluid mechanics employs a volume averaging concept which allows discrete water molecules in a body of water being visualized as a continuum—continuum assumption.
- All properties, forces, and energies which are used in fluid mechanics refer to averaged quantities over a control volume (CV), which is much larger than many molecules.
- A force is any influence that causes a volume of fluid to undergo a certain change, concerning its movement, direction, or geometrical construction. In other words, a force can cause the volume of fluid with mass to change its velocity. Energy is the measure of the work done by a force on a volume of fluid such that it moves a distance in the direction of the force. Pressure is the force per unit area.
- Energy per weight of the fluid has a unit of length and is called head, which has the dimension of length. Head is applicable to fluids with constant density only.
- The Bernoulli equation shows the conservation of energy (total head) in nonviscous fluids. The total head consists of velocity, potential, and pressure heads. The relative contribution

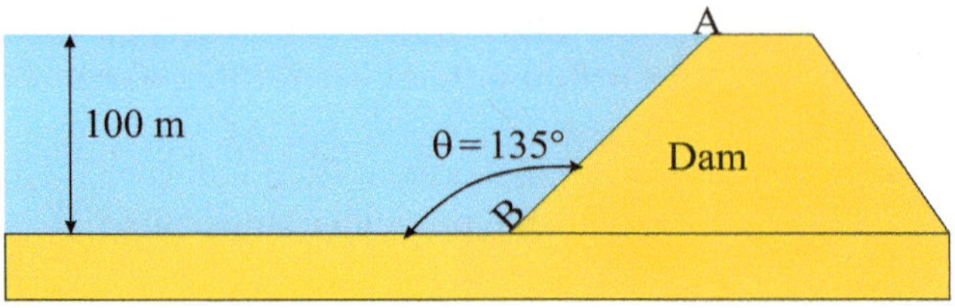

Figure 1.11.

of each component may vary in a system, but their sum must remain the same under the free-resistance situation.

- The Bernoulli equation omits the viscous effect of fluids on flow, which exhibit as a head loss: a measure of resistance to flow.
- Reynolds number is a measure of the relative strength of inertial and viscous forces in a moving fluid. Small values of Reynolds number indicate great effects of viscous forces on flow and the flow is well-behaved: a laminar flow. This is likely the case for flow in geologic media.

1.5 Exercises

1) Identify the units or dimensions for the following terms: (e.g., density = mass per length cubed [M/L^3].)
   a. acceleration
   b. force
   c. unit volume
   d. pressure
   e. dynamic viscosity
   f. kinematic viscosity
   g. shear stress
2) Determine the force acting on AB which has a width of 500 meters and the depth of water behind the dam is 100 meters (Figure 1.11). (Hint: use hydrostatic pressure concept and the fact that pressure is a scalar).
3) On a stormy day in August, a scuba diver is diving off the Mediterranean coast of Greece. A low pressure system is moving into the area and the atmospheric pressure is decreasing at a steady rate. The diver is descending at a constant vertical velocity through the water. Keep in mind that absolute pressure is defined as the sum of the gauge pressure and the atmospheric pressure. Perform the following analyses:
   a. Use the concept of the substantial derivative to develop a formula for the rate of change of the absolute pressure as a function of time experienced by the diver. Assume that the water is incompressible and has a constant density.
   b. The diver is descending at a rate of 0.5 m/sec and atmospheric pressure is decreasing by 0.0001 atmospheres/sec. The density of seawater is 1.035 g/cm^3. Determine the actual change of pressure (in Pascals) with time experienced by the diver. (1 atm = 98,690 Pa)

c. How and why would you change the assumptions from those in part a) if you were dealing with a natural ocean water system with thermal currents, etc.? Modify your equation to show these changes.

4) Consider the force balance on the triangle in a static fluid in Figure 1.3 to show pressures on all directions are the same if the triangle is infinitesimally small. That is, pressure is a scalar.

2

Darcy's Law for Saturated Porous Media

2.1 Introduction

This chapter develops and explains principles of fluid flow through porous media in terms of principles of fluid mechanics presented in Chapter 1. A **porous medium** or a porous material is defined as a solid (often called a matrix) permeated by an interconnected network of pores (voids) filled with a fluid (e.g., air or water). Usually both the solid matrix and the pore network (also known as the pore space) are assumed to be continuous, so as to form two interpenetrating continua (e.g., soils, rocks, or geologic formations). In this chapter, Poiseuille's law for flow through a capillary tube is first derived to understand effects of various factors on flow along a flow path in porous media. On the basis of Poiseuille's law, a mathematical model is subsequently developed for flow through a porous medium. Introduction of Darcy's law and the representative elementary volume (REV) concept then follow. Next, the mathematical model based on Poiseuille's law is compared to Darcy's law and is then used to elucidate the physical interpretation of Darcy's law and the hydraulic conductivity. Then, limitations of Darcy's law are discussed. Finally, laboratory methods for determining the hydraulic conductivity are presented.

2.2 Flow Through a Capillary Tube

Before discussing flow in porous media, let us examine flow through an idealized system. One of the simplest systems, which nonetheless has many characteristics of relevance to understanding flow in porous media, is flow through a capillary tube. In the following, we describe how a cylindrical capillary tube is used to approximate the actual flow in a porous medium.

As shown in Figure 2.1a, suppose we observe the path of a fluid particle at a microscopic level in a porous medium that is fully saturated with a fluid. As we focus on the fluid pathway (shown as a line with an arrow), we notice that the pore (void) spaces, between the solid particles through which flow occurs, are quite variable. We also notice that the flow path is quite tortuous, and the variable geometry of the flow path cannot be easily described. To avoid these complications, we will approximate the actual, tortuous (crooked) flow path through the voids or pore spaces as an idealized straight flow path through a cylindrical capillary tube of a very small diameter (see Figure 2.1b). As we analyze the fluid flow through an idealized capillary tube, we may postulate that because of the complicated maze of tortuous flow paths the frictional head losses must play an important role. Therefore, the viscous forces must dominate the inertial forces for flow through a porous medium. As a result, the flow regime in a porous medium will most likely be laminar in terms of our analysis of the Reynolds number in a closed-conduit or pipe flow (see Chapter 1) scenario.

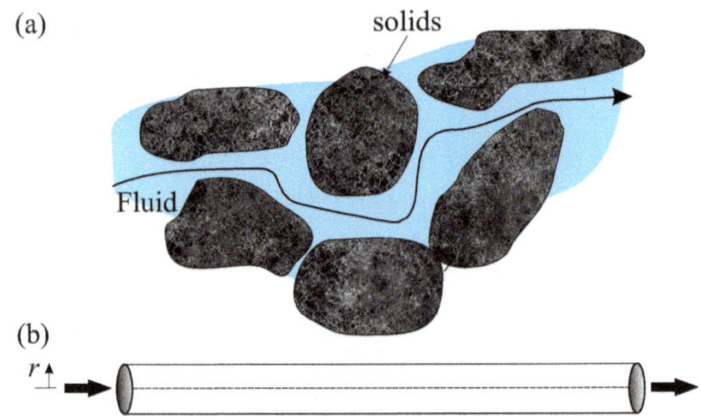

Figure 2.1. Illustration of an actual pathway for a fluid particle in a) a real porous medium and b) an idealized capillary tube with a constant radius.

The following simplifications are applied to mimic the pathway of a fluid particle in a real porous medium:

- The actual fluid pathway is stretched (i.e., the idealized flow tube is straight).
- The tube with variable radius is replaced by a tube with an "equivalent," constant, and spatially invariant radius.

Thus, a porous medium is conceptualized as a collection of stretched flow tubes of constant radius.

2.2.1 Poiseuille's Law for Flow Through a Capillary Tube

The preceding approximations for the flow paths enable us to construct a mathematical model that mimics flow complexities in a simplified sense in porous media. While complex features of flow through a porous medium are simplified, some fundamental principles that govern flow through porous media may be revealed through the analysis of this simplified mathematical model.

Before developing a mathematical model for flow through a bundle of capillary tubes, we will first study flow through one horizontal tube (Figure 2.2). Now, consider all forces (i.e., pressure and viscous) acting on a fluid in a control volume, CV (i.e., an arbitrary volume of the fluid), with a radius of r and a length of Δx along the axis of this tube. According to Newton's second law, we have

$$p(\pi r^2) - (p + \Delta p)(\pi r^2) - \tau(2\pi r\Delta x) = ma = 0 \tag{2.2.1}$$

where Δp is negative and a negative sign is used for the shear stress, which implies the opposite direction of this force; the inner surface area (A) subject to the shear stress is $2\pi r\Delta x$; m is the mass of the fluid in the control volume; a is acceleration. Notice that the viscous force is considered in the formulation because of the small

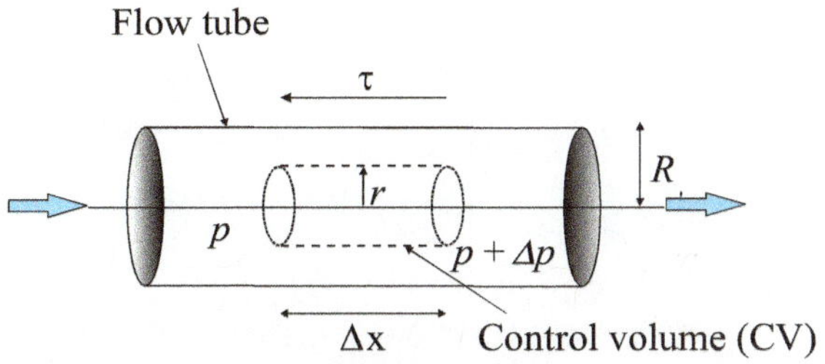

Figure 2.2. A section of a capillary tube of radius R. A force balance analysis is performed for a moving fluid in a cylindrical control volume of radius $r < R$ and length Δx in response to a pressure differential Δp. τ is shear stress.

opening of the tube. Since groundwater velocity is small, the acceleration term will be neglected. In other words, we assume the flow is steady and uniform. Thus, equation (2.2.1) can be simplified and becomes

$$\tau = -\frac{r}{2}\frac{\Delta p}{\Delta x} \tag{2.2.2}$$

If the fluid is a Newtonian fluid, then the shear stress is linearly proportional to the velocity gradient in the r direction. Under this condition, equation (2.2.2) can be expressed as

$$-\mu\frac{dv}{dr} = -\frac{r}{2}\frac{\Delta p}{\Delta x} \tag{2.2.3}$$

where μ is the dynamic viscosity and v is the velocity. Since the shear stress is positive as defined in the force balance equation and the dv/dr is a negative quantity (i.e., the velocity decreases as r, distance from the center of the tube, increases), the negative sign on the right-hand side of equation (2.2.3) is necessary.

Equation (2.2.3) is an ordinary differential equation, describing the change of v with respect to the change of r. Solving the equation by integration, that is,

$$\int dv = \left(\frac{1}{2\mu}\frac{\Delta p}{\Delta x}\right)\int rdr \tag{2.2.4}$$

we obtain a general solution:

$$v(r) = \left(\frac{1}{2\mu}\frac{\Delta p}{\Delta x}\right)\frac{r^2}{2} + C \tag{2.2.5}$$

where C is a constant of integration, which depends on the boundary condition. In our case, the boundary condition is $v = 0$ at $r = R$ (r is the variable representing the radius, measured from the center of the tube to its wall, and R is the value of the tube radius). After applying this boundary condition to equation (2.2.5), the constant then becomes

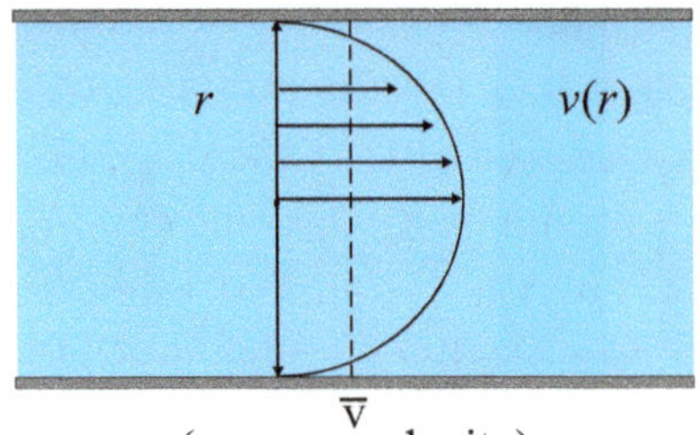

Figure 2.3. Theoretical velocity profile in a capillary tube (solid parabolic curve) and the average velocity profile (dashed line).

$$C = -\left(\frac{1}{2\mu}\frac{\Delta p}{\Delta x}\right)\frac{R^2}{2} \tag{2.2.6}$$

Substituting C from equation (2.2.6) in the general solution, that is, equation (2.2.5), we obtain a particular solution describing the velocity variation along the radius of the tube, r. That is, the velocity at any point along the radius, r, is given by

$$v(r) = -\left(\frac{1}{4\mu}\frac{\Delta p}{\Delta x}\right)(R^2 - r^2) \tag{2.2.7}$$

where $v(r)$ is positive in the direction of flow since Δp is a negative quantity. The resultant velocity profile is shown in Figure 2.3.

Total discharge from the tube is obtained by integrating the point velocity, $v(r)$, over the cross-sectional area, perpendicular to flow. That is,

$$Q = \int v dA = \int_0^R \frac{-1}{4\mu}\frac{\Delta p}{\Delta x}(R^2 - r^2)2\pi r dr \tag{2.2.8}$$

After integration, equation (2.2.8) yields

$$Q = \frac{-\pi R^4}{8\mu}\frac{\Delta p}{\Delta x} = \frac{-\pi R^4}{8\mu}\frac{dp}{dx} \tag{2.2.9}$$

The average velocity, $\overline{v}$, over the cross-sectional area, A, is then equal to

$$\overline{v} = \frac{Q}{A} = \frac{-R^2}{8\mu}\frac{dp}{dx} \tag{2.2.10}$$

Notice that the average velocity, $\overline{v}$, in equation (2.2.10) is not the same as $v(r)$ in equation (2.2.7). Furthermore, the pressure, p, is assumed constant over any cross-sectional area perpendicular to the flow path. Equation (2.2.10) is called **Poiseuille's law.** Poiseuille's law (or the **Hagen-Poiseuille law**, named after Gotthilf Heinrich Ludwig Hagen [1797–1884] for his experiments in 1839) is a physical law describing laminar stationary flow of an incompressible uniform viscous Newtonian fluid through a cylindrical tube with constant circular cross section area. This was experimentally derived in 1838 and was formulated and published in 1840 and 1846 by Jean Louis Marie Poiseuille (1797–1869).

Now, let us summarize what we have learned from Poiseuille's law for flow through a single flow tube embedded in a porous medium:

- The discharge Q varies as the fourth power of the radius of the tube ($Q \propto R^4$).
- The average velocity is proportional to the pressure gradient, dp/dx, for the horizontal flow scenario.

In other words, the simple, idealized, one-flow-tube model provides us with an explicit and quantitative relationship between the average velocity and pressure (energy) gradient in the tube. Additionally, the coefficient of the linear proportionality between the average velocity and pressure gradient is functionally related to the capillary tube properties as well as fluid properties.

2.2.2 General Form of Poiseuille's Law

Poiseuille's law, as developed in the preceding section and presented in the form of equation (2.2.10), considers only horizontal flow. For nonhorizontal flow, following a similar procedure, one can derive a generalized form of equation (2.2.10) that includes the effect of gravity (i.e., the elevation head):

$$\overline{v} = -\frac{\rho g R^2}{8\mu}\frac{dH}{dx} \tag{2.2.11}$$

where H is the total head (or hydraulic head), and it is the sum of pressure head, h, and elevation head, z, that is,

$$H = \frac{p}{\rho g} + z = h + z \tag{2.2.12}$$

Notice that the velocity head is not considered either in equation (2.2.10) or in the generalized equation (2.2.11). You may now ask why. Consider $\overline{v} = 0.1$ m/s, a rather high velocity for groundwater flow. The corresponding velocity head is

$$\frac{\overline{v}^2}{2g} = \frac{(0.1\,{}^{\mathrm{m}}/_{\mathrm{s}})^2}{2(9.81\,{}^{\mathrm{m}}/_{\mathrm{s}^2})} = 0.00051\mathrm{m} \tag{2.2.13}$$

According to equation (2.2.13), even with this fast groundwater flow velocity the velocity head is, in effect, too small to measure. Therefore, we can conclude that for most groundwater flow problems the velocity head can be ignored. Another reason to neglect velocity heads in groundwater flow is that, in general, the contribution from the velocity head is much smaller than the contributions from pressure heads and elevation heads. Notice that this is not to say that the groundwater velocities are zero. It simply means that in a moving fluid the contribution to the total energy from the kinetic energy (see Chapter 1) is much smaller than the contribution from either the potential energy or the energy due to fluid pressure.

2.3 Poiseuille's Law for Flow Through Porous Media

2.3.1 Poiseuille's Law for Flow Through a Bundle of Capillary Tubes

The next logical step following the preceding development is to extend the Poiseuille law for flow through a single capillary tube to a more realistic situation in which a porous medium is viewed as a bundle of cylindrical capillary tubes (see Figure 2.4). We therefore consider a simplified porous medium consisting of only two types of tubes: N_1 number of parallel tubes having radius a_1 and N_2 number of parallel tubes having radius a_2. The cross-sectional area of the porous medium is A (Figure 2.4). The total cross-sectional area of all the pores, A_p is

$$A_p = \pi(N_1 a_1^2 + N_2 a_2^2) \tag{2.3.1}$$

Then the porosity, n, of the entire medium is defined as

$$n = \frac{A_p}{A} = \frac{\pi(N_1 a_1^2 + N_2 a_2^2)}{A} \quad \text{and} \quad A = \frac{\pi(N_1 a_1^2 + N_2 a_2^2)}{n} \tag{2.3.2}$$

Specifically, porosity is the volume fraction of voids in a porous medium. The total discharge Q through the medium is $Q = N_1 Q_1 + N_2 Q_2$. Applying Poiseuille's law to Q_1 and Q_2, we obtain

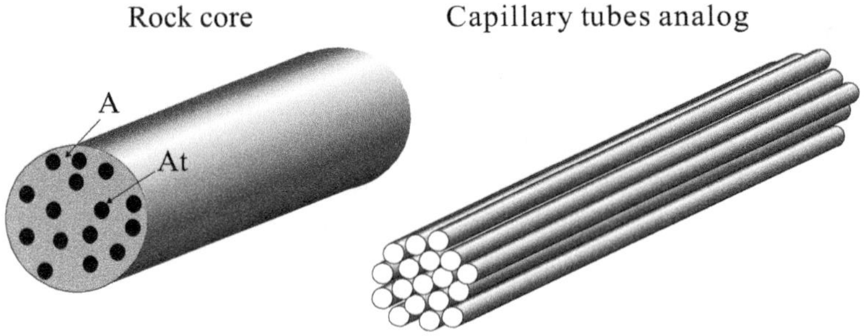

Figure 2.4. Illustrations of the conceptualization of idealized porous media. The cross-sectional area of each tube is denoted by A_t.

$$\begin{aligned} Q &= -\left[N_1\left(\rho g\frac{\pi a_1^4}{8\mu}\right) + N_2\left(\rho g\frac{\pi a_2^4}{8\mu}\right)\right]\frac{dH}{dx} \\ &= -\left(\frac{\rho g \pi}{8\mu}\right)[N_1 a_1^4 + N_2 a_2^4]\frac{dH}{dx} \end{aligned} \tag{2.3.3}$$

The discharge, q, per unit cross-sectional area of the medium then takes the form

$$q = \frac{Q}{A} = -\frac{\rho g}{8\mu}\left(\frac{\pi(N_1 a_1^2 + N_2 a_2^2)}{A}\right)\frac{[N_1 a_1^4 + N_2 a_2^4]}{[N_1 a_1^2 + N_2 a_2^2]}\frac{dH}{dx} \tag{2.3.4}$$

where q is the ***specific discharge***, the average discharge per unit cross-sectional area of the porous medium, and it has a dimension of [L/T]. If we let

$$\overline{R}^2 = \frac{[N_1\, a_1^4 + N_2 a_2^4]}{[N_1\, a_1^2 + N_2 a_2^2]} \tag{2.3.5}$$

we then have the formula for flow through the porous medium

$$q = \frac{Q}{A} = -\frac{\rho g n \overline{R}^2}{8\mu}\frac{dH}{dx} \tag{2.3.6}$$

where $\overline{R}$ denotes an average pore radius for the porous medium. Here, notice several aspects of equation (2.3.6):

- Specific discharge is "linearly" proportional to the total head gradient, dH/dx, and the constant of proportionality, $\rho g(n\,\overline{R}^2)/8\mu$, is called "conductivity."
- The conductivity depends on both fluid specific weight ($\gamma = \rho g$) and viscosity (μ) and the geometry of the fluid path, $(n\,\overline{R}^2)/8$.
- The driving force consists of two components:
  a. The component due to gravity, $\rho g dz/dx$, which exists only in the vertical direction.
  b. The component due to pressure head gradient, dp/dx.

From a comparison of equation (2.3.6) with equation (2.2.11) for a single capillary tube, it is apparent that upscaling the flow principle from one tube to an idealized porous medium yields an additional property (porosity) of the porous medium and illustrates the average nature of the pore radius in Poiseuille's law.

2.3.2 Tortuosity

While the idealized medium (a bundle of straight capillary tubes) contains many of the same characteristics of a porous medium, it lacks the tortuous ("crooked") flow paths of a porous medium Figure 2.5. Therefore, we subsequently need to introduce a correction factor (tortuosity) to Poiseuille's law to make it a more realistic representation of the tortuous flow through an actual porous medium.

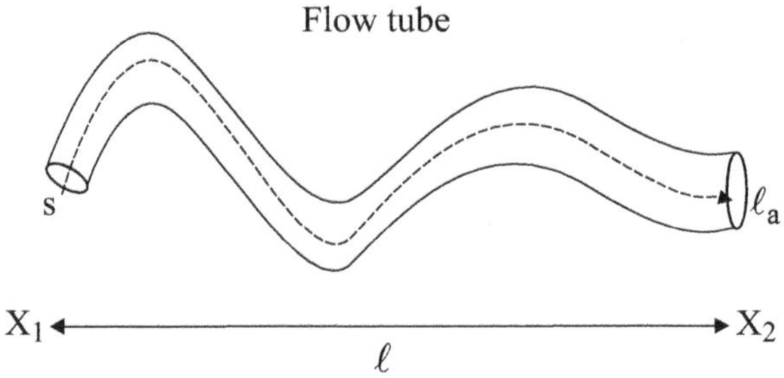

Figure 2.5. An illustration of the tortuosity concept. ℓ = straight line distance between X_1 and X_2; ℓ_a = actual distance between X_1 and X_2 along the flow path (streamline) described by a curvilinear coordinate system; X = Cartesian coordinates; S = curved streamline.

Note that while tortuosity is a term typically used to describe the tortuous connectivity of the pore space as it affects flow through a porous medium, tortuosity has no universal definition. One of the earlier definitions is by Carman (1937), who defines tortuosity (T) as the square of the ratio of the effective flow length (ℓ_a) via the pore passages to the straight path length (ℓ) in the direction of flow (see Figure 2.5), that is,

$$T = \left(\frac{\ell_a}{\ell}\right)^2 > 1 \qquad (2.3.7)$$

Consider now the flow along the streamline, which can be described by

$$\mathbf{q_s} = -\rho g \frac{nR^2}{8\mu}\frac{\mathbf{dH}}{\mathbf{ds}} \qquad (2.3.8)$$

where the specific discharge and the head gradient are vectors (bold fonts denote vectors). This equation is not useful since (a) we always use the Cartesian coordinate system for our observations and (b) we do not and cannot measure the specific discharge and the total head gradient along a streamline. For practical purposes, equation (2.3.8) should be expressed in terms of the Cartesian coordinate system. This can be accomplished by projecting the vectors in equation (2.3.8) onto the Cartesian planes. Projection of the hydraulic head gradient onto the Cartesian x-axis leads to

$$\frac{dH}{ds} = \frac{dH}{dx}\frac{dx}{ds} = \frac{dH}{dx}\left(\frac{\ell}{\ell_a}\right) \qquad (2.3.9)$$

where ℓ/ℓ_a is the "average" cosine of the angle of q_s with respect to the direction x. The average here implies that it is a mean value of the angle over the entire streamline. Similarly, the specific discharge component along the x-axis can be related to q_s.

$$q_x = q_s \frac{dx}{ds} = q_s \left(\frac{\ell}{\ell_a} \right) \tag{2.3.10}$$

Combining equations (2.3.8) and (2.3.9), we obtain

$$q_x \left(\frac{\ell_a}{\ell} \right) = -\rho g \frac{nR^2}{8\mu} \left(\frac{\ell}{\ell_a} \right) \frac{dH}{dx} \tag{2.3.11}$$

and

$$q_x = -\rho g \frac{nR^2}{8\mu T} \frac{dH}{dx}, \tag{2.3.12}$$

where $T = (\ell_a/\ell)^2$ is the tortuosity, which is always greater than or equal to 1. Equation (2.3.11) thus represents the projection of equation (2.3.8) onto the x-axis. Notice that the preceding analysis of tortuosity is merely conceptual. That is, it is practically impossible to analyze the tortuosity by using this approach or by measuring the actual length of a streamline. Nonetheless, through the tortuosity concept, we have developed an approach to incorporate effects of the tortuous flow path into the equation for flow in one straight tube. For a bundle of tortuous capillary tubes, after the application of a spatial averaging procedure similar to that in the previous section, equation (2.3.12) takes the following form:

$$q_x = -\rho g \frac{n\overline{R}^2}{8\mu \overline{T}} \frac{dH}{dx} \tag{2.3.13}$$

where $\overline{T}$ is the average tortuosity. Equation (2.3.13) is unequivocally more realistic than equation (2.3.6). According to equation (2.3.13), the specific discharge is again linearly proportional to the gradient. The constant of proportionality now can formally be called the hydraulic conductivity of a porous medium, that is,

$$K = \rho g \frac{n\overline{R}^2}{8\mu \overline{T}} \tag{2.3.14}$$

Thus, hydraulic conductivity depends on the tortuosity in addition to the other fluid and medium variables. According to this equation, a porous medium with higher tortuosity will have a lower hydraulic conductivity than for a medium with the same porosity and pore radius but smaller tortuosity.

Meanwhile, it is logical to speculate that the tortuosity of a porous medium has directional dependence. In other words, flow in a porous medium in a certain direction may have higher tortuosity than flow in other directions. This speculation is based on our analogy of flow through porous media to walking through a parking lot full of cars. As you may realize, walking through a parking lot along some specific direction is much easier than walking in other directions. This fact also suggests that the hydraulic conductivity may vary with the flow direction (see Section 2.7 on hydraulic conductivity anisotropy).

Questions:

1. Why does a well-sorted geologic material have a higher hydraulic conductivity value than a poorly sorted porous medium?
2. Why can a clayey material with a large porosity value be less conductive than a sandy material with a small value of porosity?
3. Why might the tortuosity be directionally dependent? How would it affect the hydraulic conductivity?
4. Can you design an experiment to measure the tortuosity in a core sample?

2.4 Darcy's Law for Saturated Porous Media

Darcy's law is an empirically derived constitutive relationship that describes the fluid flow through a porous medium. The law is named after Henry Darcy, a French hydraulic engineer, who performed a series of experiments on the flow of water through sand columns. Starting in late 1855, Darcy supervised a succession of experiments with the objective of determining the relationship between the volumetric flow rate of water through sand columns and hydraulic head losses. All experiments were completed in the courtyard of an unnamed hospital in Dijon, France.

Before we discuss Darcy's experiments, let us revisit some of the basic fluid mechanics concepts, namely, pressure head, elevation head, and total head, using the soil column experiment shown in Figure 2.6. Suppose that the column contains uniformly packed sand that is fully water saturated and the water flow in the column is steady – the flow rate does not vary with time. In order to know the pressure head (h), elevation head (z), and total head (H) distributions along the sand column, we have installed many water manometers along the column length. We then plot the distribution of pressure, elevation, and total head along the column length as a function of elevation (z). If we specify a datum as shown in Figure 2.6, then the distance between the water level rise in the manometer tube and the datum represents the ***total head***. The ***elevation head*** is the distance between the datum and the elevation of the center of the manometer connection with the soil column. Furthermore, the ***pressure head*** is the height of the water level measured from the center of the manometer connection on the column to the water level rise in the manometer tube. Under fully saturated conditions, all pressure heads are positive values. According to Figure 2.6, the pressure and the total head decrease as the elevation decreases. In order to explain this trend, recall Bernoulli's law: Total energy/unit

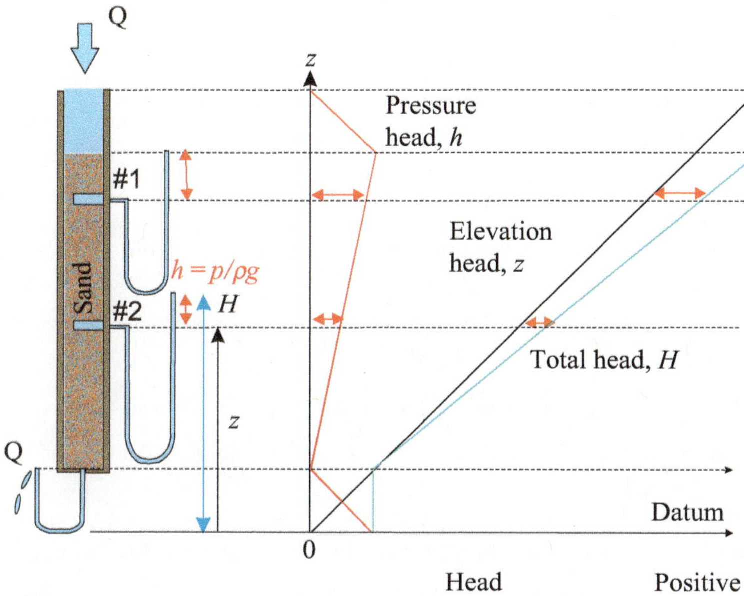

Figure 2.6. Pressure, elevation, and total head distributions in a saturated soil column under a steady flow condition.

weight (γ) at point 1 must be equal to the total energy/unit weight at point 2 in the fluid. Mathematically, it is

$$\frac{V_1^2}{2g} + \frac{P_1}{\gamma} + z_1 = \frac{V_2^2}{2g} + \frac{P_2}{\gamma} + z_2 + \Delta h \tag{2.4.1}$$

Since $(V^2/2g) << P/\gamma + z$ for flow through a porous medium, the kinetic energy terms can be omitted in the equation. Thus, we have

$$\frac{P_1}{\gamma} + z_1 = \frac{P_2}{\gamma} + z_2 + \Delta h \tag{2.4.2}$$

where

$$\Delta h = \text{headloss} = \left(\frac{P_1}{\gamma} + z_1\right) - \left(\frac{P_2}{\gamma} + z_2\right) = h_L \tag{2.4.3}$$

Therefore, the observed decrease in the total head is called head loss (h_L) or the potential loss within the sand column – energy being lost by frictional resistance due to fluid viscosity, small opening of pores, and tortuous pathways. Note that the head loss is independent of the inclination of the sand column. That is, the head loss is manifested in the pressure head difference between two points along the fluid flow path.

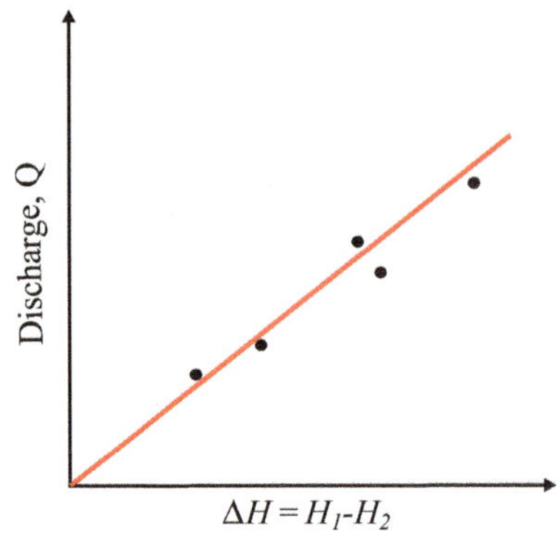

Figure 2.7. Plot of the relation between discharge and head loss in Darcy's experiments.

Besides the head loss, you should ponder why the pressure and the total head vary linearly in the sand column (see Figure 2.6). In addition, you should be curious about the hydrostatic spatial pressure head distribution in the reservoir. That is, if flow exists, why is the total head distribution in the water reservoir constant? Can you try answering these questions using Bernoulli's equation?

Next, using this soil column experiment, we can impose several different heads on the top and bottom of the soil column and obtain different discharge values. If we plot the flow or discharge, Q, versus the hydraulic head difference (head loss), ΔH, we will likely find that data points are scattered along a general linear trend as shown in Figure 2.7. This linear relationship indicates that Q is linearly proportional to the head difference. Suppose we repeat the experiment by varying the length of the sand sample, ℓ, and the size of the sample (i.e., the cross-sectional area of the column, A). Again, we would observe that the discharge is linearly proportional to the cross-sectional area of the column and inversely proportional to the column length.

On the basis of the results of these experiments, we can conclude, as Mr. Henry Darcy did, the following:

- The volumetric flow rate is proportional to hydraulic head difference: $Q \propto \Delta H$
- The volumetric flow rate is inversely proportional to the distance, ℓ, over which the flow occurs: $Q \propto 1/\ell$
- The volumetric flow is proportional to the cross-sectional area, A, contributing to flow: $Q \propto A$

Combining the preceding proportionalities, we obtain

$$Q \propto \frac{\Delta H}{\ell} A, \qquad \Delta H = H_1 - H_2 \tag{2.4.4}$$

Equation (2.4.4) can be expressed as a linear equation with a constant of proportionality (K) as

$$Q = -KA\frac{(H_1 - H_2)}{\ell} \quad \text{or} \quad Q = KA\frac{(H_2 - H_1)}{\ell} \tag{2.4.5}$$

where K is a constant of proportionality and the minus sign indicates that water flows in a down-gradient direction (i.e., from point 1 to point 2). This equation is known as **Darcy's law,** which is basically an empirical equation (i.e., obtained experimentally or as a result of a regression analysis of numerous experimental data) without a rigorous mathematical basis.

The equation can also be written as

$$q = \frac{Q}{A} = -K\left(\frac{H_1 - H_2}{\ell}\right) = -K\frac{dH}{d\ell} \tag{2.4.6}$$

where q is the Darcy velocity or specific discharge, or more simply, flux [L/T]. Since A is the total area that constitutes both solid and void spaces, while water flows only through a fraction of the total cross-sectional area, the actual average flow velocity of water in the pore space, often called **average linear velocity or pore velocity**, v, is then given by

$$v = \frac{q}{n} \tag{2.4.7}$$

where n is the porosity. Notice that the average linear velocity is an averaged water velocity over all pores in the porous medium, not the velocity of a water molecule or the water velocity within a single pore. Additionally, porosity is the ratio of void or pore volume to bulk sample volume of a porous medium. Porosity is a static property; that means that it can be measured in the absence of any fluid flow. For most naturally occurring media the porosity ranges from 0.10 to 0.40, although occasionally values outside this range have also been reported. Many times porosity is reported as a percentage, but it should always be used in calculations as a fraction. From these typical values, it is quite evident that the rock "phase" clearly occupies the largest volume in any medium. Furthermore, the pore space as well as the porosity can be divided into an interconnected or effective porosity that is available to fluid flow and a disconnected porosity that is unavailable to fluid flow. Lastly, notice that the porosity is defined over a volume that is substantially larger than a pore in that porous medium.

After all, Darcy's law simply states that specific discharge in a porous medium is linearly proportional to the magnitude of the head gradient and is in the direction of decreasing the hydraulic head.

2.5 Control Volume, Darcian Continua, and Representative Elementary Volume

2.5.1 Control Volume and Representative Elementary Volume

Figure 2.8 shows an expanded areal view of a porous medium at the pore scale. As suggested by Figure 2.8, fluid flow in a porous medium takes place through a

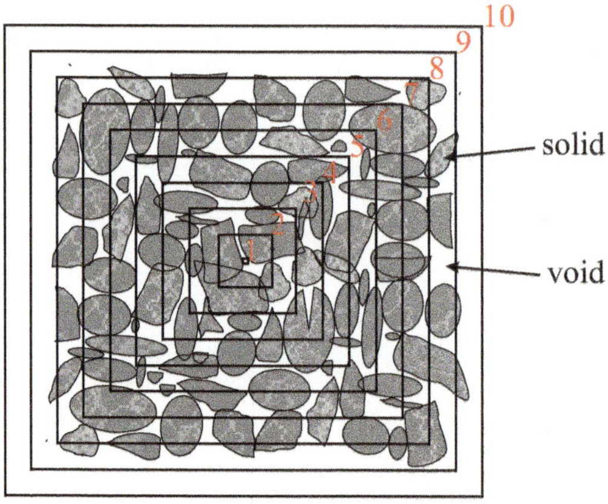

Figure 2.8. An expanded areal view of a porous medium illustrating the concept of REV for porosity (open spaces are voids and filled spaces are solids). The ten squares represent the ten CVs of different sizes.

complex network of interconnecting pores or openings. Describing such an intricate network in any exact mathematical manner is not practically feasible or desirable. Rather than describing the governing fluid flow at the pore-scale level (e.g., via the Navier-Stokes equation), we resort to the so-called Darcian continuum approach – a spatial averaging approach. Darcy's law, which we described earlier, only considers the average flow behavior over a certain volume of a porous medium that encompasses numerous pores. The volume over which the flow is averaged is then referred to as a **control volume** (CV). Using the CV approach, Darcy's law essentially bypasses both the microscopic level – the scale at which we consider what happens to each fluid molecule – and the pore level – the scale at which we consider the flow pattern within each pore and between pores. Our observations of flow in a porous medium are focused on a macroscopic level – the scale in which only average phenomena over the CV are considered. Darcy's law is therefore a macroscopic law. Any flow property defined at a point in our conceptualized porous media based on Darcy's law thus represents some averaged property over a CV; the property at every point in our conceptualized porous medium therefore varies continuously and smoothly in space such that the principles of continuous differential calculus are valid and applicable. The porous medium and the flow itself are subsequently treated as **Darcian continua**. Such a continuum concept is analogous to the continuum concept applied in fluid mechanics and other branches of science.

To illustrate the concept of the continuum, let us use the measurement of porosity of a uniformly packed sand column as an example. Suppose we select an infinitesimally small point as our CV (area 1 in Figure 2.8) to define the porosity at a given location of the soil column. The porosity value will likely have a value of either 1 (void) or 0 (solid), depending on whether the CV is located on the solid or within the pore space. As a result, the distribution of porosity along the column behaves in a

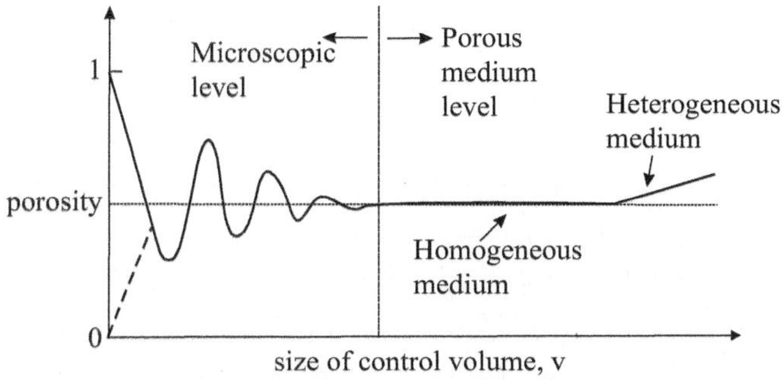

Figure 2.9. An illustration of the definition of REV using porosity as a function of volume. (after Bear, 1972)

discrete and erratic fashion, and a continuous mathematical function cannot be employed to describe its spatial variation. As the size of the CV increases to encompass several pores, the erratic behavior lessens. Since the spatial distribution of porosity becomes relatively smooth in space, a continuous mathematical function thus is suitable for representing its spatial distribution. In term of the porosity, the sand column can be treated as a Darcian continuum.

As the size of the CV increases further to a certain size, the porosity value becomes constant everywhere along the column; this porosity value defined over this volume thus is the representative porosity value everywhere in the column, except near the ends and the walls of the column. The CV of this size is then defined as a REV Bear (1972). The word "elementary" means that the size of the CV must be much smaller than the column, and it can also be thought of as a building element of the sand column. Therefore, a precise and rigorous definition of a REV should be a CV whose volume averaged property is spatially or translation invariant. This means that the averaged property over the REV must be representative of the averaged property of the entire domain (e.g., the soil column), and the REV must be much smaller than the entire domain such that the averaged property is the same regardless of where this CV is located in the entire domain, except near boundaries of the domain. We call such a REV a **spatial REV**.

The preceding definition for a REV is the same as that defined by Bear (1972). Using Figure 2.8 as an example, if we select a CV that is centered at Point 1 and greater than square 8 to define the porosity, the porosity value will be influenced by areas (heterogeneity) beyond the sample. As the size of the CV decreases, the effect of the boundary decreases (see Figure 2.9). The porosity value becomes constant as the CV size continuously decreases. At a certain size, the porosity value begins to fluctuate, as shown in Figure 2.9. The CV of this particular size is called REV. If the CV size shrinks to a mathematical point (e.g., Point 1), the porosity value can be either 1 or 0, depending on the location where the point is located.

If we apply this CV approach to a property of a porous medium, we may classify the medium as a heterogeneous or homogeneous medium in terms of that property.

If the size of the CV is small (e.g., a volume containing five grains of sand), the property is likely to vary spatially. The medium is thus called **heterogeneous**. If progressively larger CVs are used, we will find that values of the property tend toward a constant value everywhere in the medium. The smallest volume that leads to the constant value is then defined as REV, and the medium is classified as a **homogeneous** medium with respect to that property. This also means that a spatial REV exists for this medium.

This REV concept is analogous to the minimum sample volume required in any statistical analysis. In statistical analysis, a sample volume must be large enough so that the statistics of samples in the volume represents a given population. If the statistics of the minimum sample volume at a different portion of the population are the same, then the population is called homogeneous and heterogeneous otherwise. For instance, if a sand column is uniformly packed with medium sand (grain size ranging from 0.25 mm to 0.5 mm), a sample volume of 1 cm^3 (many grains) of the sand will likely be sufficient to define a mean and variance of the grain size over the entire column or a spatial REV. Note that the size of the REV is scale-dependent. For example, one may define a REV for a laboratory-scale problem, but this REV may not be applicable to field-scale problems, where heterogeneity at different scales coexists.

Lastly, a CV that encompasses the entire soil column always possesses the representative porosity value of the column. If this CV is considered as an REV, the requirement of being elementary is only met in the ensemble sense (i.e., an imaginary space, see Section 4.4 in Chapter 4). That is, the statistical distribution (i.e., mean, variance, etc.) of heterogeneity within this CV is the same as that of other possible soil columns with the same size and volume. If one adopts such an ensemble REV concept, any heterogeneous medium can always be considered as a homogeneous one. This REV, which is different from the spatial REV, is named **ensemble REV**.

2.5.2 REV Based on Flow Behaviors

It is important to recognize that both spatial and ensemble REV concepts define the homogeneity of a medium property solely on the basis of a spatial or ensemble averaging rule without accounting for flow (e.g., the fraction of total voids in a porous medium is the way we measure porosity without considering whether flow occurs in some voids or not). An appropriate definition of REV should consider the flow process. For example, if porosity is defined as the fraction of voids where flow is active, this porosity will be different from porosity that is defined as the fraction of total voids in a porous medium. Accordingly, the process-based REV will be different from that based on our spatial or ensemble averaging procedure.

Using the process-based definition of a REV, the existence of a REV for medium properties will be time-dependent. As an example, the process-based REV for hydraulic conductivity can only be defined when the REV requirement with respect to flow is met. That is to say, the flow process must take sufficient time to experience many pores of a variety of sizes and geometries in the medium. Only after sufficient time elapses does water in different parts of the medium move in a similar fashion under a given energy gradient. Another example is that the wetting front from a point source in a sand box must evolve to a certain size before the REV concept can be applied. This time-dependent phenomenon is nonetheless obscured by the fact that in a homogeneous soil column at laboratory scales, the size of heterogeneity is on the order of grain diameters and the time for flow to reach the REV is insignificant, compared with the timescale of our experiments. Therefore, the time-dependent phenomenon is often neglected.

2.5.3 Scale of Measurement (Scale Interest or Objective)

Because of the CV (or REV) embedded in Darcian continuum assumption upon which hydraulic properties (such as porosity, hydraulic conductivity, and specific storage, see section 5.4 in Chapter 5) are defined and flow processes (such as head and specific discharge) are calculated, the consistency between the observation scale of the flow process and the size of the CV must be emphasized. Specifically, if the spatial REV exists for a given medium, the measurement scale of the flow process (e.g., head) must be as large as the size of this REV such that the measurement samples the same representative pore-size distribution as does the REV. This consistency is a necessary condition for agreement between the observed and the predicted process, which is derived on the basis of properties defined over the REV. This requirement is also necessary for the situations in which the spatial REV does not exist and the ensemble REV is invoked. In these cases, the properties that are defined over the REV represent the ensemble average properties, which predict ensemble average flow processes. If the sampling space is large, the spatial average (observed) value may be close to the ensemble average one. Otherwise, the deviation between the observed and the predicted could be large although overall the predicted is unbiased.

2.6 Validity of Darcy's Law

2.6.1 Lower Limit

Darcy's law is one of the most widely used and fundamental laws in subsurface hydrology. However, Darcy's law is not universal and it is valid only for some groundwater flow regimes. In other words, it has a lower and an upper limit. Therefore, it is important to realize the limits of our assumptions, including the validity of Darcy's law, to know when they are not reasonable assumptions. First, let us examine the lower limit of Darcy's law.

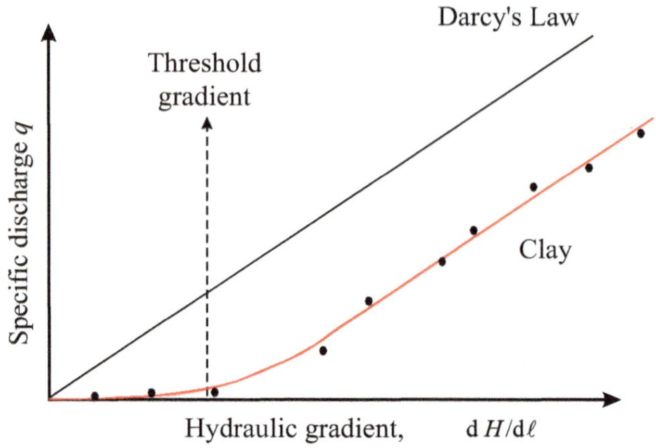

Figure 2.10. A schematic illustration of the lower limit of Darcy's law.

Darcy's law does not hold at very low head gradients in some fine-textured materials such as compacted clays. According to Darcy's law, any hydraulic head gradient, no matter how small, will cause water movement. On the contrary, many experiments have shown that the hydraulic head gradient must exceed some threshold value dh/dl, in order to cause water movement in clayey soils (Figure 2.10). The failure of Darcy's law in this case can be attributed to the fact that important electrostatic forces exist in clayey soils, which are due to electrical charge imbalances within the mineral structure of the soil. At low hydraulic head gradients, these electrostatic forces (or internal energies) cannot be neglected in the derivation of Poiseuille's law. Nonetheless, neglecting the lower limit does not have serious impact on analysis of most groundwater flow problems. At any rate, flow regimes that do not satisfy Darcy's law (i.e., the linear relationship between q and gradient) are called **non-Darcian flow**.

2.6.2 The Reynolds Number for Groundwater Flow

Before proceeding to the discussion of the upper limit of Darcy's law, let us introduce the Reynolds number for groundwater flow scenarios. In Chapter 1, we use the Reynolds number as a criterion for classifying flow in pipes as being either laminar or turbulent. Likewise, as shown in the following, a Reynolds number can be defined to categorize groundwater flow through porous media.

$$Re = \frac{q\rho d}{\mu} = \frac{qd}{\nu} \propto \frac{\text{inertial force}}{\text{viscous force}} \tag{2.6.1}$$

where q is the specific discharge; μ is the dynamic viscosity and ν is the kinematic viscosity of the fluid; d is a characteristic length, generally considered as the mean grain size, or d_{10} (i.e., the grain size for which 10 percent of the sample by weight is

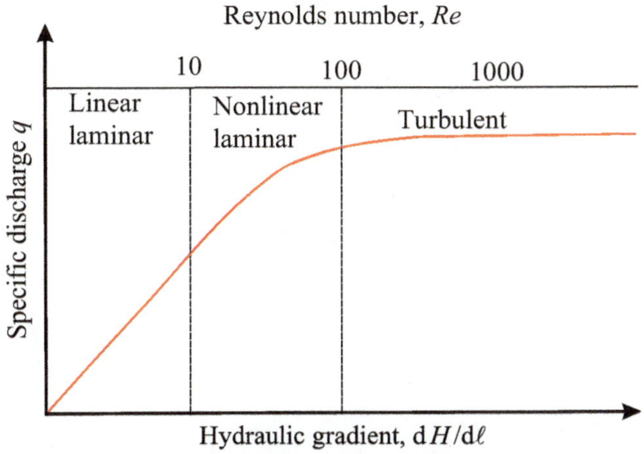

Figure 2.11. Relationships among specific discharge, the Reynolds number, and the hydraulic gradient.

finer). The term d_{50} is also sometimes used in the literature as a representative grain diameter or the characteristic length. For flow in porous media, if the Reynolds number is large (Section 2.6.3), the flow is turbulent; otherwise, it is laminar. As was mentioned previously, most groundwater flows in porous media are considered laminar because, in comparison to inertial forces, viscous forces dominate and, unlike flow in closed conduits or pipes, flow in porous media must travel through small pores and tortuous paths. Nonetheless, as described later, there are exceptions.

2.6.3 The Upper Limit

Figure 2.11 illustrates the general relationship between the specific discharge and the hydraulic head gradient as well as Reynolds number. According to Darcy's law, the specific discharge q is linearly proportional to the hydraulic head gradient. Figure 2.11 shows that such a linear relationship only exists when the hydraulic gradient is relatively small or the Reynolds number is smaller than 10. Flow behaviors deviating from the linear relationship (i.e., Darcian behavior), however, have been observed at large dh/dl values (i.e., at $10 \leq Re < 100$). Under these conditions, the relation between q and dh/dl is no longer linear, although the flow can still be in the laminar flow regime (or in a transition regime). At $Re >> 100$, turbulent flow with streamlines crossing each other takes place. Under the turbulent flow regime, q is no longer correlated with dh/dl and flow exhibits chaotic behavior (random motion). Under this scenario, describing flow behavior quantitatively becomes difficult.

The reason that flow behavior deviates from that described by Darcy's law is explained as follows: When the hydraulic gradient is large, the specific discharge is

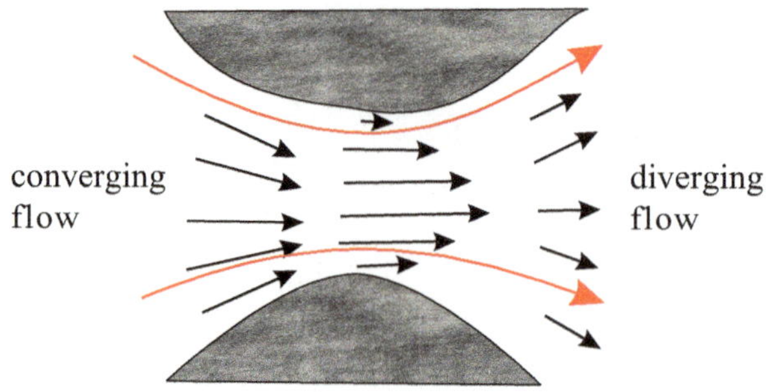

Figure 2.12. Velocity variation within a pore and converging and diverging flow.

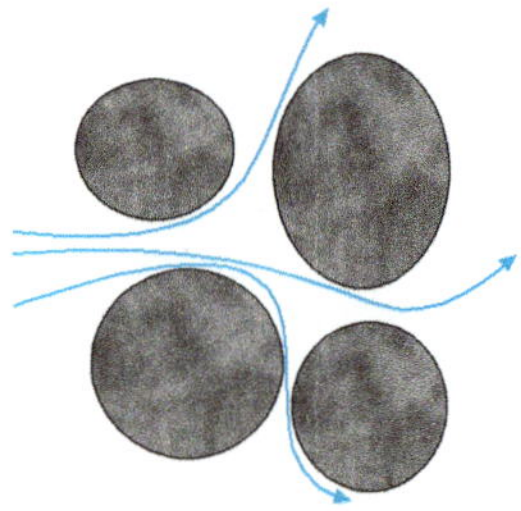

Figure 2.13. Variations in flow velocity among different pores.

large accordingly. Under this condition, equation (2.6.1) yields a large Reynolds number, indicating that the inertial forces are greater than the viscous force. Recall that during the formulation of Poiseuille's law (the surrogate of Darcy's law), we specifically ignored the inertial force that causes fluid acceleration. Therefore, ignoring the fluid acceleration in Darcy's law is the main cause of the flow deviation from that expected from Darcy's law.

The force balance from our development of Poiseuille's law is derived on a CV basis, which is moving with the flow. The acceleration is thus defined according to the substantial derivative

$$a = \frac{Dq}{Dt} = \frac{\partial q}{\partial t} + q\frac{\partial q}{\partial x} \tag{2.6.2}$$

Equation (2.6.2) consists of a local acceleration term, $\partial q/\partial t$, and a convective acceleration, $\partial q/\partial x$. Since Darcy's law is based on steady-state analysis, the local acceleration term is always 0. However, for a high Reynolds number, the convective acceleration in equation (2.6.2) becomes significant and it cannot be ignored anymore. This implies that changes in flow velocity (acceleration or deceleration) caused by variations in pore radius (see Figure 2.12) as well as in flow paths (see Figure 2.13) are significant at large flow rates. Consideration of the importance of the convective acceleration term then leads to Forcheimer's (1914–16) empirical formula:

$$J = \frac{q}{K} + bq^2 \tag{2.6.3}$$

where J is the hydraulic head gradient, K is the hydraulic conductivity, and b is a constant, depending on the porosity, average grain diameter, and shape factor. However, it is independent of the fluid viscosity. Equation (2.6.3) yields a nonlinear relationship between J and q. The first term of the equation is linear (i.e., Darcy's law) but the second term is nonlinear. This equation has been proposed to describe the non-Darcian laminar flow at $10 < Re < 100$.

Under unsteady flow regimes, the fluid acceleration due to both the local and the convective acceleration can become important. Therefore, they must be considered in the force balance analysis:

$$F_P + F_\tau = ma = m\left(\frac{\partial q}{\partial t} + q\frac{\partial q}{\partial x}\right) \tag{2.6.4}$$

where F_P and F_τ denote forces due to pressure difference and the viscous forces. A modified Darcy's law has been proposed to include these effects. That is, (Polubarinova-Kočina, 1952):

$$J = \frac{q}{K} + bq^2 + c\frac{\partial q}{\partial t} \tag{2.6.5}$$

where b and c are empirical constants. In general, the effect of the term involving $\partial q/\partial t$ is significant only during a fraction of a second. For all practical purposes, the time-dependent Darcy's law has often been ignored.

Questions:

1. Is the Darcy law valid for flow through a single fracture with large apertures? Why or why not? If not, can you make any suggestions to improve the law?
2. Do you think that classical Darcy's law is valid for flow in many coastal aquifers made of coral reefs? Why or why not?
3. Can you use Darcy's law to describe groundwater movement during an earthquake event?

2.7 Hydraulic Conductivity

Now, if we carry out many experiments similar to those by Mr. Darcy, but we only vary the grain size of the material, we will obtain results as shown in Figure 2.14.

Figure 2.14 shows that coarse-textured sand has a large slope for the linear relationship between q and *dH/dl*, and therefore a large value for hydraulic conductivity or K, the proportionality constant between specific discharge and hydraulic gradient or equation (2.4.6). On the other hand, the fine-textured sand has a small slope and therefore a small K value. These results suggest that a) the slope

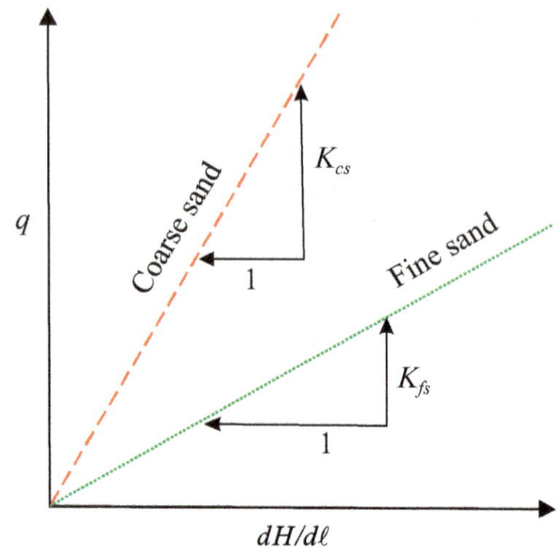

Figure 2.14. Schematic illustration of the relationship between the Darcian flux, q, and hydraulic gradient, dH/dl, for a coarse sand and a fine sand.

represents the fluid volume that a porous medium can transmit over a unit cross-sectional area under a unit hydraulic gradient condition per unit time; b) the slope reflects properties of the medium and fluid together. Therefore, the slope (or K) is usually called the hydraulic conductivity of a porous medium. These results apparently are in agreement with our previously established mathematical model for flow through porous media:

$$q_x = -\frac{n\overline{R}^2 \rho g}{8\mu\overline{T}} \frac{dH}{dx} \tag{2.7.1}$$

Comparing equation (2.7.1) with Darcy's law

$$q = -K\frac{dH}{dx} \tag{2.7.2}$$

yields

$$K = \frac{\rho g n\overline{R}^2}{8\mu\overline{T}} \tag{2.7.3}$$

Equation (2.7.3) suggests that the saturated hydraulic conductivity K is

- a function of porous media properties; that is, K is directly proportional to medium porosity, n, the second power of the average pore radius, $\overline{R}^2$, and inversely proportional to the mean tortuosity, $\overline{T}$. As we postulated in Section 2.3, the tortuosity can have preference in direction. Therefore, the hydraulic conductivity, in turn, may have directional dependence.
- a function of fluid properties; that is, K is directly proportional to fluid density (ρ) and inversely proportional to fluid dynamic viscosity (μ).

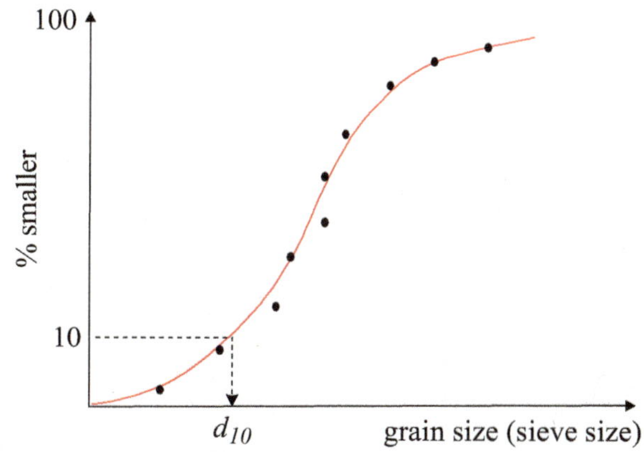

Figure 2.15. A plot of percentage of greater grain size distribution for different grain sizes. Note that d_{10} is also shown explicitly in the plot.

Note that the porosity, radius, and tortuosity in equation (2.7.3) are some type of average property of each attribute. However, qualitatively we expect that the grain size distribution (also referred as the particle-size distribution) of a porous medium gives us an indication of the average pore radius, $\overline{R}$, in equation (2.7.3). For example, we expect that a gravelly medium will have a large average pore diameter and a silty medium will have a small average pore diameter. Using this reasoning, we would anticipate a quantitative relationship between the hydraulic conductivity and an "average" grain size (or an "average" particle size). Indeed an empirical relation is frequently cited in the literature (Bear, 1972); it can be expressed as follows:

$$K = (Nd^2)\left(\frac{\rho g}{\mu}\right) = (Nd^2)\left(\frac{\gamma}{\mu}\right) \tag{2.7.4}$$

where d is the mean grain diameter [L]; $\gamma = \rho g$, which is the unit weight or weight per volume of the fluid $[M/(L^2T^2)]$; μ is the dynamic viscosity of the fluid $[M/(LT)]$; and N is a dimensionless correction factor, which is used to include effects of porosity, tortuosity, and so on. Thus, hydraulic conductivity, K, with a dimension of [L/T], depends on the properties of a porous medium and fluid properties. Thus, Darcy's law can be rewritten as

$$q = -\frac{k\rho g}{\mu}\frac{dH}{d\ell} \tag{2.7.5}$$

where $k = Nd^2$ and is called the ***intrinsic permeability***, which depends on a porous medium only and has the dimension $[L^2]$. Normally, d is assumed equal to d_{10}, which is the grain diameter such that 10 percent (by weight) of the porous matrix consists of grains smaller than it (see Figure 2.15). The parameter d_{10} is generally determined from a sieve analysis.

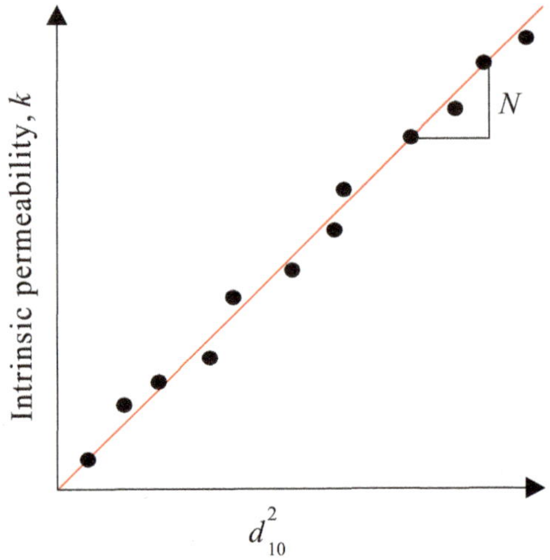

Figure 2.16. A plot showing the relationship between the intrinsic permeability and d_{10}^2.

In general, the intrinsic permeability, k, and $(d_{10})^2$ are well correlated (see Figure 2.16). If k is in square centimeters and d in centimeters, generally, N is approximately 45 for clayey material, and 100 for pure sand. Since N is related to tortuosity and porosity, soils with the same d_{10} value may have different tortuosity as a result of different grain size distributions and packing. Therefore, N may vary within the same type of soil or rock. This variation is problematic since even if we know the d_{10} value, we still do not know which value for N should be used. This implies that we still have to conduct experiments to measure hydraulic conductivity first.

The intrinsic permeability has been widely employed in the petroleum engineering field since oil reservoir problems often involve analysis of flow of fluids of multiple phases in porous media. The independence of intrinsic permeability from fluid properties allows applications to flow problems with different fluids. In petroleum engineering, the darcy is customarily used as the unit for the intrinsic permeability; it is defined as

$$1 \text{ darcy} = k = \frac{\mu(Q/A)}{dp/d\ell} = \frac{1 \text{ centipoise} \times \left(\frac{1 \text{ cm}^3/\text{sec}}{1 \text{ cm}^2}\right)}{1 \text{ atm/cm}}$$

where p is pressure. Specifically,

$$1 \text{ darcy} = 0.987 \times 10^{-8} \text{ cm}^2 = 1.062 \times 10^{-11} \text{ ft}^2$$

One darcy corresponds to 9.613×10^{-4} cm/sec in terms of hydraulic conductivity if water is at 20°C.

As we pointed out earlier, the intrinsic permeability is a function of grain size or porosity, as well as tortuosity, which may vary with the pore-size distribution,

Table 2.1. *Grain size values of geologic materials*

Name		**Size Range (mm)**
Large boulder		>630
Boulder		200–630
Cobble		63–200
Gravel	Coarse gravel	20–63
	Medium gravel	6.3–20
	Fine gravel	2.0–6.3
Sand	Coarse sand	0.63–2.0
	Medium sand	0. 2–0.63
	Fine sand	0.063–0.2
Silt	Coarse silt	0.02–0.063
	Medium silt	0.0063–0.02
	Fine silt	0.002–0.0063
Clay		≤0.002

packing, and other factors. Nevertheless, for many practical purposes an explicit relationship between the intrinsic permeability and grain size, and so forth, is useful. For example, the best-known formula (the Kozeny-Carman equation) is

$$k = [n^3/(1-n)^2]/5M^2 \tag{2.7.6}$$

where n is the porosity and M is the surface area per unit volume of the solid exposed to the fluid flow. If M can be related to some mean particle size, d_m, (e.g., $M = 6/d_m$), one can rewrite the Kozeny-Carman equation (Bear, 1972) as

$$k = \frac{{d_m}^2}{180} \frac{n^3}{(1-n)^2} \tag{2.7.7}$$

Nevertheless, suppose this formula is true and one can indirectly estimate the porosity distribution in a field site using a geophysical survey, and we may be able to estimate the mean particle size and then the intrinsic permeability distribution estimated. While this approach may not lead to accurate estimates as compared to direct permeability measurements, it will reveal the heterogeneity pattern at the field site, which can be useful.

Tables 2.1 and 2.2 list the range of grain size and porosity values for some geologic materials, while Table 2.3 tabulates possible ranges of saturated hydraulic conductivity values for different geologic materials.

2.7.1 Pore-Scale Anisotropy and Isotropy

If the hydraulic conductivity defined over a CV of a geologic medium is higher for flow in one direction than in other directions, the hydraulic conductivity is

Table 2.2. *Range in values of porosity for geologic materials*

Material	Porosity (percentage)
Gravel, coarse	24–36
Gravel, fine	25–38
Sand, coarse	31–46
Sand, fine	26–53
Silt	34–61
Clay	34–60
Sandstone	5–30
Siltstone	21–41
Limestone, dolomite	0–40
Karst limestone	0–40
Shale	0–40

Table 2.3. *Range of saturated hydraulic conductivity of various geologic materials*

Soil Type	Saturated Hydraulic Conductivity, *K* (m/yr)
Unconsolidated Deposits	
Gravel	1×10^{4}–1×10^{7}
Clean sand	1×10^{2}–1×10^{5}
Silty sand	1×10^{1}–1×10^{4}
Silt, loess	1×10^{-2}–1×10^{2}
Glacial till	1×10^{-5}–1×10^{1}
Unweathered marine clay	1×10^{-5}–1×10^{-2}
Rocks	
Shale	1×10^{-6}–1×10^{-2}
Unfractured metamorphic and igneous rocks	1×10^{-7}–1×10^{-3}
Sandstone	1×10^{-3}–1×10^{1}
Limestone and dolomite	1×10^{-2}–1×10^{1}
Fractured metamorphic and igneous rocks	1×10^{-1}–1×10^{3}

Source: Modified from Freeze and Cherry (1979).

called **anisotropic.** On the other hand, an **isotropic** hydraulic conductivity means that the hydraulic conductivity of the medium within the CV is the same in all directions.

First of all, we must recognize that the anisotropy of the hydraulic conductivity of a porous medium is a by-product of the use of the CV (volume average) for defining the hydraulic conductivity, which homogenizes structured heterogeneities at scales smaller than the CV. Anisotropy therefore is a macroscopic property of a medium and can be scale-dependent. That is, the anisotropy may depend on the size of the CV and the scale and type of heterogeneity within the CV. In the following section, we examine the cause of the anisotropy at pore-scale CVs.

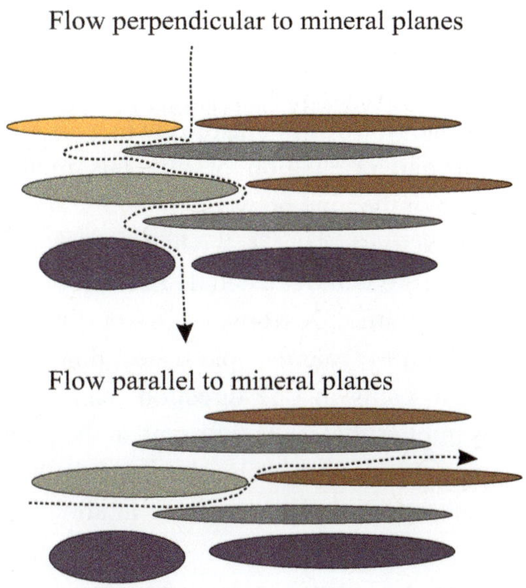

Figure 2.17. Flow perpendicular (top) to the mineral's longest axis and flow parallel (bottom) to the mineral's longest axis.

Examining a small volume of a geologic medium (e.g., a soil core) in detail, we will conclude that natural sedimentation processes and pressure of overlying materials often cause flat grains (or minerals) to orient with the longest dimension parallel to the plane on which they settle (Figure 2.17). This sedimentation process produces different flow channels parallel to the bedding plane from those oriented normal to this plane. The flow moving in the direction perpendicular to the longest dimension of minerals thus has a greater probability of being detoured, so that it can find alternative paths to flow through the medium. On the contrary, flow in the direction parallel to the longest dimension of the minerals is less likely to be detoured. In other words, the shape of minerals and their orientation can create greater tortuosity for flow moving in the direction perpendicular to the longest dimension than for flow moving in other directions. This means that the tortuosity will depend on flow direction or the tortuosity must be anisotropic or the tortuosity value changes with direction.

Recall the hydraulic conductivity model we presented in equation (2.3.12): The greater the tortuosity, the smaller the hydraulic conductivity. Therefore, the hydraulic conductivity in the direction parallel to the longest dimension of minerals should be greater than that in the direction perpendicular to it. As a result, the hydraulic conductivity of this soil core must be anisotropic. The hydraulic-conductivity anisotropy due to anisotropic tortuosity caused by the mineral orientation is referred to as the pore-scale anisotropy.

Conversely, if a porous medium consists of many rounded minerals of different sizes and packed randomly without any particular order, the tortuosity of flow paths in a direction is likely to be the same as that in other directions. The hydraulic conductivity is thus likely to be the same in all directions. In this case, the hydraulic conductivity of the porous medium is isotropic.

Accordingly, the pore-scale isotropy or anisotropy in hydraulic conductivity can be attributed to a) the shape and orientation of minerals in the soil core or, more precisely, the shape and size of the void between minerals, and b) the volume average nature of the hydraulic conductivity. That is, the hydraulic conductivity is defined for the entire volume of the soil core without a detailed description of tortuous flow paths at scales smaller than the volume. Isotropy or anisotropy is a macroscopic-scale property of a geologic medium. At microscopic scales, fluid particles merely migrate through more tortuous flow paths in one direction than in other directions in an anisotropic medium. This implies that if the flow path at the pore scale can be mapped in detail, and flow is simulated at this scale, the term or the concept of hydraulic conductivity isotropy or anisotropy is not necessary at all.

2.7.2 Hydraulic Conductivity for Fractures

A **fracture** (or fault) is a large planar void in a geologic formation such as a joint or fault. Fractures can provide either greater or lower permeability for fluids such as water or hydrocarbons as compared to the bulk soil or rock matrix. Laminar flow through a fracture (Figure 2.18) is often described by the cubic law, that is,

$$Q = -b\frac{e^3 \rho g}{12\mu}\frac{dH}{dx} = -be\frac{e^2 \rho g}{12\mu}\frac{dH}{dx} \tag{2.7.8}$$

where b is the width of the fracture perpendicular to the flow and e is the aperture opening of the fracture, both with dimensions of [L]. This equation can be derived by considering force balance in a CV similar to the procedure we used during the development of Poiseuille's law. Instead of flow through a capillary tube, we can consider flow through

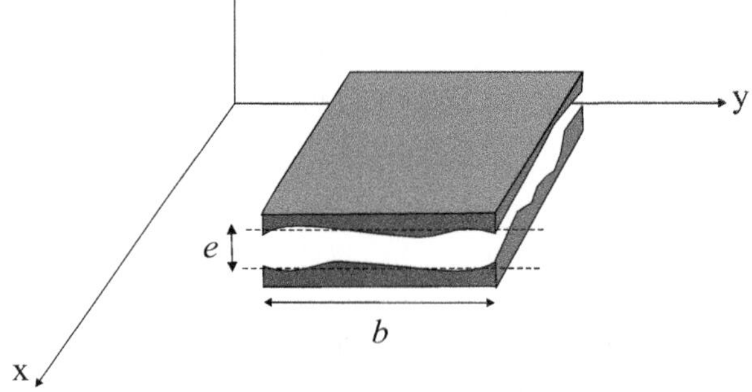

Figure 2.18. Illustration of an idealized fracture.

fractures as flow through space bounded by two parallel plates (Figure 2.18). Note that e is the average aperture opening, similar to the average pore radius used in Poiseuille's law. This cubic law has a similar form to Poiseuille's law or Darcy's law as well, showing that the discharge is linearly proportional to the hydraulic gradient.

Similar to porous media, hydraulic conductivity can be defined for fractures $K=e^2\rho g/12\mu$. However, sometimes it is more convenient to use the term: fracture transmissivity, instead, that is, the hydraulic conductivity multiplied by the average aperture opening, $T=e^3\rho g/12\mu$. The term: transmissivity, is used often because the fracture aperture is often too small to be accounted for explicitly. As a result, the flow in a fracture is generally treated as sheet flow. The usage of the transmissivity is also attributed to the fact that we are not interested in flow or pressure variation in the direction of the aperture. Of course, a fracture surface is not as smooth as the surface of a parallel plate, and the actual fracture aperture opening will vary along the fracture surface. How would these complications affect the cubic law?

Question:

Why does the cubic law have a linear relationship between gradient and discharge?

2.8 Laboratory Measurements of Saturated Hydraulic Conductivity

The constant of proportionality in Darcy's law can be determined in a laboratory by two different methods: (a) a constant (prescribed) head permeameter and (b) a falling head permeameter. The latter is used for fine-textured materials having a K value lower than 10^{-5} cm/sec whereas, coarse-textured materials are analyzed by using constant head permeameters.

2.8.1 Constant-Head Permeameter

The hydraulic conductivity of a porous medium is generally measured by using a constant head permeameter, an apparatus illustrated in Figure 2.19. Generally, it consists of a column containing the porous medium whose hydraulic conductivity is to be measured. If the soil consists of unconsolidated sediments, the medium is confined with an upper and a lower porous plate that prevents (a) upward movement of the medium and (b) flushing out of the soil material. Using various techniques (Klute et al., 1965) to maintain constant head conditions at the top and the bottom of the flow, the steady outflow rate through the soil column can be measured. That is,

$$Q = \frac{V}{t} \tag{2.8.1}$$

where V is the discharged volume of water collected from the column in a beaker at Δt. Then, using Darcy's law,

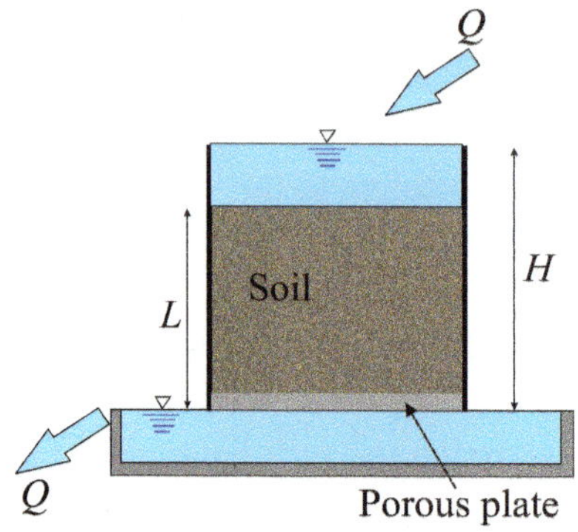

Figure 2.19. A schematic illustration of a constant head permeameter.

$$Q = KA\frac{H}{L} \tag{2.8.2}$$

to determine the hydraulic conductivity. That is,

$$K = \frac{QL}{AH} \tag{2.8.3}$$

While the apparatus and procedures are straightforward, the following precautions must be taken: a) use deaired water to prevent entrapping air in the system, b) fully saturate the sample from below as they are emplaced, and c) always take many different Q measurements under different gradients and then carry out a linear regression between Qs and gradients to obtain the slope that is the average hydraulic conductivity.

Questions:

1. How does a fine porous plate, used to support the porous medium in a soil column during a hydraulic conductivity test, affect your conductivity measurements?

2.8.2 Falling-Head Permeameter

An illustrative setup for a falling-head permeameter is shown in Figure 2.20. It consists of a soil column; a flexible tube is connected to the bottom of the column and the top of the column is confined by a porous plate.

Before the experiment is started, the soil and the tube are saturated with water. Initially, the water levels in the tube and the soil column are set to the same height by manipulating the flexible tube such that there is no flow in the system. Afterward, at any time t_0, the tube is raised so that the water level in the tube is higher than that in the soil column. Meanwhile, the difference in the water level of the tube and that at the top of

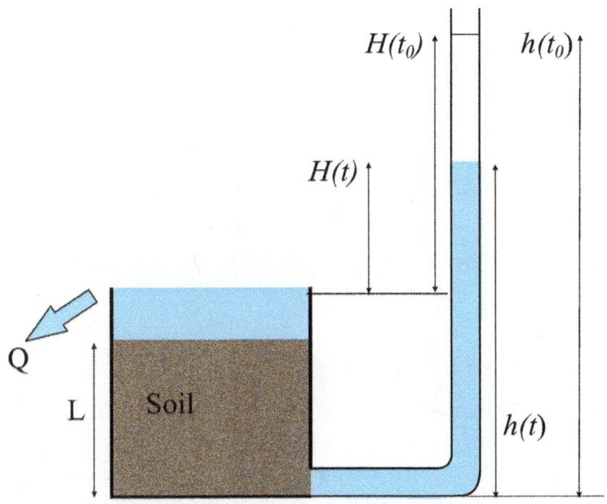

Figure 2.20. A schematic illustration of a falling-head permeameter.

the soil column is measured at several time intervals. Then, the differences in head at different times can be used to determine the saturated hydraulic conductivity of the soil. A mathematical formulation of the theory for this approach is given in the following.

Consider the system at a given time, t. The discharge at the top of the soil column is assumed to follow Darcy's law:

$$Q_{top} = KA\frac{(H_{bottom} - H_{top})}{L} \tag{2.8.4}$$

where $H_{top} = \frac{p_{top}}{\rho g} + z_{top} = L$ since $p_{top} = p_a = 0$ (p_a is the atmospheric pressure) and A is the cross-sectional area of the permeameter. If the datum is chosen at the bottom of the column, the total head at the bottom of the column is

$$H_{bottom} = \frac{p_{bottom}}{\rho g} + z_{bottom} = \frac{p_{bottom}}{\rho g} = h(t) \tag{2.8.5}$$

where $h(t)$ is the water level in the tube at any time t. Equation (2.8.4) thus becomes

$$Q_{top} = KA\left(\frac{h(t) - L}{L}\right) \tag{2.8.6}$$

Now consider the discharge from the manometer into the soil column at the bottom of the column. It is given by

$$Q_{bottom} = -a\frac{dh(t)}{dt} \tag{2.8.7}$$

The negative sign denotes that discharge decreases with time. Here we assume a linear relation between Q and $\frac{dh}{dt}$, where a is the cross-sectional area of the manometer. If storage effects of the column are neglected or an instantaneous outflow at the top of the column is assumed, we have

$$Q_{bottom} = Q_{top} \tag{2.8.8}$$

Therefore, using equations (2.8.6) and (2.8.7), we have

$$-a\frac{dh(t)}{dt} = KA\frac{\left(h(t){-}L\right)}{L} \tag{2.8.9}$$

This is the governing equation describing the change of water level in the tube as a function of time. Now, solve equation (2.8.9) for $h(t)$ by the method of integration. That is,

$$\int_{h_o}^{h(t)} \frac{dh}{(h-L)} = \frac{KA}{aL}\int_0^t dt \tag{2.8.10}$$

This yields the following results:

$$\ln\left[\frac{(h(t){-}L)}{(h_o{-}L)}\right] = \frac{-KA}{aL}t \tag{2.8.11}$$

Or in terms of the water level in the tube relative to the top of the soil column (a shifted datumn),

$$\ln\left(\frac{H(t)}{H_o}\right) = \frac{-KA}{aL}t \tag{2.8.12}$$

Utilizing the fact that $\ln = 2.3\log_{10}$, equation (2.8.12) can be rewritten as

$$\log_{10}\left(\frac{H(t)}{H_o}\right) = -\frac{1}{2.3}\frac{KA}{aL}t \tag{2.8.13}$$

According to the preceding theory, if we plot $\frac{H(t)}{H_o}$ on the log scale and t on the linear scale, a linear relation appears (see Figure 2.21). If we choose 1 log cycle on the $\frac{H(t)}{H_o}$ axis and determine Δt corresponding to this chosen log cycle, the K is determined as

$$K = \frac{2.3aL}{A\Delta t} \tag{2.8.14}$$

Questions:

The falling-head permeameter assumes an instantaneous outflow, thus neglecting the storage effects. How would you design your apparatus to prevent the impacts of this assumption? Will this analysis be much different from the mathematical analysis of a slug test result?

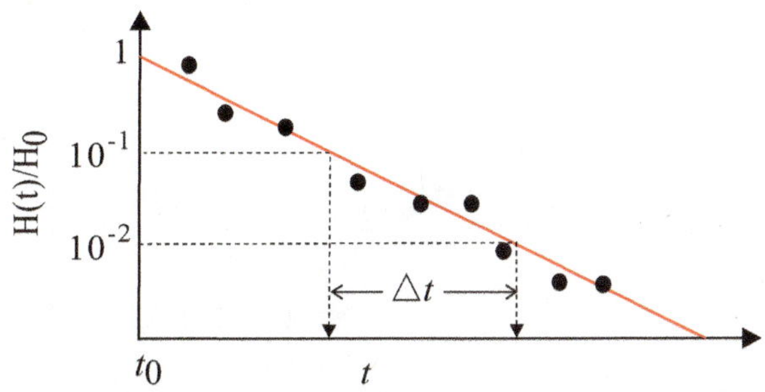

Figure 2.21. Relationship between observed log normalized water level and time in a falling-head permeameter.

2.9 Summary

This chapter has been focused on the development of Darcy's law and some of the porous media and fluid properties such as the hydraulic conductivity and porosity. Salient points include the following:

- Capillary tube flow follows Poiseuille's law, where $Q \propto R^4$, and average velocity is proportional to the pressure gradient, dp/dx, for the case of horizontal flow.
- Flow through porous media is analogous to flow through a bundle of capillary tubes. We show that specific discharge (q) is "linearly" proportional to the driving force, and the constant of proportionality, called "conductivity, (K)." K is directly proportional to medium porosity, n, the second power of the average pore radius, $\overline{R}^2$, and inversely proportional to the mean tortuosity, $\overline{T}$.
- Darcy's law is an empirical equation describing flow through porous media where $Q \propto \frac{\Delta H}{\ell} A$ and hydraulic conductivity (K) is a constant of proportionality.
- Deviations from Darcian behavior have been observed at both small and large dh/dl values.
- Hydraulic conductivity is directly proportional to the intrinsic permeability (k) of the medium and the fluid density (ρ) and inversely proportional to fluid dynamic viscosity (μ).

Reading Assignment:

(1) Darcy, Henri, 1856. les fontaines publiques de la ville de Dijon: Paris (*Groundwater*, 23 (1), p. 260. Jan.–Feb. 1985).
(2) Hubbert, M. K., 1940. The theory of groundwater motion, *Journal of Geology*, Vol. 48 (8), pp. 785–944.
(3) Hubbert, M. K., 1956. Darcy's law and the field equations for the flow of underground fluids. (*J. of Petro. Tech.*, Oct.).

2.10 Exercises

1. Consider the force balance in a volume of fluid in a circular pipe as shown in Figure 2.22 and derive the general form of the Poiseuille equation. That is, you are requested to show that

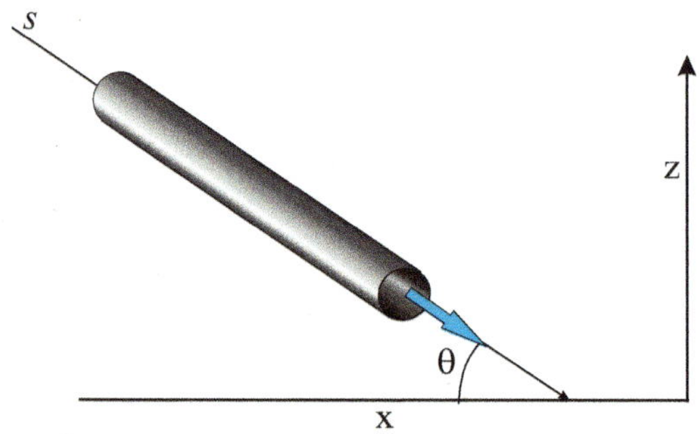

Figure 2.22. Illustration for problem 1.

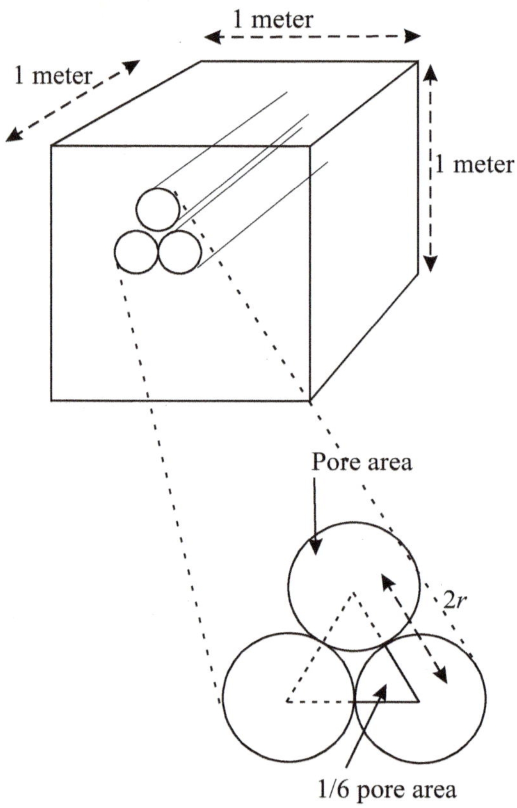

Figure 2.23. Illustration for problem 2.

$$q = -\frac{\rho g R^2}{8\mu}\frac{dH}{ds}$$

where $H = p/\rho g + z$, p is the pressure, s is the streamline along the tube, θ is the angle between the pipe and the horizontal surface, and z is the elevation.

2. Assume a large region composed of horizontal, equal-diameter tubes packed as shown in Figure 2.23. Assume also that water is flowing through the tubes at a hydraulic gradient,

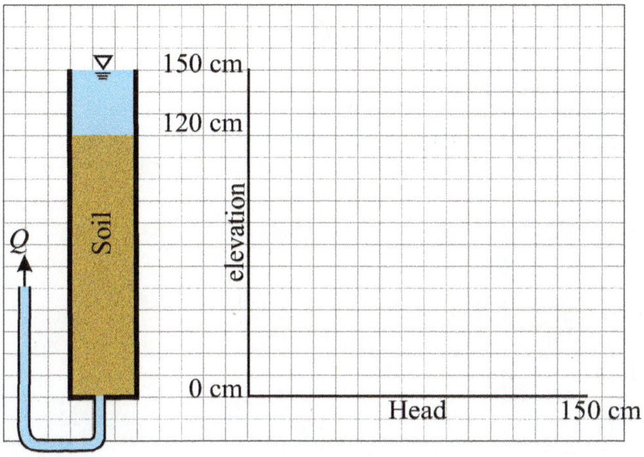

Figure 2.24a. Illustration for problem 3a.

dh/dx, of 0.01 m/m. Let the kinematic viscosity, υ, of the water be equal to 0.01 cm^2/sec, and *g* equal to 980 cm/sec^2.

a) Assuming Poiseuille's equation applies; find the value of the hydraulic conductivity, *K*; porosity, *n*; and discharge, Q, when the number of tubes per unit cross section (number/$meter^2$) is i) 10^2 and ii) 10^6. Note that the cross section may contain some fractional tubes as well.
b) Is the flow laminar in both cases? If not, is it valid to use the Poiseuille equation?
c) According to this model, how does i) *n* vary as a function of *r*, ii) *K* vary as a function of *n*, and iii) *K* vary as a function of *r*?

3. a) For the following steady-state flow experiment, plot pressure head, elevation head, and total head versus elevation. If the hydraulic conductivity of the soil is 10 cm/min, the inside diameter of the column is 10 cm, and the length of the soil sample is 120 cm, calculate the specific discharge at the outlet (see Figure 2.24a).
   b) If the soil column in (a) is uniformly packed with medium sand with grain sizes ranging from 0.2 mm to 0.6 mm, roughly what is the size of the spatial REV for this sand column?
   c) Suppose eleven horizontal piezometers are inserted into the soil column at one-foot intervals to measure the pressure profile along the column during the steady flow experiment. Discuss possible differences between the theoretical pressure profile you have drawn in (a) and any possible observed pressure profile under the following conditions: the screen interval of piezometer is 1) less than the size of the REV and 2) larger than the size of the REV. No measurement errors will be assumed.
   d) Supposing the sand column also contains some localized aggregates of pebbles with grain sizes ranging from 4 mm to 60 mm, do you expect a spatial REV to exist? Speculate how the observed pressure profile is different from the theoretical one.
   e) What is the pressure head at the bottom of the column if the outlet is plugged and the water level at the top is maintained constant (see Figure 2.24b)?

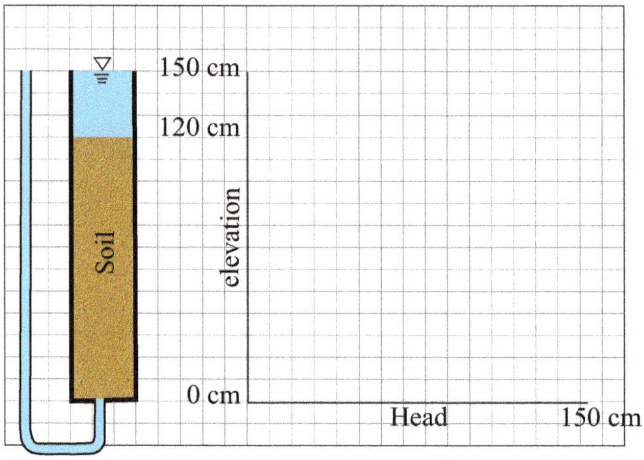

Figure 2.24b. Illustration for problem 3e.

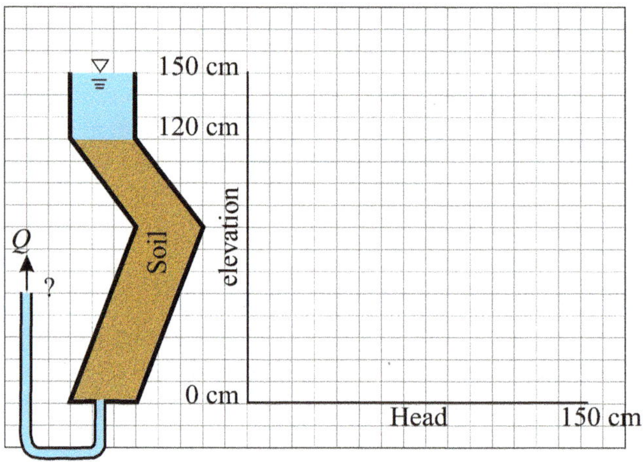

Figure 2.25. Illustration for problem 4.

4. The steady-state flow experiment shown in Figure 2.25 has the same soil as in Figure 2.24.
   a) If the outflow rate (specific discharge) is the same as in Problem 3a, plot the pressure, the elevation, and the total head along the column. What is the total head at the bottom?
   b) If the head at the top and the head at the bottom of the soil are the same as in Problem 3a, determine the specific discharge (and plot the pressure and the total heads along the column).
5. Flow through a horizontal fracture shown in Figure 2.18 is often described by the cubic law, that is, $Q = -(be^3)/(12\mu)\ dp/dx$, where e is the aperture, b is the width in the direction perpendicular to the flow, and μ is the dynamic viscosity of the fluid. Carry out the force balance on a CV in Figure 2.18 and then derive the cubic law. State all the assumptions clearly. If the plane surface inside the fracture is not smooth and there are meandering flow channels, how would you improve your cubic law?

3

Darcy's Law for Unsaturated Porous Media

3.1 Introduction

This chapter first introduces the concept of negative pressure head (a measure of the energy state of a fluid in an unsaturated porous medium). The relationships among this energy state and the amount of moisture in the media, the transmissibility of moisture in it, and its ability to store moisture are then discussed by introducing the moisture retention, unsaturated hydraulic conductivity, and moisture capacity characteristic curves. Extensions of Darcy's law for flow through unsaturated media and its limitations are discussed next. Lastly, some laboratory methods are presented for determining unsaturated hydraulic properties.

3.2 Concept of the Negative Pressure Head

3.2.1 Capillary Rise

Consider a capillary rise experiment as shown in Figure 3.1. Capillary rise is a well-known phenomenon that is caused by adhesive forces between fluids and material as well as cohesive forces of the fluids. A more in-depth discussion of these microscopic forces can be found in Bear (1972). At macroscopic scales, capillary rise, L, can be related to the surface tension of the fluid; σ, the contact angle; ω, which is defined as the angle between the intersection of the liquid-solid interface and the liquid-vapor interface; density of the fluid, ρ; and radius of the capillary tube, R. Specifically,

$$L = \frac{2\sigma \cos \omega}{\rho g R} \tag{3.2.1}$$

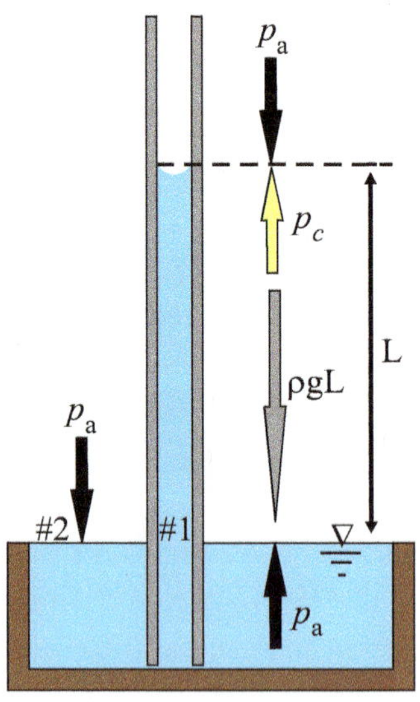

Figure 3.1. An illustration of the capillary rise experiment.

where g denotes gravity. According to equation (3.2.1), for a given fluid and a capillary tube, the rise (or the capillary force pulling the water upward) is inversely related to the radius of the tube: the smaller the tube, the higher will be the rise.

This capillary rise can also be described by considering the force balance at a macroscopic level. Consider the force balance at point 1 in the capillary rise experiment shown in Figure 3.1. The upward pressure at point 1 must be equal to the pressure at point 2, which is the atmospheric pressure, p_a. If the cross-sectional area of the tube is denoted by A, the total upward forces at point 1 then consist of the capillary force that pulls the water up into the tube, p_cA, and the upward force due to atmospheric pressure at point 1, p_aA. On the other hand, the total downward force includes the downward force, p_aA, due to atmospheric pressure, and the weight of water column, ρgLA. The force balance equation at point 1 thus becomes

$$p_aA + \rho gLA = p_cA + p_aA, \text{ and } p_c = \rho gL \qquad (3.2.2)$$

According to Equation (3.2.2), the capillary pressure (p_c) at the top the capillary rise equals the unit weight of the fluid multiplied by the capillary rise, L. Therefore, the capillary rise can be attributed to either the upward pulling of the positive capillary force in the tube or the negative water pressure in the tube (i.e., pressure smaller than the atmosphere pressure, which is assumed zero).

3.2.2 Tensiometer

In the preceding section, we introduced the concept of capillary pressure in a tube, and we now are ready to apply this concept to unsaturated porous media. Consider a sand column filled with water and equipped with a manometer on one side of the column (Figure 3.2a). If the column is sealed at the bottom and there is no flow in the column, the water levels at points 1 and 2 or the total head at points 3 and 4 must be equal. At point 3, the elevation head, z, is the distance between points 4 and 3, if a datum is chosen as the location 4 where z is zero. The pressure head (which is positive) is the distance between points 3 and 2. Therefore, the total head at point 3 is the distance between points 2 and 4. Since the elevation head at point 4 is zero (assumed datum), the total head at this point is the pressure head, which is the distance between point 4 and point 2.

Suppose now the water level in the column is below the location of the water manometer (point 3). The sand above the water level (indicated by ∇ in Figure 3.2b) is unsaturated. Air from the unsaturated zone thus may enter the manometer and be trapped in the water column in the manometer. Therefore, the water in the manometer would become disconnected from the sand column (Figure 3.2b) and the manometer would no longer function as a pressure measurement device to register the pressure in the unsaturated sand.

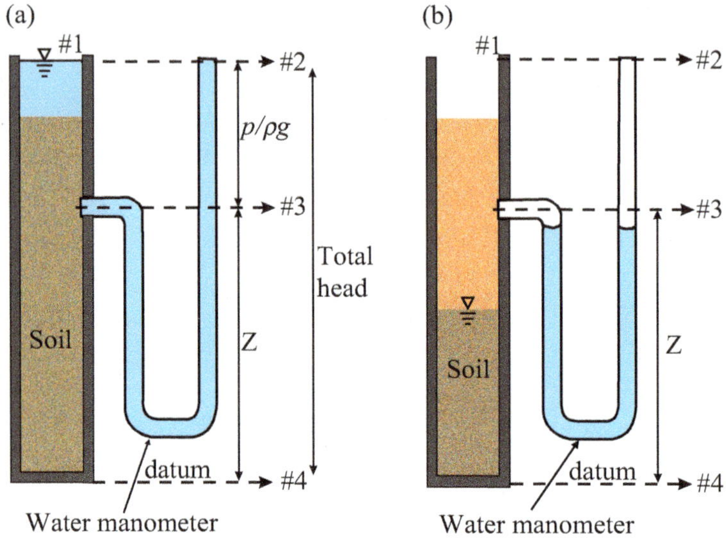

Figure 3.2. Pressure head and total head at point 3 a) when the column is fully saturated and b) when the water level is below point 3.

In order to prevent this problem while measuring the pressure in the unsaturated sand, the end of the water manometer in the sand column must be equipped with a ceramic cup (or a membrane). This ceramic cup has extremely fine pores, which are always water-saturated as a result of strong capillary forces of the fine pores. The saturated cup, therefore, prevents invasion of air into the manometer as the surrounding sand becomes unsaturated. The continuity of the water phase in the manometer is thus maintained and the water level in the manometer can be used as an indicator of the pore-water pressure in the sand. Such a manometer equipped with a ceramic cup is called a tensiometer.

Suppose a tensiometer is used in the preceding column experiment. If the water table drops to the position of the ceramic cup, the water level in the manometer tube should align with the water table position in the column. Under this situation, the pressure head in the sand column around the ceramic cup is zero or equal to the atmospheric pressure head. The total head at the ceramic cup (or point 3) is equal to the elevation head, z (see Figure 3.3a).

As the water table drops below the ceramic cup (Figure 3.3b), water in the sand around the cup will drain. Large pores of the sand drain first, and the remaining water is confined to small pores. Specifically, the adhesive forces between sand and water and the cohesive forces between water particles in the small pores create a capillary force in the sand – a force that resists the external forces, which cause drainage of the small pores. After this partial drainage, the sand around the cup becomes relatively dry while the ceramic cup remains wet. Consequently, the capillary forces in the sand

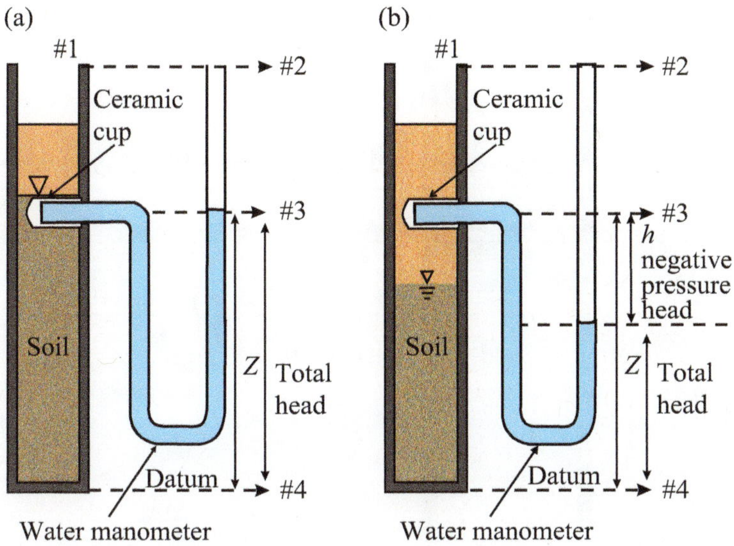

Figure 3.3. Pressure and total heads at the location of the tensiometer: when the water table is a) at the tensiometer location and b) below the tensiometer location.

around the cup start pulling water from the ceramic cup and, in turn, from the manometer to compensate for its loss of water. Therefore, this flow of water into the unsaturated sand results in lowering the water level in the manometer. After the capillary forces in the unsaturated sand reach equilibrium with the capillary force in the ceramic cup, the flow of water from the tensiometer to its surrounding sand ceases. The water level in the tensiometer then represents the capillary pressure head or pore-water pressure head in the unsaturated sand around the cup. Since the water level is now below the cup, the pore-water head of the unsaturated sand near the cup is considered to be negative, indicating that it is less than the atmospheric pressure, which is zero. Because the pore-water pressure head is negative, the total head (hydraulic head) becomes less than the elevation head (Figure 3.3b).

To describe this negative pore-water pressure (or water pressure) concept quantitatively, consider the force balance in a tensiometer (see Figure 3.4) when the tensiometer registers a steady pressure. The pressure at point #1 equals the atmospheric pressure, p_a, which must be equal to the pressure at point #2. At point #2, the upward force ought to be balanced by the downward force. The upward pressure is p_a and the downward pressure is the sum of the pressure due to the height of water column, ρgL, and the water pressure, p_w, in the soil registered by the ceramic cup. That is,

$$p_w + \rho gL = p_a \qquad (3.2.3)$$

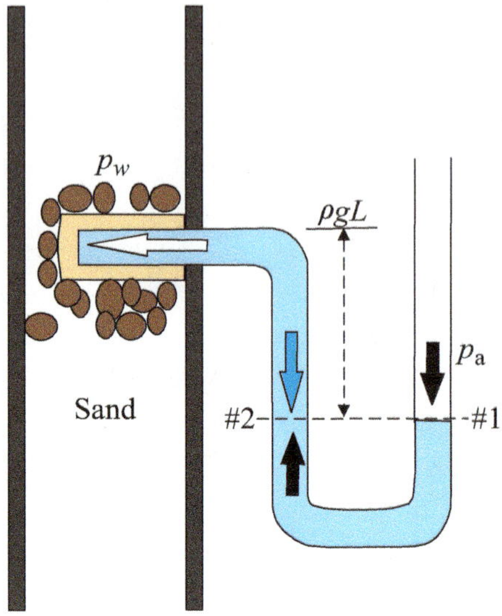

Figure 3.4. Force balance in a tensiometer.

Equation (3.2.3) leads to the following definition of the water pressure in the soil around the ceramic cup:

$$p_w = p_a - \rho gL \tag{3.2.4}$$

If we use p_a as the reference pressure, $p_a = 0$, then $p_w = -\rho gL$. That is, the water in the soil around the ceramic cup is under tension and has a negative value (i.e., less than the atmospheric pressure). Then, the capillary pressure, defined as the pressure of air (or the non-wetting fluid) minus the pressure of water (the wetting fluid), is:

$$p_c = p_a - p_w = \rho gL \tag{3.2.5}$$

This capillary pressure (positive) represents the resultant capillary forces in all of the pores adjacent to the ceramic cup due to the cohesive and adhesive forces between water, air, and soil particles. Again, both capillary and pore-water or water pressure are defined at a macroscopic level. That is, they are macroscopic quantities that represent the capillary force in an average sense over the surface of the ceramic cup, instead of the pressure within any single pore. A variety of terminologies have been used for negative water pressure, namely, soil-water pressure, and matric potential. To the contrary, capillary pressure, suction, and tension represent the absolute value of the negative pressure. Note that we will use pressure head to represent the pore-water pressure for rest of this book.

3.2.3 Pressure, Elevation, and Total Heads

Next, we consider a uniformly packed sand column experiment as shown in Figure 3.5. If we first apply a constant flow rate of water to the top of the column that equals the saturated hydraulic conductivity of the sand such that no ponding occurs at the top the column and flow reaches a steady flow condition, then the sand is fully saturated and the flow is driven by gravity only. Because the gravity is the only driving force, the pressure distribution along the column must be uniform and the pressure equals the atmospheric pressure, $p_a = 0$. Under this condition, the total head gradient is unity and we say that the flow is under a unit gradient condition.

If we reduce the flow rate such that it is less than the saturated hydraulic conductivity of the sand, the sand column becomes unsaturated because the amount of water supplied is not sufficient to fill all of the pores in the sand column. This unsaturated condition creates capillary forces in the sand. That is, the pressure head of water in the sand pores is less than the atmospheric pressure. If a tensiometer is installed to register the pressure head in the unsaturated sand column, the water level in the tensiometer should be lower than the elevation of the ceramic cup (Figure 3.5) (i.e., a negative pressure head). Assuming that many tensiometers have been installed along the column length, a plot of the tensiometer readings yields the pressure head distribution along the column (Figure 3.5). Notice that in Figure 3.5 the pressure at the outlet at the bottom of the column is equal to atmospheric pressure, which is zero, and the height of the outlet thus represents the water table position.

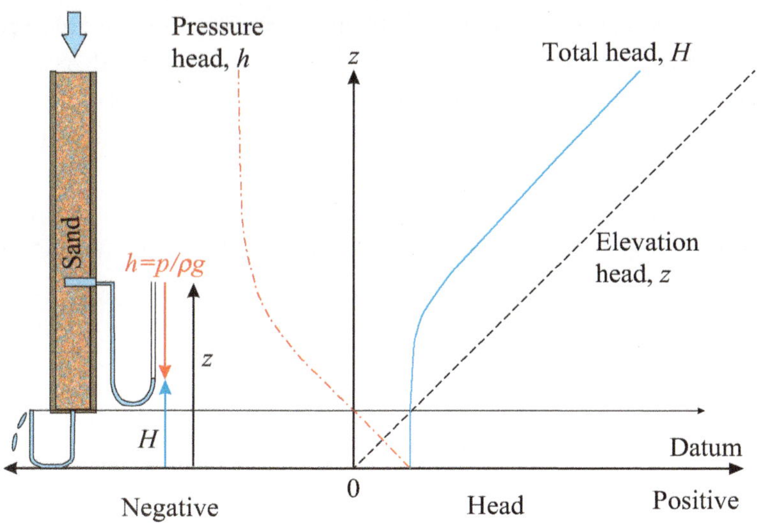

Figure 3.5. Pressure, elevation, and total head distributions along a soil column during a steady-state, unsaturated flow experiment.

Comparing this figure with Figure 2.6 in Chapter 2 for the head profile along a sand column under saturated conditions, you should notice some differences and should ask yourself, "Why is the pressure head profile a curve, not a straight line as in the saturated flow experiment even though the sand is homogeneous?" This question will be addressed in Section 3.8 after the introduction of some of the properties of an unsaturated medium.

Questions:

a) The water level in a manometer takes time to reach an equilibrium at which you can have a steady measurement of pressure. Speculate about the factors controlling the time required to reach the new equilibrium (the response time of a tensiometer).
b) If the response time is large, how can you design a better tensiometer so that the response time is minimal?
c) Can a tensiometer measure positive pressures?

3.3 Moisture Retention Curve

3.3.1 Moisture Content and Degree of Saturation

Previous sections have shown that the pressure in porous media is negative when they are unsaturated, and is positive when they are saturated. How the negative pressure head is related to the degree of saturation or moisture content of the medium is the next logical question. However, before answering this question, we must define some terms. Figure 3.6a illustrates the presence of three phases, that is, air, water, and solid phases, in a representative unsaturated medium sample. Figure 3.6b illustrates the mass and volume relations among the three phases for the hypothetical sample. Let the subscripts s, w, and a represent the solid, water, and air phases, respectively; the subscript t represents the total for either mass, M, or volume, V. The subscript *pore* or *fluid* represents the pore volume, which is also equal to the fluid (air plus water) volume in unsaturated media. The mass-volume relationships among the three phases are

$$\begin{aligned} M_T &= M_s + M_w + M_a \\ V_T &= V_s + V_w + V_a \\ V_{pore} &= V_{fluid} = V_w + V_a \end{aligned} \tag{3.3.1}$$

The solid particle density, ρ_s (g cm^{-3}), and the dry bulk density, ρ_b (g cm^{-3}), for the soil are defined as

$$\rho_s = \frac{M_s}{V_s} \text{ and } \rho_b = \frac{M_s}{V_T} \tag{3.3.2}$$

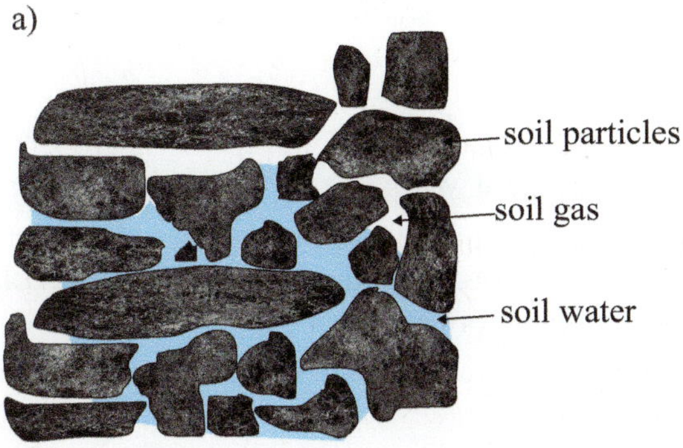

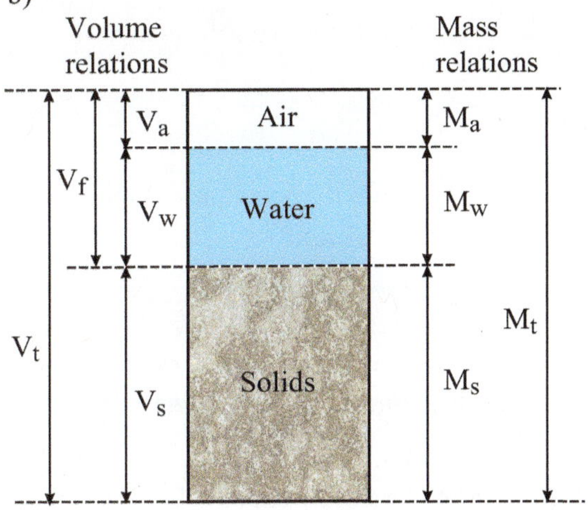

Figure 3.6. a) an illustration of air, water, and solid phases in an unsaturated porous medium and b) mass and volume relations of the three phases.

The amount of water in a volume of soil can be defined in terms of gravimetric water content or volumetric water content. The gravimetric water content is defined as

$$\theta_g = \frac{M_{wet\ soil} - M_{dry\ soil}}{M_{dry\ soil}} = \frac{M_w}{M_s} \tag{3.3.3}$$

On the other hand, the volumetric water content is defined as the volume of moisture per unit bulk volume of the porous medium. This can be represented as

$$\theta_v = \frac{V_w}{V_T} \tag{3.3.4}$$

The relationship between the gravimetric water content and the volumetric water content is

$$\theta_g = \frac{M_w}{M_s} = \frac{\rho_w V_w}{\rho_s V_s} = \frac{\rho_w V_w}{\rho_b V_T} = \frac{\rho_w}{\rho_b}\theta_v \tag{3.3.5}$$

Notice that the water content and the moisture content are synonymous; we will use the term "moisture content," and we will use θ to denote θ_v throughout the book, unless otherwise noted. Here we must emphasize the fact that moisture content is a spatially averaged state of the porous medium over a control volume (CV). Such an averaged state implies that some pores in the CV may be fully saturated even though the moisture content of the entire medium is less than its value for a completely saturated condition.

In addition to moisture content, another term frequently used is "degree of saturation," *S*, which is defined as the percentage of the pore volume that is saturated. This can be represented as

$$S = \frac{\theta}{n} = \frac{\theta}{\theta_s} \tag{3.3.6}$$

where n is the porosity, and it is equal to the saturated moisture content of the porous medium, θ_s, only if entrapped air is absent.

3.3.2 Analysis of Pressure Head-Moisture Content Relationship Using a Bundle of Capillary Tubes

Using these newly defined terms, we are ready to introduce the soil moisture retention curve. A soil moisture retention curve (also known as a moisture retention curve) is a functional relationship between the energy state of the soil water (as represented by matric potential or pressure head, h) and the corresponding soil moisture content, θ. A moisture retention curve defines the storage or moisture-holding characteristics of an unsaturated soil or rock sample under different matric potentials, θ (h). More specifically, this relationship defines how much moisture a geologic medium can hold against a suction that is applied to the medium. To understand the characteristic of the function that describes the moisture retention and pressure head relation, we will consider an idealized porous medium having porosity n (same as θ_s) and a cross-sectional area A, which consists of a bundle of straight, parallel capillary tubes of different radii of a given length, L (Figure 3.7).

If we plot the number of tubes of a specified radius on the vertical axis and all possible radii on the horizontal, we then have a frequency distribution of the radius of the bundle of capillary tubes. If the number of tubes of a given radius, r, is defined as $f(r)$, the total number of tubes in this bundle of capillary tubes then becomes

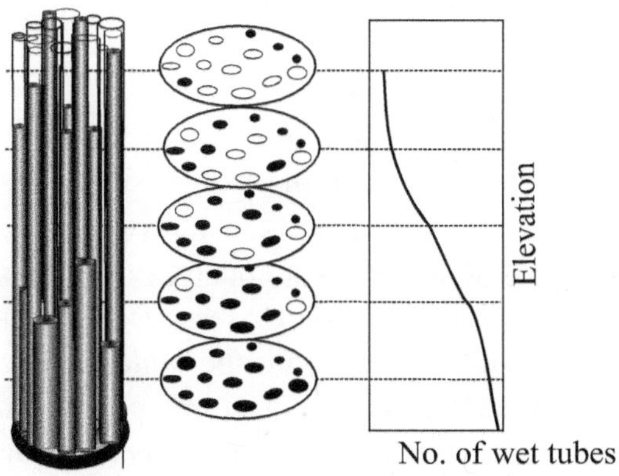

Figure 3.7. A capillary tube analogue that explains the moisture retention curve of a porous medium. (modified from Tuller and Or, 2005)

$$\int_{r_{\min}}^{r_{\max}} f(r)dr = \text{total number of tubes.} \tag{3.3.7}$$

where $r_{\max}$ and $r_{\min}$ are the largest and the smallest radius in the bundle of capillary tubes, respectively. The porosity of the bundle of capillary tubes is then given by

$$n = \frac{V_v}{V_T} = \frac{\left(\int_{r_{\min}}^{r_{\max}} \pi r^2 f(r)dr\right)L}{AL} = \frac{\left(\int_{r_{\min}}^{r_{\max}} \pi r^2 f(r)dr\right)}{A} = \theta_s \tag{3.3.8}$$

That is, the porosity (or saturated water content) of the bundle of tubes is the ratio of the total cross-sectional area of the tubes to the cross-sectional area of the bundle.

If we dip this bundle of capillary tubes into a shallow water reservoir, we will see that water rises to a certain height in a given tube, which is inversely proportional to its radius (Equation 3.2.1). At the surface of the reservoir, all the tubes are fully saturated with water, and thus the water content is θ_s. As we move our observation up along the bundle to a new elevation, we see some of these tubes may be empty because their radii are greater than the radius that can produce a capillary rise to this elevation. Therefore, the water content at this elevation (say, z_1) is given by

$$\theta(z_1) = \left(\int_{r_{\min}}^{r_1} \pi r^2 f(r)dr\right) \bigg/ A \tag{3.3.9}$$

where r_1 is smaller than the r_{max} and is given by the capillary rise, equation (3.2.1):

$$r_1 = \frac{2\sigma \cos \omega}{\rho g z_1} \tag{3.3.10}$$

This water content $\theta(z_1)$ is the ratio of the total amount of water in capillary tubes that remain saturated at the elevation z_1 to the amount of water in all capillary tubes when they all fully saturated, and this water content should be smaller than θ_s. As we continuously move up our observation along the bundle, we will find the moisture content decreases. Notice that the elevation is the same as the capillary rise, which is equal to the capillary pressure head. Thus, if we plot the moisture content as a function of the elevation, we obtain a moisture retention curve for this bundle of capillary tubes (right-hand plot in Figure 3.7).

While the idealized bundle of capillary tubes provides a conceptual and theoretical basis for a moisture retention curve for a porous medium, a real porous medium sample is considerably more complex than the bundle of capillary tubes. To represent a porous medium properly, we must consider any variation in the tube radii with elevation, and possibly any spatial discontinuity of tubes and other such factors. These factors make mathematical analysis of the moisture retention curve of a geologic medium a very complicated task. In the following paragraphs, we thus will only provide qualitative description of the moisture retention behavior of any geologic medium.

Consider a long column of an initially dry geologic medium that has been dipped into a shallow water reservoir for a period. Assuming that there is no evaporative water loss, one would expect water will continuously rise along the column up to the height that is inversely proportional to the radius of the smallest pore as a result of small sizes of pores and capillary forces in the medium. Analogous to the bundle of capillary tubes of different radii, the medium will have different pore sizes in the cross section of the column at any given elevation. Some of these pores will remain fully saturated with water, while others will still be empty because the capillary force that keeps water within the pores is less than the gravitational force. If moisture content is defined as the percentage of pore spaces that are saturated with water over the cross section at a given elevation, we will observe that the moisture content decreases with the elevation. Notice that a given elevation can be directly related to the capillary rise (or pressure head) of the largest pore that remains saturated at this elevation. A plot of the moisture content versus elevation is therefore a plot of the moisture retention curve (moisture content–pressure head relation) of the geologic medium.

The moisture retention curve concept can also be elucidated by using a hanging column experiment, which is one simple method for measuring the moisture retention curve of a porous medium. We will discuss this method next.

3.3.3 Hanging Column Experiment

Figure 3.8 illustrates a possible setup of a hanging column experiment. The apparatus comprises a funnel, a porous plate, and a water manometer. If we place a soil sample

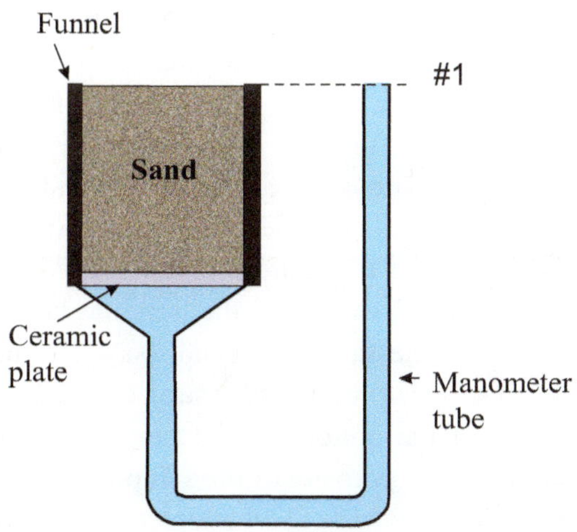

Figure 3.8. A typical setup for a hanging column experiment.

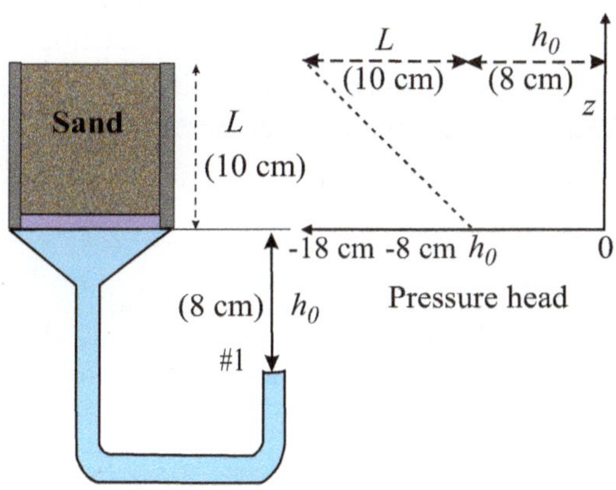

Figure 3.9. An illustration of the lowering of the pressure head in a hanging column experiment.

of 10 cm in height on the porous plate in the funnel, which is connected to a manometer, and the soil is initially fully saturated from the bottom to the top, the water level in the manometer should be at the same level as the water level at the top of the soil column (location 1 in Figure 3.8).

If we lower the open end of the manometer to a new elevation (point 1 in Figure 3.9), water will flow out from the manometer and the soil sample starts to drain. After a period of time, the flow will cease because capillary forces of the soil balance out the gravitational force due to lowering the manometer. Since the flow ceases, the pressure along the vertical direction of the soil sample must be negatively hydrostatic such that the hydraulic head everywhere in the soil sample is the same.

That is, $h = -z + h_O$, where h is the pressure head, z is the elevation measured from the top of the porous plate in the funnel, and h_O, is the pressure head at $z = 0$ (Figure 3.9). If we neglect the thickness of the porous plate, the pressure head at the bottom of the soil sample, h_O, must be equal to the distance between the water level in the manometer and the porous plate, which is considered to be negative. The pressure head distribution in the sand becomes more negative as the elevation increases. In terms of the drawing in Figure 3.9, the pressure head at the bottom of the porous plate will be −8 cm of water and the pressure head at the top of the soil sample is −18 cm of water. If the sample is small (say, less than 10 centimeters in height), we may assume that the entire sample is subject to the same negative pressure head as the pressure head at the bottom of the plate (the shorter the sample, the smaller the error). If we know the porosity of the sample at saturation and the amount of water being discharged after lowering the manometer the moisture remaining in the sample can be determined. A pair of moisture content and pressure head (θ, h) values is thus obtained.

If we successively lower the manometer and determine the moisture content of the sample, then a plot of the relationship between the negative pressure and the moisture content of the sample (i.e., moisture retention curve) can be made (Figure 3.10). Knowledge of this curve permits us to determine the amount of water being retained in a porous medium for a given suction.

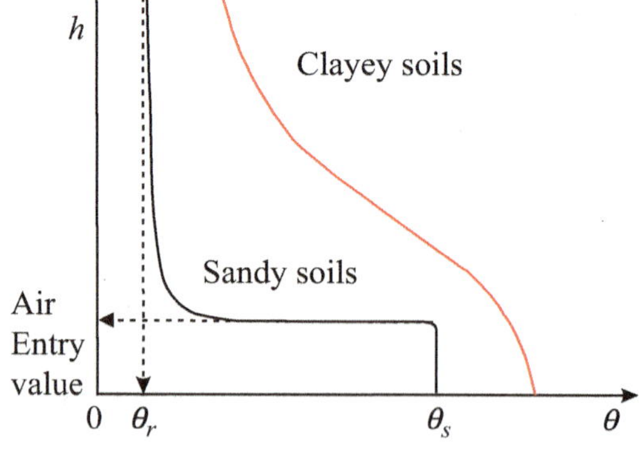

Figure 3.10. Typical moisture retention curves for sandy soils and clayey soils, with residual and saturated water content identified for the sandy soils curve.

Figure 3.10 also illustrates typical moisture release behavior for two types of porous media (sandy soils and clayey soils) as pressure becomes more negative. For both media at saturation, the pressure head is zero and the moisture content is equal to the porosity (typically, less than the porosity because of entrapped air). This water content is called the saturated water content and is denoted by θ_s. As the pressure head becomes negative, but less than the air entry value (h_e), the soil remains saturated. The air entry value is the pressure at which air starts entering the pores and forcing out the water in the largest pores. Once the pressure head exceeds the air entry value, the moisture content of the sample decreases rapidly because of drainage of large pores, which have low capillary forces. After drainage of the large pores, an increase in the negative value of h does not reduce the moisture content significantly, because of presence of strong capillary forces within the small pores. As a result, the moisture retention curve of a porous medium often exhibits an *S*-shape curve, as shown in Figure 3.10. The moisture content, which does not change significantly for very dry conditions, regardless of the increase in negative pressure, is called the residual moisture content, θ_r.

As demonstrated by our analysis of the moisture retention curve using the bundle of capillary tubes, the moisture retention curve is related to the distribution of tube radii of the bundle. By the same token, the rate of reduction in moisture content of a porous medium, as the pressure head becomes more negative, depends on the pore size distribution of the medium. For instance, well-sorted sands tend to have a narrow pore-size distribution (i.e., a large number of large pores and only a few small pores). Therefore, sandy soils tend to have a rapid reduction in moisture content as suction increases. Conversely, fine-textured materials such as clayey soils or rocks have a widespread pore-size distribution and the reduction in moisture content is thus gradual (see Figure 3.10).

The moisture retention curve of a porous medium is generally nonunique. In other words, the moisture-pressure head relationship measured during the wetting of an initially dry medium will be different from the relationship during the draining of an initially saturated medium (Figure 3.11). This nonuniqueness or history dependence is called hysteresis. It has been attributed to variations in pore radius (the ink-bottle effect), and entrapment of air in some pores, as well as the raindrop effect, that is, the dependence of the contact angle on the direction of fluid displacement at microscopic scales (Bear, 1972). Generally, the moisture content of a porous medium is higher at any given pressure head for drying than wetting (Warrick, 2003). Models have been developed for describing this hysteresis phenomenon (e.g., Parker and Lenhard 1987; Lenhard and Parker 1987), but the data for quantifying the hysteresis phenomenon are often not available. While the hysteresis phenomenon has been observed in laboratory experiments, it has rarely been emphasized in field scale problems, largely because of the relatively greater importance of spatial variations of the moisture retention curve in the field due to heterogeneity.

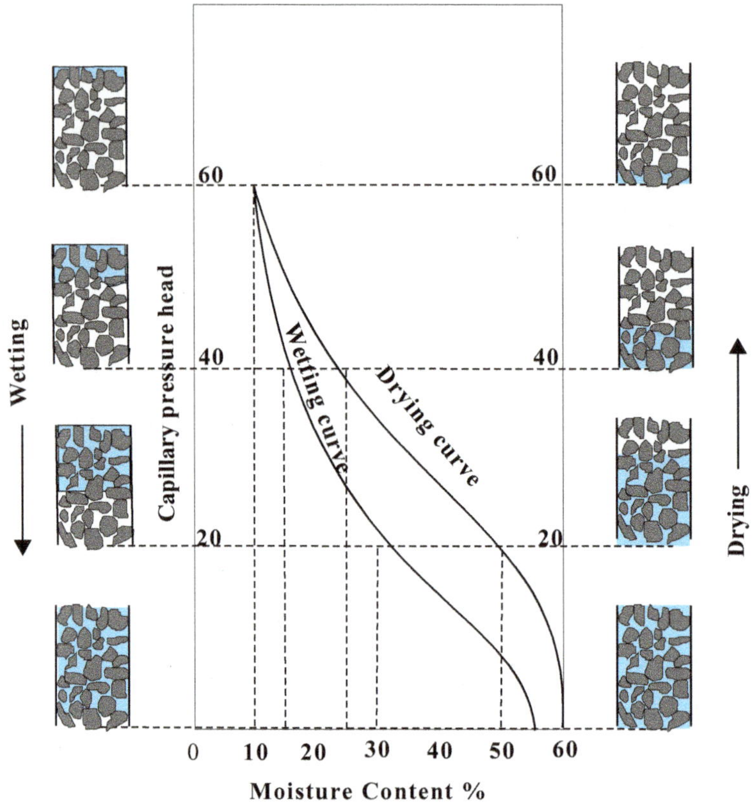

Figure 3.11. A hysteresis loop of a moisture retention curve: hysteresis effect in soils subjected to wetting and drying processes. (modified from Tuller and Or, 2005)

Finally, a related term called effective saturation curve, S_e (h), (ranging from 0.0 to 1.0), is defined as

$$S_e(h) = \frac{\theta - \theta_r}{\theta_s - \theta_r} = \frac{S - S_r}{1 - S_r} \tag{3.3.11}$$

where θ_r and θ_s are the residual and saturated moisture contents, respectively, and S_r is the residual saturation.

3.4 Unsaturated Hydraulic Conductivity Curve

Unlike the saturated hydraulic conductivity, the unsaturated hydraulic conductivity is not a single value for a given porous medium, but a functional relationship between the energy states of the unsaturated porous medium, as represented by the associated pressure head, h, and the associated transmissive characteristics, as represented by unsaturated hydraulic conductivity, K.

3.4.1 Analysis of Pressure Head – Unsaturated Hydraulic Conductivity Relationship Using a Bundle of Capillary Tubes

According to our discussion about Poiseuille's law for flow through porous media under fully saturated conditions (see Chapter 2), we learned that the saturated hydraulic conductivity is a function of the average radius of saturated pores, porosity, and tortuosity. In order to envision how the hydraulic conductivity for an unsaturated porous medium would be different from its saturated hydraulic conductivity, we ask ourselves the question, "How would the degree of saturation of a porous medium affect the average radius of saturated pores, porosity, and tortuosity of the medium?" To answer this question quantitatively, we again use the capillary tube bundle analogy. That is, we approximate a real porous medium by an idealized medium that is made of a bundle of straight capillary tubes. The bundle of capillary tubes will consist of three classes of tubes for simplicity – namely, N_1 tubes of the radius of r_1, N_2 tubes of the radius of r_2, and N_3 tubes of the radius of r_3. We will assume that $r_1 > r_2 > r_3$. As shown in Chapter 2, Poiseuille's law for flow through this bundle of tubes under fully saturated condition can be formulated as

$$
\begin{aligned}
q = \frac{Q}{A} &= -K\frac{dH}{dz} = -\left[\frac{\rho g n \overline{R}^2}{8\mu}\right]\frac{dH}{dz} \\
&= -\left[\frac{\rho g}{8\mu}\left(\frac{\pi(N_1r_1^2 + N_2r_2^2 + N_3r_3^2)}{A}\right)\left(\frac{N_1r_1^4 + N_2r_2^4 + N_3r_3^4}{N_1r_1^2 + N_2r_2^2 + N_3r_3^2}\right)\right]\frac{dH}{dz}
\end{aligned}
\tag{3.4.1}
$$

where the porosity and the average (i.e., effective) radius for the K term for the bundle of capillary tubes are

$$
n = \left(\frac{\pi\left(N_1r_1^2 + N_2r_2^2 + N_3r_3^2\right)}{A}\right) \text{ and } \overline{R} = \sqrt{\left(\frac{N_1r_1^4 + N_2r_2^4 + N_3r_3^4}{N_1r_1^2 + N_2r_2^2 + N_3r_3^2}\right)}, \tag{3.4.2}
$$

respectively. The porosity (saturated porosity) in equation (3.4.2) is referred to as the ratio of total area of the saturated tubes to that cross-sectional area of the bundle of the capillary tubes, and it is equal to the saturated moisture content of the bundle of capillary tubes. As this fully saturated bundle starts to drain, the tubes of radius r_1 will empty first because they have the largest radius and the smallest capillary forces, whereas tubes of the other two radii will remain to be saturated and flow will take place only in the tubes of radius r_2 and r_3. Thus, equation (3.4.1) becomes

$$
q = -K\frac{dH}{dz} = -\left(\frac{\rho g}{8\mu}\left(\frac{\pi(N_2r_2^2 + N_3r_3^2)}{A}\right)\left(\frac{N_2r_2^4 + N_3r_3^4}{N_2r_2^2 + N_3r_3^2}\right)\right)\frac{dH}{dz} \tag{3.4.3}
$$

The saturated porosity and the effective radius in the K term for the remaining fully saturated tubes are

$$n = \left(\frac{\pi\left(N_2r_2^2 + N_3r_3^2\right)}{A}\right) \text{ and } \overline{R} = \sqrt{\left(\frac{N_2r_2^4 + N_3r_3^4}{N_2r_2^2 + N_3r_3^2}\right)}, \text{ respectively} \qquad (3.4.4)$$

Notice that this saturated porosity is less than the saturated moisture content of the bundle of capillary tubes. After drainage of the tubes of the radius of r_2, the flow in the remaining tubes can be determined by

$$q = \frac{Q}{A} = -K\frac{dH}{dz} = -\left(\frac{\rho g n\overline{R}^2}{8\mu}\right)\frac{dH}{dz} = -\left(\frac{\rho g}{8\mu}\left(\frac{\pi N_3 r_3^2}{A}\right)\left(r_3^2\right)\right)\frac{dH}{dz} \qquad (3.4.5)$$

Likewise, the saturated porosity and the effective radius of the K term for the remaining fully saturated tubes are

$$n = \left(\frac{\pi\left(N_3r_3^2\right)}{A}\right) \text{ and } \overline{R} = r_3, \text{ respectively.} \qquad (3.4.6)$$

Based on the above analysis, we could draw the following conclusions. When the capillary tubes drain progressively according to the size of their radius, the saturated porosity (i.e., moisture content) and the effective radius of the entire bundle of capillary tubes become smaller, so does the hydraulic conductivity, compared to those under previous saturation levels. As the effective radius is smaller, the capillary force becomes greater, and the pressure head is more negative. We then can state that under non–fully saturated conditions, the hydraulic conductivity of the bundle of capillary tubes will decrease as the pressure head becomes more negative or the moisture content becomes less. If an ideal porous medium consists of tubes of a variety of radii, instead of tubes of three different classes of radius, we further state that the hydraulic conductivity and fully saturated porosity of the bundle will decrease smoothly as the pressure head continuously becomes more negative.

We now are ready to extend the findings from the idealized porous media to a geologic porous medium. Consider a fully saturated porous medium that is being desaturated. As in the case of the bundle of capillary tubes, water in the largest pores will drain and be emptied first and the remaining water will be confined to and flow in smaller pores. As a result, the number of fully saturated pores (the saturated porosity) will decrease; the average radius of the pores that are saturated becomes smaller. Unlike in the capillary tube bundle analogy, flow channels in geologic media are tortuous, some of them are interconnected, and opening of any flow channel also varies along its length. If the medium is being desaturated, connectivity of flow channels will differ from that under saturated conditions. Some of the channels will likely be blocked by air and become dead ends and water thus will have to detour. Therefore, active flow

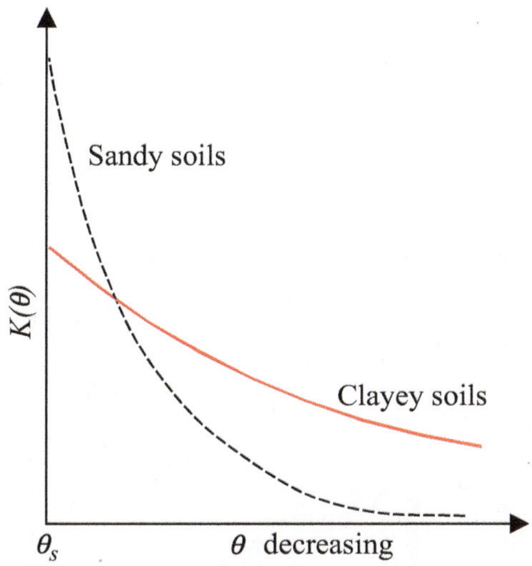

Figure 3.12. A typical unsaturated hydraulic conductivity and moisture content relation.

channels will likely become more tortuous than those when the same medium is fully saturated. If all this is true, the hydraulic conductivity term in equation (3.4.5) then should also include the tortuosity term in the denominator as in the analysis of the hydraulic conductivity of saturated porous media (see Chapter 2). This tortuosity, however, varies with the degree of saturation of the medium as opposed to that under saturated conditions. Notice that this relationship is generally difficult to define. In fact, the tortuosity may increase or decrease depending upon the pore-size distribution, as the medium is being desaturated. The tortuosity of the medium under unsaturated conditions will vary with the direction of flow as does the tortuosity of the medium under saturated conditions. Furthermore, its variation with the flow direction may also depend on the averaged moisture content of the medium.

These reasons lead us to conclude that the unsaturated hydraulic conductivity of a porous medium will decrease with a decrease in moisture content (see Figure 3.12). It will asymptotically approach zero after a threshold value of moisture content (i.e., residual water content) is reached. That is, porous media with moisture content less than the threshold value virtually cannot transmit any significant amount of water, because remaining water (perhaps in a form of films) becomes isolated and disconnected from each other. Again, we remind you that the preceding arguments invoke the CV concept in describing the unsaturated hydraulic conductivity. That is, the hydraulic conductivity is a volume-averaged property.

Since moisture content is related to the pressure head, the unsaturated hydraulic conductivity is also a function of pressure head (see Figure 3.13). As shown in this figure, the shape of the unsaturated hydraulic conductivity curve depends on the

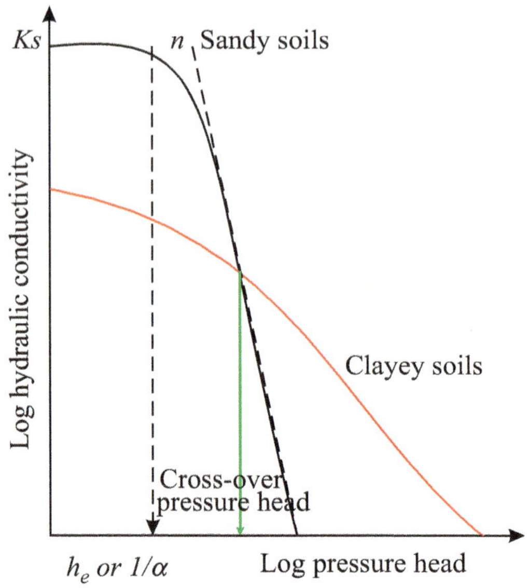

Figure 3.13. Typical unsaturated hydraulic conductivity and pressure head relationship for sand soils and clayey soils on a log-log plot. The physical meanings of the parameters in the MVG model (see Section 3.5) are also illustrated.

texture of the medium. That is, near saturation, the hydraulic conductivity of coarse-textured materials (e.g., sandy soils) is higher than that of fine-textured materials (e.g., clayey soils). The hydraulic conductivity of coarse-textured materials decreases much more rapidly than does that of fine-textured materials, as the pressure head becomes more negative. The hydraulic conductivity curves of the two types of materials thus intersect each other at some pressure head (i.e., crossover pressure head). At the cross-over pressure head, the hydraulic conductivities of the two materials are the same. Beyond the cross-over pressure head, the fine-textured material has a higher hydraulic conductivity than the coarse-textured material. This can be attributed to the fact that more air exists in the coarse-textured material at this range of pressure and air impedes the flow of water. This is a unique character of unsaturated hydraulic conductivity of porous media – a rather counterintuitive one.

Because of this unique character of unsaturated media, consider the case where a fine-textured material overlies a coarse-textured material. Suppose the pressures in the two materials are more negative than the cross-over pressure. In such a case, water in the fine material will not be able to flow into the coarse material below because the latter has a lower hydraulic conductivity. In other words, there exists a large amount of air in the coarse material below, and water from the overlying fine material cannot enter the coarse material unless the pressure of water is built up high enough to expel the air in the coarse material. This phenomenon is called the capillary barrier effect, and it has been used as a fundamental principle in the design of earth liners for landfills and "umbrellas" for waste storage facilities to prevent surface infiltration (Yeh and Gelhar, 1983; Ross, 1990; Yeh et al. 1993; 1995; Selker, 1997).

Lastly, if the unsaturated hydraulic conductivity curve of a medium is normalized by its saturated hydraulic conductivity value, a new curve is formed and this curve is called the relative hydraulic conductivity curve for the medium:

$$K_r(h) = \frac{K(h)}{K_s} \tag{3.4.7}$$

Similar to the moisture retention curve of a porous medium, the relative hydraulic conductivity curve also tends to exhibit the hysteresis effect.

3.5 Mathematical Models for Moisture Retention and Hydraulic Conductivity Curves

3.5.1 Gardner and Mualem-van Genuchten Models

Certainly, one could measure the unsaturated conductivity of a porous medium at every pressure head value, and subsequently tabulate the $K(h)$ relationship. However, for many practical purposes and convenience, mathematical models are often used to describe this relationship. One such mathematical model, frequently used to depict the unsaturated hydraulic conductivity and moisture retention curves, is the exponential model (Gardner, 1958),

$$\begin{aligned} K(h) &= K_s \exp(\beta h) \\ \theta(h) &= (\theta_s - \theta_r)\exp(\beta h) + \theta_r \end{aligned} \tag{3.5.1}$$

where K_s is the saturated hydraulic conductivity; β is the pore-size distribution parameter $[1/L]$, representing the rate of reduction in conductivity as the soil desaturates; θ_s is the saturated moisture content; and θ_r is the residual moisture content. The exponential model has been very popular because of its simplicity and convenience in mathematical analyses. It, nevertheless, fits the observed $K(h)$ or $\theta(h)$ data over only a limited range of pressure heads. Other widely used models for $K(h)$, and $\theta(h)$ are those by Mualem (1976) and van Genuchten (1980), also known as the MVG model. This model can be expressed as

$$\begin{aligned} K(h) &= K_s\left(1 - (\alpha|h|)^{n-1}\left[1 + (\alpha|h|)^n\right]^{-m}\right)^2\left[1 + (\alpha|h|)^n\right]^{(-m/2)} \\ \theta(h) &= (\theta_s - \theta_r)[1 + (\alpha|h|)^n]^{-m} + \theta_r \end{aligned} \tag{3.5.2}$$

in which | | represents the absolute value; α $[1/L]$, n [-], and m [-] are soil parameters; and $m = 1 - 1/n$. In general, the air entry value (h_e), that is, the threshold pressure head at which air starts displacing water in the largest pores of the porous medium, is related to $1/\alpha$. The slope of the hydraulic conductivity curve is controlled by the value of n (see Figure 3.13). Equation (3.5.2) is valid over a broader range of pressure head values than the exponential model (van Genuchten and Nielson, 1985). However, several investigators (Khaleel et al., 1995; Yates et al., 1992) have found that the van Genuchten-Mualem model does not provide accurate estimates of unsaturated hydraulic conductivity for soil sediments at relatively low moisture contents when these estimates are solely based on moisture retention data and a saturated hydraulic conductivity measurement. This condition may be explained by the fact that the saturated hydraulic conductivity for the relatively coarse material is dominated by large pores, whereas the unsaturated hydraulic conductivity appears to be dominated by small pores. Khaleel et al. (1995) recommended the use of at least one direct measurement of hydraulic conductivity, as a match point, for the moisture regime of interest.

Because of the use of these mathematical models to describe the relationships among unsaturated hydraulic conductivity, pressure head, and moisture content, soils or geologic media can be categorized by the magnitude of these parameters such as α, β, n, θ_s, θ_r, and K_s. For example, coarse-textured soils are reported to have large values of α, n, β , and K_s, and fine-textured soils have small values (e.g., Stephens et al., 1987 and Table 3.1) of the same parameters. However, values of some of these parameters are not necessarily unique for a given geologic medium because of hysteretic behavior in $K(h)$ and $\theta(h)$ relations; these values can be different according to the wetting and drying histories of the medium.

Generally speaking, moisture retention curves of soils can be measured with less cost and effort than unsaturated hydraulic conductivity. Parameters, α, β, and n for unsaturated hydraulic conductivity are often conveniently assumed to be the same as those of moisture retention curves, even though they are different (e.g., Yeh and

Table 3.1. *Values of the parameter of MVG model for soil textural groups*

Texture	θ_r	θ_s	**α 1/cm**	n	**K_s cm/d**
Sand	0.020	0.417	0.138	1.592	504.0
Loamy sand	0.035	0.401	0.115	1.474	146.6
Sandy loam	0.041	0.412	0.068	1.322	62.16
Loam	0.027	0.434	0.090	1.220	16.32
Silt loam	0.015	0.486	0.048	1.211	31.68
Sandy clay loam	0.068	0.330	0.036	1.250	10.32
Clay loam	0.075	0.390	0.039	1.194	5.52
Silty clay loam	0.040	0.432	0.031	1.151	3.60
Sandy clay	0.109	0.321	0.034	1.168	2.88
Silty clay	0.056	0.423	0.029	1.127	2.16
Clay	0.090	0.385	0.027	1.131	1.44

Source: Rawls et al. 1982.

Harvey, 1990). Therefore, instead of directly measuring the $K(h)$ curve of a medium, this curve is often conveniently generated by using the measured saturated hydraulic conductivity, K_s, and the parameter α, β, and n values obtained from the measured $\theta(h)$ curve of the same medium. In addition, equations (3.5.1) and (3.5.2) are frequently used to extrapolate the unsaturated hydraulic conductivity to very dry conditions where a direct measurement of unsaturated hydraulic conductivity is typically not available.

Overall, the MVG model can fit field data better than the exponential model, since the MVG model has more parameters that can be adjusted. However, this model often leads to numerical convergence problems during numerical simulations. For simulations of flow through heterogeneous, unsaturated media, a simpler model such as the exponential model for the unsaturated hydraulic conductivity coupled with the moisture retention function of the MVG model would be quite adequate for analysis of bulk behaviors of water movement in the vadose zone.

3.5.2 Parameter Values for Soils

Tables 3.1 and 3.2 list average values of the α, n, θ_s, θ_r, and K_s parameters for soil textural groups according to the USDA classification (SCS, 1975) as estimated by Rawls et al. (1982) and Carsel and Parrish (1988), respectively, from analyses of a large number of samples. These tables may serve as guides for making initial estimates of the parameters, which may be refined by using or by collecting the site specific data (Rawls et al., 1982).

Khaleel and Freeman (1995) examined a database that composed of six soil categories and used it as the basis for characterizing the variability in parameters θ_r, θ_s, α, n, and K_s. Descriptive statistics are given in Table 3.3, and the variability of the moisture

Table 3.2. *Values of the parameters of MVG model for soil textural groups*

Texture	θ_r	θ_s	**α 1/cm**	n	**K_s cm/d**
Sand	0.045	0.43	0.145	2.68	712.8
Loamy sand	0.057	0.41	0.124	2.28	350.2
Sandy loam	0.065	0.41	0.075	1.89	106.1
Loam	0.078	0.43	0.036	1.56	24.96
Silt	0.034	0.46	0.016	1.37	6.00
Silt loam	0.067	0.45	0.020	1.41	10.80
Sandy clay loam	0.100	0.39	0.059	1.48	31.44
Clay loam	0.095	0.41	0.019	1.31	6.24
Silty clay loam	0.089	0.43	0.010	1.23	1.68
Sandy clay	0.100	0.38	0.027	1.23	2.88
Silty clay	0.070	0.36	0.005	1.09	0.48
Clay	0.068	0.38	0.008	1.09	4.80

Source: Carsel and Parrish, 1988.

Table 3.3. *Descriptive statistics for van Genuchten parameters and saturated hydraulic conductivity Ks for six soil categories*

Soil Category	Hydraulic Property	Lower Limit	Upper Limit	Transform	Statistics		
					Mean	Standard Deviation	D_{max}
SS	θ_s	0.321	0.566	NO	0.438	0.059	0.103
	θ_r	0.000	0.881	SN	0.458	0.255	0.148
	α	−7.131	−0.949	LN	−4.489	1.352	0.164
	n	1.262	2.894	NO	1.824	0.344	0.064
	K_s	−12.058	−4.057	LN	−8.487	1.813	0.144
S	θ_s	0.197	0.519	NO	0.346	0.073	0.050
	θ_r	0.000	0.881	SN	0.189	0.146	0.134
	α	−5.547	−0.149	LN	−3.097	1.347	0.057
	n	−5.756	4.330	LR	−1.459	1.523	0.080
	K_s	−11.191	−2.847	LN	−6.849	2.129	0.089
SSG	θ_s	0.187	0.375	NO	0.262	0.072	0.178
	θ_r	0.000	0.064	NO	0.030	0.029	0.202
	α	−5.843	−2.276	LN	−3.957	1.166	0.153
	n	1.256	1.629	NO	1.400	0.131	0.233
	K_s	−10.854	2.995	LR	−5.262	5.499	0.198
GS	θ_s	0.203	0.334	NO	0.272	0.048	0.182
	θ_r	0.010	0.069	NO	0.040	0.019	0.226
	α	0.004	0.074	NO	0.027	0.023	0.191
	n	1.529	2.537	NO	1.994	0.315	0.176
	K_s	−7.966	2.989	LR	−1.569	3.582	0.159
SG1	θ_s	0.113	0.260	NO	0.166	0.036	0.071
	θ_r	0.000	0.062	NO	0.023	0.015	0.134
	α	−6.075	−0.084	LN	−4.086	1.550	0.232
	n	0.233	1.081	LN	0.489	0.184	0.169
	K_s	−15.476	−3.297	LN	−7.932	3.322	0.150
SG2	θ_s	−2.888	−2.234	LN	−2.590	0.216	0.226
	θ_r	0.000	0.0197	NO	0.010	0.007	0.171
	α	−5.952	−3.590	LN	−5.008	0.882	0.187
	n	1.347	1.885	NO	1.621	0.178	0.162
	K_s	−10.473	−2.040	LN	−7.137	2.332	0.145

Note: SS, sand mixed with finer fraction; S, sand; SSG, sand and gravel mixed with finer fraction; GS, gravelly sand; SG1, sandy gravel with gravel fraction < 60 percent; SG2, sandy gravel with gravel fraction > 60 percent. NO (no transform); LN (natural log transform); LR (log transform). *Source:* Khaleel and Freeman,1995.

retention data for the six soils is shown in Figure 3.14. Among the five parameters, the largest variability is in K_s, while the smallest variability is observed in n. The coefficient of variation (CV) for K_s ranges from about 104 to 293 percent, whereas the CV for n ranges from 3 to 39 percent. The large K_s variability is not unexpected, given the highly heterogeneous nature of vadose zone sediments. Carsel and Parrish (1988) also noted

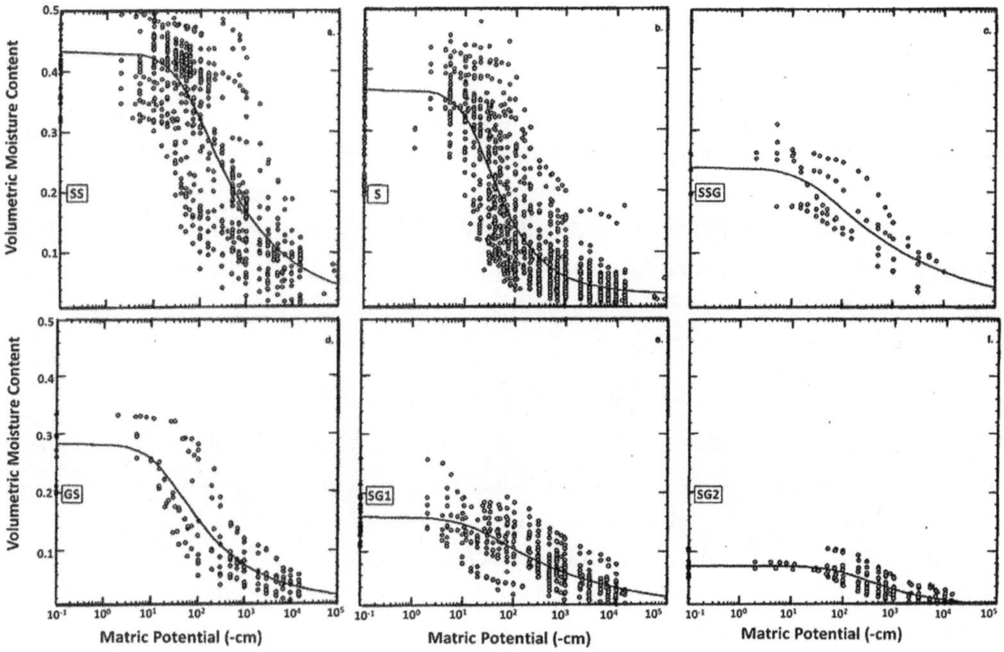

Figure 3.14. Moisture retention data and best-fit van Genuchten curves for the six soil categories (SS, sand mixed with finer fraction; S, sand; SSG, sand and gravel mixed with finer fraction; GS, gravelly sand; SG1, sandy gravel with gravel content approximately < 60 percent; SG2, sandy gravel with gravel content approximately > 60 percent). *Source:* Khaleel and Freeman, 1995.

similar variability. The variability in θ_s was minimal; the CV for θ_s was less than 25 percent for most soil categories. This is also consistent with data reported by other investigators (Carsel and Parrish, 1988). The CV for θ_r ranged from 43 to 95 percent, whereas the CV for α ranged from 95 to 247 percent.

Gravels can make up large fractions of the subsurface, including the vadose zones. The extensiveness of some gravel deposits is evident from the fact that they serve as major sources of groundwater for agricultural, industrial, and municipal usage. Some gravel deposits are mined for use in construction. Because some gravel deposits do occur in the vadose zone, understanding unsaturated flow and transport in such settings requires knowledge of their hydraulic properties. Some gravel deposits occur in heavily contaminated vadose zones such as the Hanford Site, Washington State, in the United States, where radioactive wastes have leaked. Gravels are also an important component in engineered capillary barriers for subsurface waste isolation. However, relatively little information is available on the unsaturated hydraulic properties of gravels. Please note that measurements of moisture characteristics on gravelly media have been reported by Khaleel and Relyea (1997), Conca et al. (1998), Khaleel and Heller (2003), and Tokunaga et al. (2002), while unsaturated hydraulic conductivities of gravels have been reported by Mehuys et al. (1975), Conca et al. (1998), and Khaleel and Relyea (2001).

Because of their large size, water retention in unsaturated gravels is expected to be mainly associated with grains rather than with the macroscopic pore network. Thus, water retention depends on intragranular and grain surface morphology. Some previous studies have examined aspects of water retention in gravel, especially as it relates to availability for plant uptake (e.g., Coile, 1953; Flint and Childs, 1984; Montagne et al., 1992; Brouwer and Anderson, 2000).

3.6 Moisture Capacity Curve

As illustrated earlier, moisture content is a function of pressure head, and this relationship is nonlinear. If we differentiate θ of the relationship with respect to h, we will obtain a curve that represents the amount of change in moisture content per unit change in pressure head at different pressure head values. This curve is important because it represents the change in storage of the porous medium per change in pressure head at a given pressure head. This curve is called the moisture capacity curve of the porous medium. That is,

$$C(h) = \left.\frac{d\theta}{dh}\right|_h = \left.\frac{\Delta\theta}{\Delta h}\right|_h \tag{3.6.1}$$

The vertical line in equation (3.6.1) indicates that the term is evaluated at a given h. The unit of the moisture capacity term, $C(h)$, is $[1/L]$. The sign of this term is always positive when h is negative. The moisture capacity term is analogous to the specific storage, a term used in saturated porous media (to be discussed in Chapter 5). Please note that even with such similarities, the physical mechanisms represented by the moisture capacity and specific storage are quite different. Whereas the former represents the strength of capillary pressure in porous media to hold or release water, the latter is related to the compressibility of both porous media and water.

3.7 Isotropy and Anisotropy

Few studies have investigated the unsaturated hydraulic conductivity anisotropy, especially at the laboratory scale. Unsaturated hydraulic conductivity at this scale is commonly considered to be isotropic. However, shapes and orientations of minerals in a soil core certainly affect the unsaturated hydraulic conductivity as they do the saturated hydraulic conductivity. That is, unsaturated flow paths perpendicular to the longest dimension of the minerals are more tortuous than those parallel to the longest dimension. Hence, the unsaturated hydraulic conductivity can be anisotropic. In Section 3.4, we deduce from the increase in the tortuosity due to a decrease in water saturation that the unsaturated hydraulic conductivity should also decrease. Therefore, one would expect that the anisotropy in unsaturated hydraulic conductivity would depend on the moisture content of a medium as well. The anisotropy in the unsaturated hydraulic conductivity nevertheless has often been considered as being an intrinsic property of a porous medium, the same as the anisotropy in saturated hydraulic conductivity. As a result, any anisotropy in the unsaturated hydraulic conductivity has been treated by scaling the unsaturated hydraulic conductivity in different directions in accordance with the anisotropy of the saturated hydraulic conductivity. This anisotropy thus remains constant over the full range of saturation or pressure head – a constant anisotropy. That is,

$$K_x(h) = K_{sx} \cdot K_r(h),\ K_y(h) = K_{sy} \cdot K_r(h),\ \text{and}\ K_z(h) = K_{sz} \cdot K_r(h) \qquad (3.7.1)$$

where $K_x(h)$, $K_y(h)$, and $K_z(h)$ are the unsaturated hydraulic conductivity values in the x, y, and z directions, respectively; K_{sx}, K_{sy}, and K_{sz} are the saturated hydraulic conductivity values in the x, y, and z directions, respectively; $K_r(h)$ is the relative unsaturated hydraulic conductivity (see equation 3.4.7). While this assumption may be counterintuitive, the anisotropy at this scale generally is insignificant.

Finally, we emphasize that the moisture content (and, in turn, the moisture retention curve) and moisture capacity terms are volumetric properties of a geologic medium. They are scalars, independent of directions, as is the specific storage term for a saturated medium.

Questions:

1. Draw some hypothetical $C(h)$ curves for sand and clay. If they are different, why are they different?
2. What is the equivalent term for moisture capacity in saturated porous media?
3. Since all the aforementioned unsaturated hydraulic properties represent the average properties over a volume of a geologic medium, can you apply the concepts of these properties to a pile of boulders or pebbles with pores as large as the boulders or pebbles?
4. Would the moisture retention curve or the moisture capacity curve exhibit anisotropy?

3.8 Darcy-Buckingham Law for Unsaturated Porous Media

3.8.1 Extension of Darcy's Law for Unsaturated Porous Media

Darcy's law was developed for flow in saturated porous media, and Edgar Buckingham (1907) extended it to moisture movement in unsaturated media. We will now introduce the so-called Darcy-Buckingham law for unsaturated porous media. Again, we will accomplish this by using reason and intuition, without resorting to detailed mathematics. Detailed mathematical derivation of the law can however be found in

Corey (1994). Up to now, we have shown that under saturated conditions, a linear relationship exists between specific discharge and the hydraulic gradient for flow through porous media. This is attributed to the nature of laminar flow or the fact that viscous force is greater than the inertial force for flow through a saturated porous medium (see Chapter 2, Section 2.6.2). In Section 3.4, we demonstrate that in an unsaturated porous medium, the wetted radius of pores and wetted porosity are smaller than those for fully saturated medium. We also suggest that the tortuosity of the medium under unsaturated conditions is much greater than that for saturated conditions. These results imply that flow in unsaturated media must be laminar, and a linear relationship should therefore exist between specific discharge and the hydraulic gradient. That is, Darcy's law will hold for flow through unsaturated media, and the governing equation for one-dimensional vertical unsaturated flow is

$$q = -K(h)\frac{dH}{dz} = -K(h)\frac{d(h+z)}{dz} = -K(h)\left[\frac{dh}{dz} + 1\right] \tag{3.8.1}$$

where q is the specific discharge, and H is the hydraulic head, which is equal to the sum of the pressure head, h, and the elevation head, z, positive upward. The term dh/dz denotes the pressure gradient and $dz/dz = 1$ represents the gravity (or elevation gradient). Equation (3.8.1) is Darcy's law for unsaturated flow and in soil physics literature it is often called Darcy-Buckingham law. For three-dimensional unsaturated flow, Darcy's law is given as

$$\mathbf{q} = -K(h)\nabla H \tag{3.8.2}$$

where $K(h)$ is a scalar (assuming that unsaturated hydraulic conductivity at the laboratory scale is isotropic) and ∇H is the hydraulic gradient vector. Equation (3.8.2) can be written in terms of the Cartesian coordinates as

$$\begin{gathered} q_x = -K(h)\left[\frac{\partial h}{\partial x}\right], \quad q_y = -K(h)\left[\frac{\partial h}{\partial y}\right], \text{ and} \\ q_z = -K(h)\left[\frac{\partial h}{\partial z} + 1\right] \end{gathered} \tag{3.8.3}$$

Equations (3.8.1), (3.8.2), and (3.8.3) also are macroscopic equations that employ the Darician continuum concept, as is the Darcy's law for saturated media. The variables in the equation represent spatial averages over a CV that consists of many pores. Therefore, these equations will not hold for processes within a volume of pores that is smaller than the CV. For example, these equations will not describe dripping of water in large voids of a rock pile consisting of large pebbles or boulders.

Thus, we can state that the major difference between Darcy's law for flow in saturated versus unsaturated porous media stems from the difference in the hydraulic conductivity. Under saturated conditions, hydraulic conductivity is solely a function of the medium and fluid properties (i.e., a constant for a given medium and fluid). But under unsaturated conditions, it varies as a function of pressure head (or moisture content), which changes with flow in both space and time. This dependence of hydraulic conductivity upon pressure head makes equation (3.8.1) nonlinear.

Lastly, at a given pressure head, the specific discharge in the unsaturated medium is assumed to be linearly proportional to the hydraulic gradient, as articulated previously. Note that this linear relationship does not imply that q is a linear function of the head. Although the relationship may be intuitive after our discussion, no experiment has been conducted explicitly to verify this assumption. This may be attributed to the difficulty in designing specific experiments for this purpose. Regardless, this convenient assumption has been taken for granted and seems to work well for many practical purposes, in particular those that involve granular porous media.

3.8.2 Pressure Head Profile during Steady Unsaturated Flow

Having introduced Darcy's law for unsaturated flow, we are ready to explain how the dependence of hydraulic conductivity on pressure head affects head distributions in a soil during unsaturated flow. Consider the steady downward unsaturated flow situation in Figure 3.15 (one-dimensional steady infiltration toward the water table). Recall

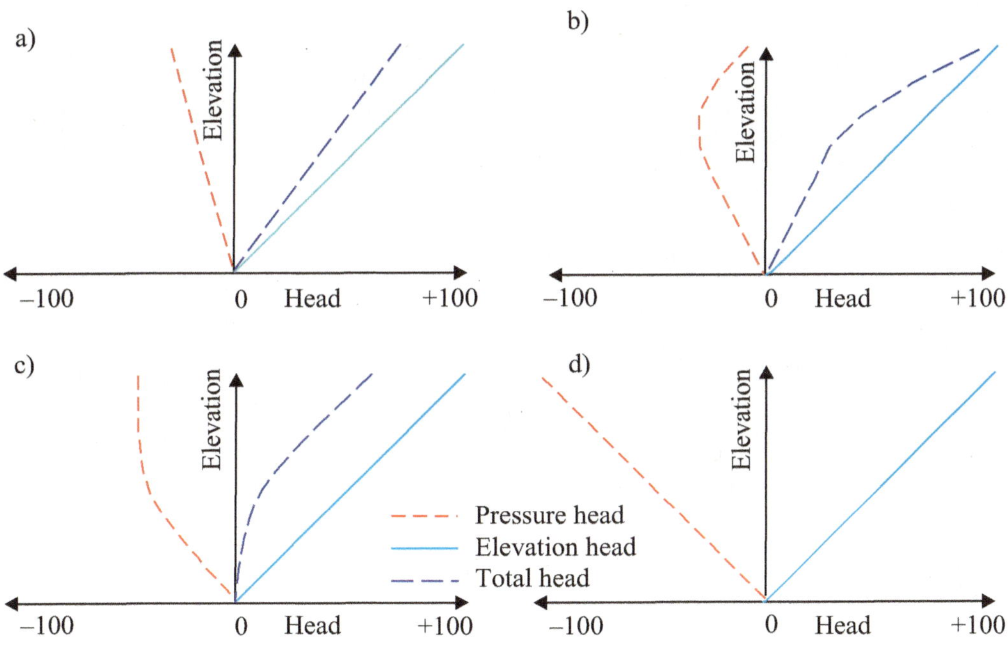

Figure 3.15. Pressure, elevation, and hydraulic head profiles during one-dimensional steady-state infiltration; a) and b) are physically incorrect pressure profiles, and c) and d) (no flow) are the likely pressure head profiles.

that in an early section, we asked the reason why the pressure head distribution along a soil column during steady infiltration is a curve (as shown in Figure 3.5 in Section 3.2), as opposed to the linear pressure head distribution in saturated flow (Chapter 2). To explain this nonlinear pressure head profile, we will use different possible head profiles presented in Figure 3.15a, b, c, and d. In these figures, the vertical axis is the elevation (z) and the horizontal axis is the head ranging from −100 to +100 unit of water. The elevation head is a 45° line increasing from 0.

Consider Figure 3.15a, where the negative pressure head is assumed to be linearly increasing with elevation from the zero pressure at the water table. The hydraulic head corresponding to this pressure profile thus is the straight line on the right, indicating that the hydraulic gradient must be a constant along the profile. Such a linearly increasing negative pressure head with elevation implies that the unsaturated hydraulic conductivity value of the soil becomes smaller with the elevation. On the basis of these facts and Darcy's law, the specific discharge must decrease with elevation. This decreasing specific discharge with elevation, however, contradicts the principle of a one-dimensional (1-D) steady flow in a homogeneous soil, where the specific discharge must be the same throughout the length of the column. Such a linear pressure head profile therefore cannot be physically correct during 1-D steady infiltration in an unsaturated homogeneous medium.

Figure 3.15b illustrates another possible pressure profile, in which the pressure head becomes more negative first then less negative as the elevation increases. This profile implies that the unsaturated hydraulic conductivity first decreases and then increases with an increase in elevation. Therefore, the unsaturated hydraulic conductivity values at two elevations could be the same. The hydraulic head, however, indicates that the hydraulic gradient continuously increases with the elevation. The calculated specific discharge based on Darcy's law therefore will not remain constant throughout the length of the column and this variable discharge violates the steady-uniform-flow principle, as stated previously for Figure 3.15a. This pressure profile is also physically incorrect.

One possible pressure profile that obeys the constant specific discharge principle is illustrated in Figure 3.15c, where the pressure head profile becomes increasingly negative as the elevation increases and then remains constant at higher elevations. As a consequence, the unsaturated hydraulic conductivity behaves in the same way. The hydraulic head shown on the right side of the figure suggests that the hydraulic gradient initially increases with the elevation and then becomes unity (that is, unit gradient: gravitational flow). Such a combination of unsaturated hydraulic conductivity and hydraulic gradient profiles thus may likely yield a constant specific discharge throughout the column.

Figure 3.15d shows another possible pressure head distribution: a straight line that becomes more negative from the water table upward with a slope of −1. This pressure head profile is in essence a hydrostatic pressure head profile for unsaturated media.

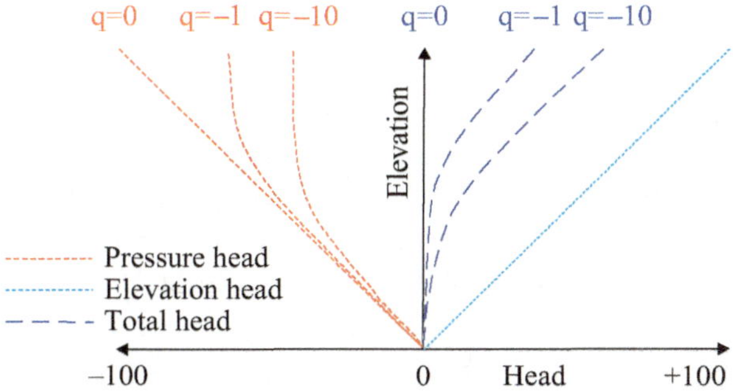

Figure 3.16. Pressure head, elevation head, and total head profiles during a one-dimensional steady infiltration (downward flow) toward a water table.

After combining it with the elevation head, the hydraulic (or total) head is uniformly zero, as are the hydraulic gradient and the specific discharge. Therefore, this pressure head profile is physically possible.

Once the effects of the dependence of the unsaturated hydraulic conductivity on the pressure head are understood, general head profiles for one-dimensional steady infiltration toward a water table can be evaluated; they are also illustrated in Figure 3.16 for different infiltration rates (i.e., q = 0, −1, and −10 cm/min). The negative sign of the infiltration rate denotes the downward flow. As shown in the figure, when there is no upward and downward flow (q = 0), the pressure head profile is negatively hydrostatic, and, hence, the total head is uniformly zero. As the infiltration rate increases (q = −1 cm/min), the profile exhibits some uniform pressure head condition near the top and gradually curves toward the water table. If the infiltration rate increases further (q = −10 cm/min), the steady pressure profile is similar to that of q = −1 cm/min, but the overall pressure head becomes less negative and the uniform pressure region increases. The total head profiles for these cases indicate that the hydraulic gradients gradually increase from the water table and reach unity at the uniform pressure region of the pressure head profiles.

Pressure head and total head profiles for steady-state evaporation (upward flow) at different rates are depicted in Figure 3.17. Opposite to the infiltration rate (q), which has a negative value in Figure 3.16, the evaporation rate (q) is denoted by a positive value, indicating that the hydraulic gradient must always be negative (i.e., upward flow). Therefore, the capillary pressure head gradient must be negative and has magnitudes greater than 1, representing the gradient due to gravity. In other words, the upward capillary force must be greater than the gravity, which pulls water downward. Similar to the infiltration cases, the pressure head profiles become more negative as the elevation increases from the water table. The pressure head gradient in the evaporation case, however, continues to increase and increases rapidly with the

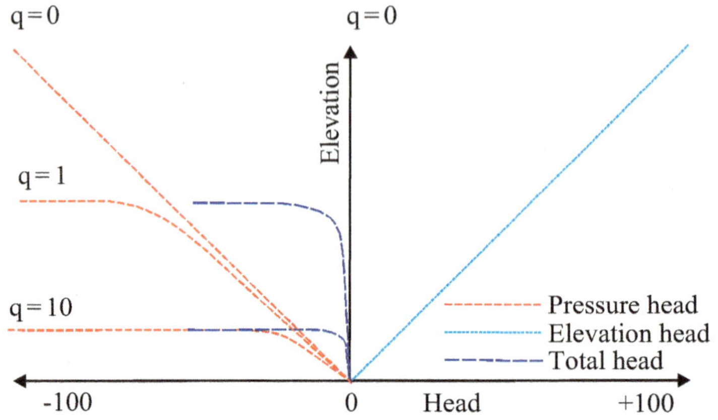

Figure 3.17. Head distributions during evaporation in a one-dimensional unsaturated medium with a water table at the bottom.

elevation. This suggests that the deeper the water table is, the greater the capillary force that is required to lift a given amount of water from the water table.

For one-dimensional vertical infiltration ($q < 0$) to or evaporation ($q > 0$) from zero elevation ($z = 0$) where the pressure head (h_0) is prescribed, equation (3.8.1) can be used to derive an analytical solution for the head profile. The solution is given in the following if the exponential model is used for the unsaturated hydraulic conductivity function:

$$\int_{z=0}^{z=z} dz = \int_{h_0}^{h(z)} \frac{-dh}{\left(\frac{q}{K_s} e^{\beta h} + 1\right)} \tag{3.8.4}$$

The result of the integration is

$$h(z) = \frac{1}{\beta} \ln \left[\exp[-\beta(z - h_0)] + \frac{q}{K_s} \exp(-\beta z) - \frac{q}{K_s} \right] \tag{3.8.5}$$

Note that this solution is only valid for unsaturated conditions (i.e., pressure head less than zero). Yeh (1989)) extends this solution to layers of different K_s and β parameters.

Lastly, flow through unsaturated media is a very complex process that involves air, water, and solids. However, the discussion of unsaturated flow presented in this chapter ignores possible air movement in the media. This assumption generally is valid as long as the media have any outlets for air to exit. In addition, when the media become dry, the flow in an unsaturated medium can be highly discontinuous at a local scale because the propagation of the negative pressure head is generally not as diffusive as that in the positive pressure head in saturated media. This can be attributed to the presence of air in unsaturated media. More importantly, the movement of moisture often involves advection due to gravity, in addition to diffusion induced by the pressure head gradient. It is highly sensitive to heterogeneities at different scales. As a result, prediction using the continuum assumption (Darcian continua) likely will fail for flow through initially very dry media, or media with a high degree of heterogeneity. More specifically, the Darcy-Buckingham law or equation is strictly valid for media with a continuous pressure head field. These complications do not necessarily preclude the use of our continuum assumption, however. Nevertheless, we must bear in mind that such an application of the Darcian continua assumption may lead to gross misrepresentations of the actual processes in unsaturated media.

Questions:

1. While flow through a pile of boulders or pebbles may be visualized as flow through unsaturated geologic media all over, is the Darcy-Buckingham law valid for flow through this type of environment? Please explain. (Hint: The existence of the pressure gradient.)
2. Can you design an experiment to test Darcy-Buckingham law for unsaturated flow?

3.9 Measurements of Unsaturated Hydraulic Conductivity

Several methods for determining the unsaturated hydraulic conductivity are described in this section. They will be categorized as a) steady flow and b) transient flow approaches.

3.9.1 Steady Flow Approach

The steady flow approach is solely based on Darcy-Buckingham law for unsaturated flow (see equation 3.8.1). That is,

$$q = -K(h)\frac{d(h+z)}{dz} = -K(h)\left[\frac{dh}{dz}+1\right] \tag{3.9.1}$$

Generally, the steady flow approach uses either a non-unit-gradient or unit-gradient method.

1. Non-unit-gradient method. Consider steady unsaturated flow in a soil column enclosed by two ceramic plates (Figure 3.18). The bottom plate is connected to a closed reservoir, under a specified vacuum. The top plate is connected to a water supply source that is also under a specified vacuum. Two tensiometers, separated by a short distance L, are installed along the column at locations #1 and #2 (see Figure 3.18) to measure the pressure heads h_1 and h_2, at the top and the bottom, respectively.

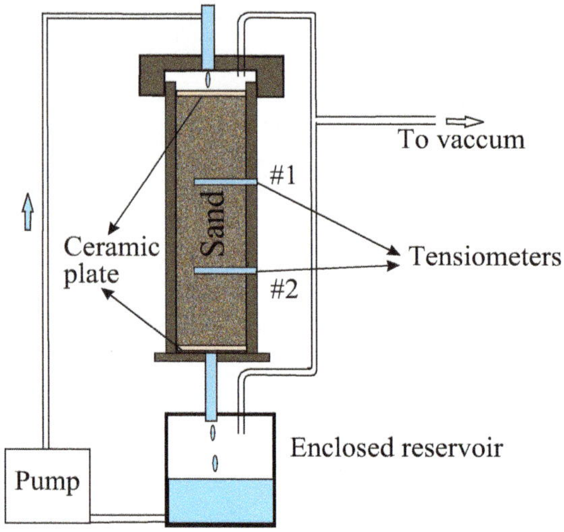

Figure 3.18. Illustration of an experimental setup for measuring unsaturated hydraulic conductivity.

If the specific discharge q is known or measured, the unsaturated hydraulic conductivity can then be calculated by

$$K(\bar{h}) = -\frac{q}{\left(\frac{h_1 - h_2}{L}\right) + 1} \tag{3.9.2}$$

where $\bar{h} = (h_1 + h_2)/2$. By varying the specific flux, this approach yields many pairs of K and h values, and, thus, a conductivity/pressure head relationship can be established. In this approach, the two tensiometers must be installed close to each other, so that the pressure head variation between the two points is approximately linear, and the pressure head difference is small such that the average of the two pressure heads is a good approximation of the pressure head for the corresponding conductivity.

2. Unit-gradient method. This method generally uses either a) a short soil column or b) a long soil column.

A short soil column approach. This method can use the apparatus shown in Figure 3.18. If the pressure values at points 1 and 2 reach the same value and the flow is steady, according to Darcy's law we have $q = -K(h)$. By varying the q at the surface of the column to reach different steady-state conditions, one can obtain many $K(h)$ values and, in turn, the $K(h)$ over a large range of h values. Please note that while the method is conceptually simple, it is difficult to adjust fluxes and pressures at the bottom and top of the soil column to reach a unit gradient and steady flow conditions.

A long column approach. An alternative to the short column approach is the long column approach. That is, it is possible to obtain the unit gradient condition in

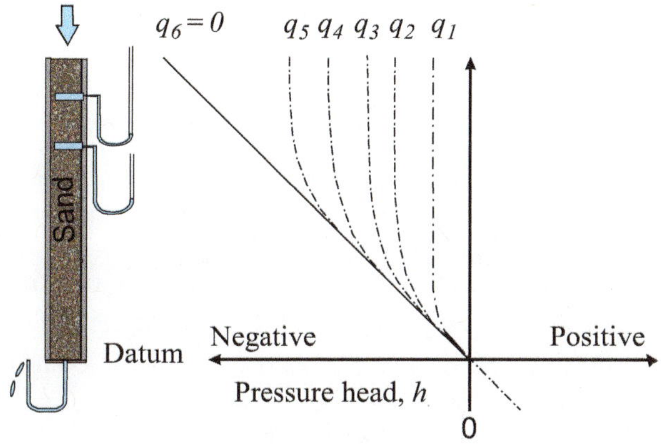

Figure 3.19. A long column approach for measuring the unsaturated hydraulic conductivity and pressure head relationship.

the upper part of a long column, during steady downward flow to a "water table" near the base of the column. This can be done without any ceramic plates at the top and bottom, and without any adjustment other than that necessary to control the inflow rates at the top of the column. Figure 3.19 illustrates the approach, showing a long soil column instrumented with several tensiometers for monitoring the pressure profile along the column. First, a constant flux that is slightly smaller than the saturated hydraulic conductivity is applied at the top of the column. Once the steady flow corresponding to the applied flux is established, the pressure head profile is then recorded. By reducing the inflow rate, several pressure head profiles corresponding to different steady flows can be obtained, as illustrated in Figure 3.19. Notice that all the pressure head profiles near the upper portion of the column exhibit essentially uniform pressure head distributions, indicating a unit gradient condition. Therefore, the uniform pressure head and the known inflow rate constitute a pair of unsaturated hydraulic conductivity and pressure head values. Many pairs of hydraulic conductivity and pressure head data sets will yield an unsaturated conductivity curve during a drying cycle for the material in the column.

Similarly, one can obtain an unsaturated conductivity curve during a wetting cycle for the material in the column if the flow experiment starts with the smallest flow rate and the rate is successively increased.

3.9.2 Transient Approach

The transient approach to determine the unsaturated hydraulic conductivity curve relies on the assumption that a one-dimensional unsaturated flow equation (see the

Richards equation in Chapter 5) adequately describes unsaturated flow in a soil column. The equation is given as

$$\frac{\partial}{\partial z}\left[K(h)\frac{\partial(h+z)}{\partial z}\right] = \frac{\partial\theta}{\partial t} \tag{3.9.3}$$

where $K(h)$ is the unsaturated hydraulic conductivity at a pressure head of h, z denotes the elevation head (positive upward), and θ is the moisture content.

After integrating the Richards equation (3.9.3) over the length of the soil column (from $z = 0$, the bottom of a column, to $z = L$, the top of the column), we have

$$-K(h)\left[\frac{\partial h}{\partial z}+1\right]\bigg|_{z=0} = \frac{\partial}{\partial t}\left[\int_{z=0}^{z=L}\theta(z,t)\right] - K(h)\left[\frac{\partial h}{\partial z}+1\right]\bigg|_{z=L} \tag{3.9.4}$$

The term on the left-hand side of equation (3.9.4) is the outflow at $z = 0$. The first term on the right-hand side of equation (3.9.4) denotes the change in storage while the second term is the specific discharge applied to the top of the soil column ($z = L$). The sum of the two terms on the right-hand side of equation (3.9.4) yields the specific discharge at the end of the column, $q(0, t)$.

$$-K(h)\left[\frac{\partial h}{\partial z}+1\right]\bigg|_{z=0} = q(0,\ t) \tag{3.9.5}$$

Equation (3.9.5) suggests that the pressure head, the pressure gradient, and the discharge at $z = 0$ at a specified time must be known in order to define $K(h)$ uniquely at the specified pressure head. Generally, the change in moisture content along the column can be determined either by a weighting method or by a neutron probe or by a time domain reflectometer or by an electric method, and so forth. Additionally, the pressure head and its gradient are determined by tensiometers. This scenario actually forms the basis of the well-known instantaneous profile method (Rose et al., 1965; Watson, 1966). Determination of unsaturated hydraulic conductivity is a rather complicated and difficult inverse problem, and it will be dealt with in further detail in Chapter 9.

3.10 Summary

- The pressure head in unsaturated media is negative, less than the atmospheric pressure head, representing the capillary pressure or suction head. If the pressure head is equal to or greater than zero, the geologic medium is fully saturated.
- The moisture content of a medium decreases as a function of its negative pressure. This relationship is the moisture release or retention curve characteristic of a medium.

It represents the amount of water that can be released from a medium if a given suction is applied to the medium.

- Hydraulic conductivity of the medium also decreases as the pressure becomes more negative or the medium becomes less saturated.
- The shapes of the moisture retention curve and hydraulic conductivity curve are related to the cumulative pore size distribution of the medium.
- The linear relationship between the flux and hydraulic head (or energy) gradient remains valid for flow through unsaturated media. The medium becomes more resistant to the water flow than under saturated conditions, and therefore the average flow must be laminar.
- The governing equation for flow under unsaturated conditions is a nonlinear equation because of the dependence of the hydraulic conductivity on the pressure head or the moisture content.
- In a vadose zone, the pressure head gradient can spread water horizontally, it can assist gravity to move water downward, or it can work against gravity, by pulling water upward.

3.11 Exercises

1. Two tensiometers placed in a soil profile at locations A and B indicate matric potentials of $h_A = -200$ cm and $h_B = -300$ cm, respectively; location B is 10 cm below location A. The unsaturated conductivity for the soil is given by $K(h) = K_s \exp(\beta h)$, where K_s = saturated hydraulic conductivity = 0.01 cm/day and $\beta = 0.01$ cm^{-1}. Assume location A is 100 cm below the land surface and flow in the soil is at a steady state. Answer the following questions.
   a. What are the hydraulic heads (in centimeters) at A and B?
   b. What is the hydraulic gradient based on the linear approximation?
   c. In which direction is flow occurring?
   d. For an average h within the profile, what is the soil water flux (centimeters/day)?
   e. Determine the flow through an area of 10 cm^2.
   f. What is the average moisture content in the soil profile? Assume that the relationship between matric potential and moisture content is given by $\theta = a\,|h|^n$, where $a = 0.8$, $n = -0.2$, and | | denotes absolute value.
   g. What is the average linear velocity through the soil profile?
   h. How long (days) will a contaminant take to travel from A to B by advection alone?
   i. What would be soil water flux (centimeters/day) through the profile under a unit hydraulic gradient condition?
2. A soil column contains 50 cm of sand over 50 cm of clay. A piezometer at the sand-clay interface measures a pressure head of 50 cm. The entire column is saturated and 20 cm of water is ponded on the top while the bottom is the water table. The steady measured flux rate at the clay bottom is 20 cm/h. Calculate the saturated hydraulic conductivity of the sand and the clay materials.
3. A vertical cylindrical soil column of 100 cm^2 cross-sectional area and 50 cm height is filled with homogeneous soil; 10 cm of water is continuously ponded on the top of the column to reach a steady-state condition, with a volumetric flow rate Q through the soil equal to 1000 cm^3/h (downward). (a) Calculate the steady-state flux (discharge per unit area or specific discharge) through the column. (b) Calculate the saturated hydraulic conductivity of the soil. A 1-mm-diameter tube is pushed through the column and hollowed out (representing

macropores). Under steady state, water flows through the soil in accordance with Darcy's law and through the tube in accordance with Poiseulle's law. (c) Calculate the volumetric flow rate Q through the tube. (d) Calculate the flux through the column-tube system.

4. A bundle of capillaries composed of glass tubes having the following diameters and numbers was dipped (vertically) into a water reservoir. Compute and plot the relative saturation of the capillaries at different elevations above the free water.

No. of capillaries	300	400	325	250	150	90	50	10	5	2
Diameter (mm)	0.001	0.005	0.01	0.02	0.05	0.1	0.2	0.5	1	2

5. Assume that 1-D steady infiltration of water into a porous medium can be described by Darcy's law,

$$q = -K(h)\left[\frac{dh}{dz} + 1\right]$$

where q is the infiltration rate, z is the elevation (positive upward), $K(h) = K_s \exp(\beta h)$, and K_s is the saturated hydraulic conductivity and β is the pore-size parameter. The analytical solution to the equation for a given infiltration rate, q, and a bottom pressure head boundary condition, h_0, is given as

$$h(z) = \frac{1}{\beta}\ln\left\{\exp[-\beta(z - h_0)] + \frac{q}{K_s}\exp(-\beta z) - \frac{q}{K_s}\right\}$$

when $h(z) < 0$. Figure 3.20 shows a possible pressure profile in a soil during steady infiltration. If $q > K_s$ and the pressure head at the bottom h_0 is negative (unsaturated condition), please determine the elevation, z^*, where $h(z^*) = 0$ (saturation) and the ponding depth (h_p) at the surface if the distance between the bottom boundary and the surface is d. Express the solution in a dimensionless form, using d, and plot the result.

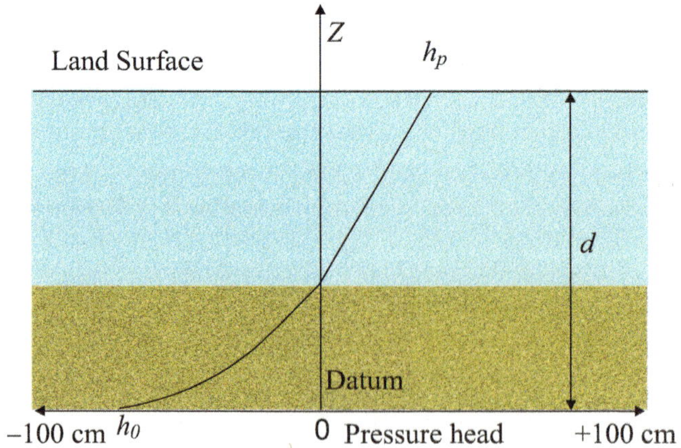

Figure 3.20. A possible pressure profile in a soil during steady infiltration.

6. a) Use the following table to explain the 1-D steady flow phenomenon associated with each pressure head profile illustrated in Figure 3.21. Note that q is negative downward.

	dH/dz > or < 0 where $H = h + z$	**Downward or Upward Flow**	**q <, =, or > K_s, Saturated K**	**Driving Forces of the Flow (dh/dz, dz/dz, or both)**
A				
B				
C				
D				
E				

b) Which one of the profiles does not follow the principle of one-dimensional steady flow in a homogeneous soil under either saturated or unsaturated conditions? Please explain.

7. A capillary pressure/degree of saturation relationship for water in a soil is shown in Figure 3.22. Draw the relationship curve if a different fluid, say, alcohol, is used and explain your logic.

8. If a tensiometer (see Figure 3.23) is filled with water to the top and a pressure gauge reads the water pressure head at the top, what is the pressure head around the ceramic cup of the tensiometer?

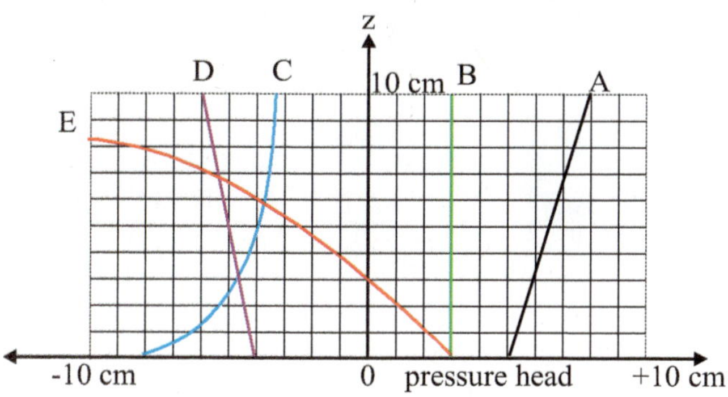

Figure 3.21. A variety of possible pressure profiles along a vertical line of a soil during steady flows.

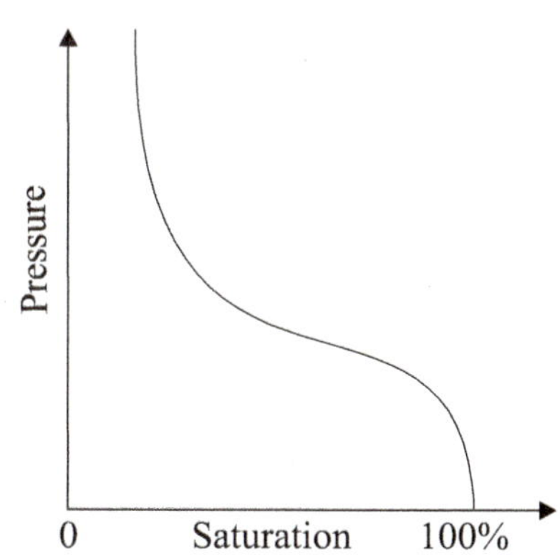

Figure 3.22. A capillary pressure/degree of saturation relationship for water in a soil.

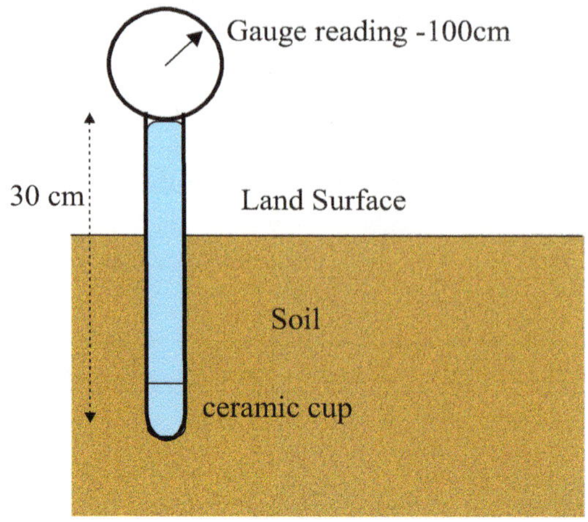

Figure 3.23. A tensiometer in a soil for Problem 8.

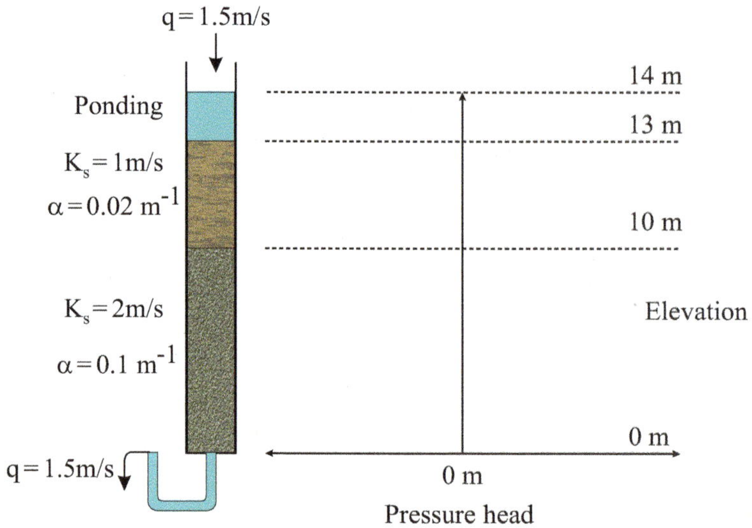

Figure 3.24. A steady-state infiltration in a soil column setup.

9. Sketch the pressure head distribution along the soil column during a steady infiltration experiment, as shown in Figure 3.24. Show your calculation, or state the rationale behind your sketch. Specifics: 1) determine the elevation where the pressure head is zero and 2) determine the pressure head at the interface between the upper soil and lower soil. We assume that the soil hydraulic conductivity–pressure head relationship can be described by Gardner's exponential model.
10. You are required first to solve Problem 9 using a numerical model and plot the profile, and then derive the analytical solution to the problem and compare the results with those from a numerical model.

4

Stochastic Conceptualization of Heterogeneity

4.1 Introduction

In Chapters 2 and 3, we defined hydraulic conductivity and explained Darcy's law for porous media over some control volume (CV), which must be large enough to include many pores (i.e., a laboratory scale CV). As we apply this CV approach to a field site or a groundwater basin, the hydraulic conductivity value defined over such a CV exhibits significant variability at various scales. In order to conceptualize and quantify this spatial variability, in this chapter we first define various scales of variations (i.e., heterogeneities) and then introduce stochastic random field concepts and approaches for describing heterogeneities. Field applications of such stochastic approaches and concepts are presented in the end.

4.2 Scales of Heterogeneity

Hydrologic properties of geologic media always exhibit some degree of spatial variability at various scales (see Figures 4.1 and 4.2). This variability is a rule rather than an exception! For laboratory-scale problems (i.e., small cores, soil columns, and sandboxes), variations in pore size, pore geometry, and tortuosity of pore channels are the major sources of heterogeneity. These types of heterogeneities will be referred to as the **laboratory-scale heterogeneity**. Microstratification, foliation, cracks, and plant roots are also possible sources of heterogeneity at this scale. As the scale of a problem increases, stratification or layering in a geologic formation becomes the dominant heterogeneity, which will be classified as the **field-scale heterogeneity**. At an even larger observation scale, the **mesoscale heterogeneity** represents the variations of the geologic formations, or facies, as well as variations in depositional environments.

Figure 4.1. An outcrop of a glacial outwash sand deposit at Hanford Site, Washington. Figure on the left shows stratifications of the deposit; figure on the right shows a zoom-in view of the small-scale cross-bedding within one of the strata.

Figure 4.2. Figure on the left shows small-scale heterogeneity within stratifications; figure on the right shows large-scale heterogeneity (stratifications). (photos provided by Brian A. Hampton)

In Chapters 2 and 3, we adopted the CV and representative elementary volume (REV) approaches to overcome problems associated with the laboratory-scale heterogeneity (this CV and REV will be referred to as the laboratory-scale CV and REV). As a result, fundamental theory for flow through porous media (i.e., Darcy's law) was derived, and associated hydraulic properties were defined. These hydraulic properties will hereafter be referred to as the laboratory-scale hydraulic properties.

4.3 Need for a Probabilistic Description of Geologic Heterogeneities

Geologic heterogeneities at multiple scales are a well-known fact (e.g, Scheibe et al., 2014). For example, Nielsen et al. (1973) and others reported that even seemingly uniform field sites manifest large spatial variations in the saturated hydraulic conductivity for laboratory-scale samples. As we analyze the flow processes in a groundwater basin, we can adopt the laboratory-scale CV, as well as Darcy's law, and then determine the laboratory-scale properties at every location across the entire basin. An immediate question confronting this approach is, "Can we determine these properties at this scale everywhere over the entire basin?" If we can, how to quantify the variability or spatial distribution of the properties becomes the next major question. On the other hand, if we can only take measurements at a limited number of locations, as is more likely, then how we estimate the property at other locations and assess the uncertainty associated with the estimate become the next set of questions. The stochastic conceptualization of heterogeneity, to be discussed below, provides the basis to answer these questions in a meaningful way (Yeh, 1992; 1998).

To introduce stochastic concepts to subsurface hydrology, we will consider spatial distribution of the laboratory-scale saturated hydraulic conductivity along a transect line at a field site (see Figure 4.3). Conventional wisdom suggests that if the hydraulic

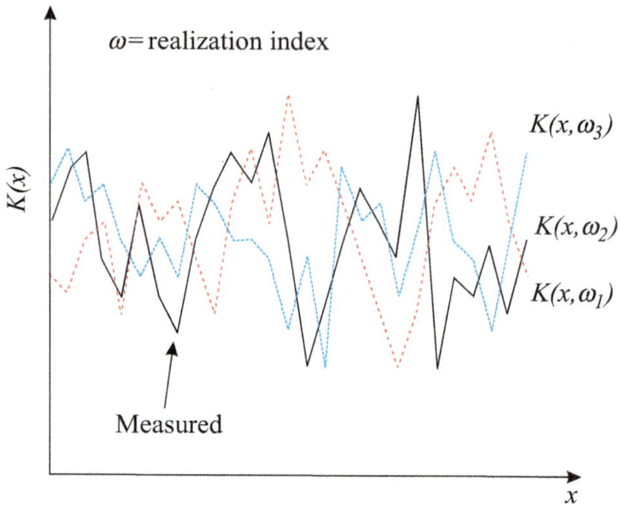

Figure 4.3. Schematic illustration of the stochastic process of the hydraulic conductivity distribution along a transect line. The solid line represents the measured hydraulic conductivity values; the dashed lines denote some possible values (different realizations) along the transect line.

conductivity, $K(x_i)$, at a location, x_i, along this transect line is measured exactly, then there is no uncertainty about the K value at that location, x_i, and the hydraulic conductivity should be a deterministic parameter (i.e., a parameter known with certainty). Similarly, had the hydraulic conductivity at every location of the transect line been measured precisely, the K field of this transect (a collection of all Ks) should be treated as a deterministic parameter field. Therefore, probabilistic conceptualization of the hydraulic conductivity field is unnecessary.

On the other hand, if there is no measurement of K, or its measurement involves errors whose magnitudes are unknown, we have to estimate or guess its value at a location. Our guess or estimate in essence makes us unknowingly conceptualize the K at the location as a random variable, characterized by a probability distribution.

While the probabilistic conceptualization of a K field is comprehensible as discussed, the fact is that the K at a given location, even if it has been determined exactly, always can be treated as a random variable. This is equivalent to saying that the value of K determined at that location represents just one of the many possible geologic materials that may have been deposited at this location when the environment came into existence. This is analogous to an experiment in which a coin is flipped, where the outcome is always deterministic since it has already occurred. However, this outcome is just one of the two possibilities (a head or tail). To determine the likelihood of occurrence of head or tail, the outcome must be considered as a random variable. Therefore, a hydraulic property of a geologic formation always can be considered as a random variable regardless of whether the property is known or unknown or involves measurement errors.

4.4 Stochastic Processes and Ensemble Moments

According to the preceding discussion, the hydraulic conductivity at a given location, $K(x_i)$, can be considered as a random variable, with a range of possible values, some of which may be more likely to occur than others. Specifically, at a given location, x_1, along the transect line there exists a probability density distribution $f(k(x_1))$ for hydraulic conductivity. Similarly, at other locations, $x_2, \ldots x_n,$ the hydraulic conductivity values, $K(x_2), \ldots K(x_n)$, are random variables with the distributions $f(k(x_1)), \ldots$ and $f(k(x_n))$ respectively. As a consequence, the hydraulic conductivity field along the transect line can be considered as a collection of random variables. This collection of an infinite number of random variables is then called a **stochastic process**. For geologic or hydrologic heterogeneity, the term **random field** is often used instead of the term "stochastic process."

Since the hydraulic conductivity distribution along the transect line is a stochastic process, there must exist many possible sequences of $K(x)$ values along this transect line, some of which may be more likely to occur than others. The actual hydraulic conductivity values, if they are measured along the transect line (for example, the solid black line, $K(x, \omega_2)$, in Figure 4.3.) are then just one possible sequence of K values out of all the possible sequences of $K(x, \omega_i)$. Note that ω_i is an ensemble index indicating the sequence number, and it ranges from ω_1 to ω_∞. For convenience, we will drop this index in the notation. In addition, it is the convention that the capital letter, K, is used to denote the hydraulic conductivity random variable, and the lowercase letter to designate its value (a number). The **ensemble** is then defined as a collection of an infinite number of space series measuring the same variable. A **realization**, then, is a single space series in an ensemble. For example, a set of porosity values measured in an aquifer is considered as a realization of all the possible porosity values of the aquifer. Note that this stochastic concept is also applicable to variables in the time domain (i.e., time series). For example, any temporal water level fluctuation in a well can be considered as a single realization, out of all the possible temporal series, or an ensemble.

Since there are a large number of possible sequences of $k(x)$ values along the transect line, we then ask, "What is the probability of finding a particular sequence of the $k(x)$ value?" In order to answer this question, a joint probability distribution (JPD) of these random variables must be used. This is because, in general, knowledge of the probability density function of the hydraulic conductivity at each location reveals nothing about the joint variation of these random variables, or the likelihood of concurrence of all the hydraulic conductivity values in any sequence. Therefore, a joint probability density function is necessary for this purpose.

4.4.1 Joint Probability Distribution

Joint probability distribution is defined as a distribution of the probability of occurrence of many random variables. More precisely, if $K(x_1), \ldots K(x_n)$ are n random variables, we will refer to a function, f, that is, $f(K(x_1), \ldots, K(x_n))$, as the joint probability density of these random variables, if the probability that $a_1 < k(x_1) < b_1, \ldots$ and $a_n < k(x_n) < b_n$ is given by

$$\int_{a_1}^{b_1}\int_{a_2}^{b_2}\cdots\int_{a_n}^{b_n} f[K(x_1),\, K(x_2), \ldots K(x_n)]dk_1 \cdot dk_2 \cdot \ldots dk_n \qquad (4.4.1)$$

where the density function f must be greater than or equal to zero for all values of $K(x_1), \ldots K(x_n)$ for which the function is defined, and

$$\int_{0}^{\infty}\int_{0}^{\infty}\cdots\int_{0}^{\infty} f[K(x_1),\, K(x_2), \ldots K(x_n)]dk_1 \cdot dk_2 \cdot \ldots dk_n = 1 \qquad (4.4.2)$$

If $K(x1), \ldots K(x_n)$ are statistically independent (i.e., the likelihood of occurrence of a $k(x_1)$ value at x_1 is unrelated to the likelihood of occurrence of $k(x_2)$ at x_2, then the joint probability density function is given by

$$f[K(x_1), K(x_2), \ldots K(x_n)] = f[K(x_1)] \cdot f[K(x_2)], \ldots f[K(x_n)] \qquad (4.4.3)$$

Once a joint probability density function or JPD is determined, the probability of the occurrence of a particular realization of the k values along the transect line can then be calculated.

4.4.2 Statistical Moments

While the joint probability density function of a stochastic process can be generally described as normal (Gaussian), or log-normal, and so on, specification of its ensemble statistical parameters (i.e., ensemble moments) is necessary to characterize the density function fully. These ensemble statistics are discussed in the following.

First moment. The first moment of a JPD is defined as

$$E[K] = \int_{0}^{\infty} kf(k)dk = \mu \qquad (4.4.4)$$

where K denotes the stochastic process, and k represents its value. The operator, E[], denotes the mean or ensemble average and $f(k)$ is the joint probability density function. The first moment defines the ensemble mean of the process, representing the most likely value of the hydraulic conductivity at any location.

Second moment. The second moment, covariance function, or autocovariance function is given as

$$C(\xi) = E[(K(x) - \mu)(K(x + \xi) - \mu)] \tag{4.4.5}$$

where ξ (pronounced as "xi") is the separation distance, or lag, or shift. When $\xi = 0$, the covariance function becomes the variance:

$$C(0) = E[(K(x) - \mu)^2] = \sigma^2 \tag{4.4.6}$$

The variance characterizes the spread of the hydraulic conductivity around its mean value.

An autocorrelation function is defined as the ratio of the autocovariance function to the variance

$$\rho(\xi) = \frac{C(\xi)}{\sigma^2} \tag{4.4.7}$$

where $-1 \leq \rho(\xi) \leq 1$. It is a function showing the correlation of the series to itself at different separation distances. In other words, the autocorrelation function describes the spatial relation between random variables at different locations. Higher order moments (such as skewness and kurtosis) certainly can be defined, but they are often ignored because of mathematical difficulties in analyzing them.

Next, we will introduce stationary, nonstationary, and second-order stationary stochastic processes, which are assumptions we commonly evoke when estimating the JDP.

4.4.3 Stationarity and Nonstationarity

Under this condition, statistical properties of a stochastic process remain stationary or constant in space. A more precise definition of stationarity can be given as follows: A process, $K(x)$, is said to be strictly stationary (or **statistically homogeneous or translational invariant**) if, for any admissible $x_1, x_2, \ldots x_n$ and any separation, ξ, the JPD of $\{K(x_1), K(x_2), \ldots, K(x_n)\}$ is identical to the JPD of $\{K(x_1 + \xi), K(x_2 + \xi), \ldots, K(x_n + \xi)\}$. Any stochastic process that does not meet the requirement of stationarity is called a **nonstationary (statistically heterogeneous)** process. That is, if the moments of a stochastic process fail to be stationary, the process is nonstationary. Strict stationarity is a rather strong requirement. To relax this requirement to some extent, the concept of a second-order stationary process is introduced next. Notice that the stationarity has a similar definition to that of the spatial homogeneity discussed in Chapter 2. That is, if a spatial REV exists for a geologic formation (i.e., the property defined in the CV is spatially invariant), this geologic formation is considered as a homogeneous medium, and it is a heterogeneous medium if the REV does not exist.

Second-order stationarity: A stochastic process is said to have second-order stationarity if the first and second moments of the process have the following properties:

1) $E\,[K(x)] = \mu$, a constant, invariant in x
2) $E\,[(K(x) - \mu)^2] = \sigma^2$, a constant, invariant in x
3) Cov $[K(x), K(x+\xi)] = C(\xi)$

Specifically, the covariance function of the process depends only on the separation distance, ξ. That is, the covariance of any pair of random variables will be the same as long as the distance between the variables in the pair is the same, regardless of the locations of the pair. The second-order stationarity means that a stochastic process is stationary in terms of the first and second statistical moments only, and anything other than these two moments is ignored. As mentioned previously, it is difficult to cope with moments of orders higher than 2 mathematically. Therefore, second-order stationarity, or implicitly a Gaussian (multivariate normal) process, is merely an assumption for mathematical convenience. If the process is Gaussian, then the second-order stationary assumption is optimal. Otherwise, the second-order stationary assumption is the best we can do.

4.5 Ergodicity and Moment Estimates

Any application of the second-order stationarity concept to a field problem requires the **ergodicity assumption**. Defining the first and second moments of a second-order stationary process demands knowledge of its joint probability density function (JDP). However, this JPD is seldom known in reality, since the determination of the JPD requires that all the possible sequences of K values of a field site (e.g., along the transect line) be known in advance. Furthermore, any hydraulic conductivity data set collected from a field site presents merely a single realization out of the ensemble. Hence, there is no way we can determine the JPD of the hydraulic conductivity field, and in turn, its first and second ensemble moments.

To alleviate this difficulty, we must invoke the ergodicity assumption. This assumption allows us to estimate the first and second moments of the stochastic process from our samples, which are only single realizations of the ensemble. **Ergodicity** implies that any realization of a stochastic process in space has the same probability distribution – the first and second moments in the case of the second-order stationary process – as the ensemble of possible realizations. This suggests that by observing the spatial variation of a realization of a stochastic process, it is possible to obtain statistical properties of the process for all realizations. In other words, this is equivalent to saying that if we see one, then we have seen them all. Oops! An unrealistic assumption! Yes, your common

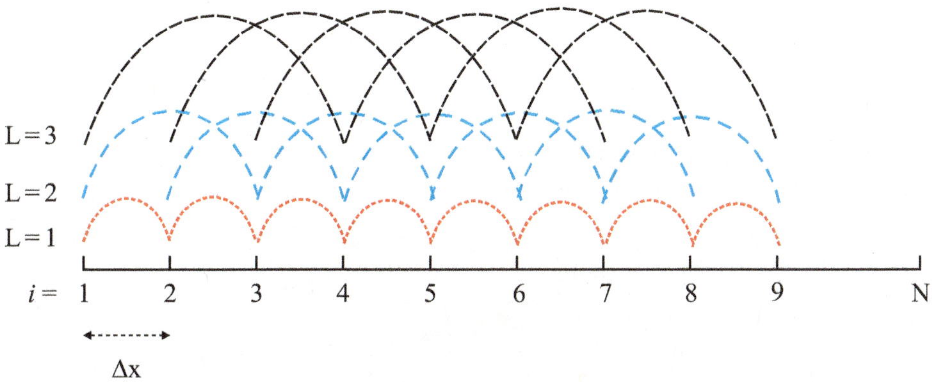

Figure 4.4. A schematic illustration of the sampling locations along a transect line and pairs and lags used in the autocovariance function estimates.

sense is correct! However, we have no choice but to accept this rather unrealistic assumption since we can only sample one realization at a given field site. Note that ergodicity can neither be verified nor tested. This gross assumption, of course, will impact our analysis of uncertainty. But **the objective of the uncertainty analysis is not to seek the absolute uncertainty but rather the relative uncertainty among different operational strategies, based on the same simplified assumptions.** We therefore may as well accept ergodicity as a working hypothesis.

The definitions of mean, variance, autocovariance, and autocorrelation functions [equations (4.4.4) to (4.4.7)] are based on the assumption that we know the joint probability density function of the stochastic process. When these definitions are applied to a field data set, the previously mentioned ergodicity assumption must be invoked so that the moments of the stochastic process can be estimated from the samples we have collected. The following example illustrates how estimates of the mean and covariance (more specifically, sample mean and sample covariance) are calculated from a spatial data set.

Suppose we take N measurements of the saturated hydraulic conductivity at equal intervals Δx along a transect line (Figure 4.4). The measured value of the saturated hydraulic conductivity at the ith location is denoted by $K(i)$. Suppose we consider that the hydraulic conductivity is a stochastic process; then the sample mean, covariance, and autocorrelation functions of the process are defined as follows.

The sample mean of a process is determined by using

$$\overline{K} = \frac{1}{N}\sum_{i=1}^{N} K(i) \tag{4.5.1}$$

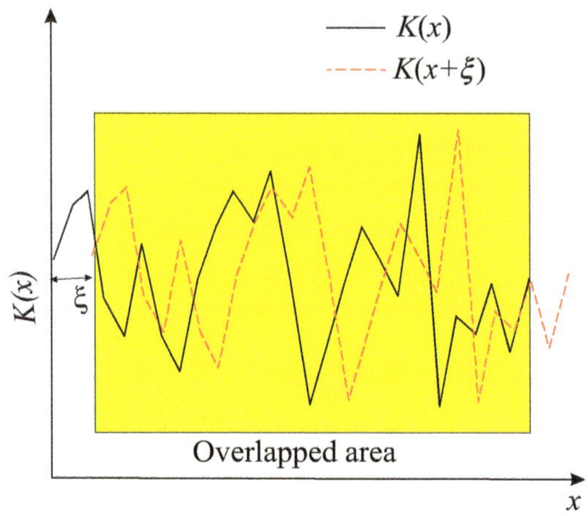

Figure 4.5. An illustration of the concept of autocovariance and autocorrelation analysis.

where N is the total number of the hydraulic conductivity samples. Sample covariance of the stochastic process, $K(i)$, can be estimated by

$$\mathrm{Cov}(L) = \frac{1}{(N-L)}\sum_{i=1}^{N-L}[K(i)-\overline{K}]\,[K(i+L)-\overline{K}] \tag{4.5.2}$$

where $L = 1, \ldots, N-1$, which is the number of lags. Each lag and the separation (offset) distance, ξ, is equal to $\Delta x \times L$. When $L = 0$, equation (4.5.2) becomes

$$S^2 = \mathrm{Cov}(0) = \frac{1}{N}\sum_{i=1}^{N}[K(i)-\overline{K}]\,[K(i)-\overline{K}] \tag{4.5.3}$$

where S^2 is the variance estimate. In other words, at zero offset, the sample covariance becomes the sample variance. It is the arithmetic average of the sum of the square of the hydraulic conductivity perturbation (i.e., the difference between the measured hydraulic conductivity value and its mean) at each location over the N sample locations. On the other hand, the covariance of the data at a given lag, L, (or a separation distance, $\xi = \Delta x \times L$), as defined by equation (4.5.2), is the sum of the product of the hydraulic conductivity perturbation pair separated by the given lag over the total number of pairs (see Figure 4.4). Figure 4.5 offers another way to explain the covariance at a given separation distance, ξ. The black solid line in Figure 4.5 denotes the original K space series; the red dashed line represents the K series shifted by ξ; the yellow region denotes the portions of the two series overlapping each other. The covariance at this given shift is then the sum of the product of the mean removed hydraulic conductivity values of the two series at each location along the overlapped region, divided by the total number of K values within the overlapped portion. As the number of lags increases, the overlapped portion of the data will become smaller. At $L = N - 1$, only one pair of K values is multiplied.

Sample autocorrelation function: A sample autocorrelation function is defined as the covariance function divided by the variance, and it represents the correlation between the spatial series and itself offset at different separation distances. At zero separation (offset) distance, the autocorrelation must be 1.

$$\rho(0) = 1 \tag{4.5.4}$$

When the separation distance is not equal to 0, the autocorrelation function is given as

$$\rho(L) = \frac{\mathrm{Cov}(L)}{S^2} = \frac{1}{S^2(N-L)} \sum_{i=1}^{N-L} [K(i) - \overline{K}]\,[K(i+L) - \overline{K}] \tag{4.5.5}$$

In fact, for each lag, equation (4.5.5) calculates a linear correlation coefficient between the spatial series, $K(i)$, and the series itself at a shift, $\Delta x \cdot L$, that is, $K(i + L)$. Therefore, the correlation analysis discussed here is strictly a linear statistical analysis. Finally, notice that hereafter the word "sample," will be dropped from the sample mean, covariance function, and autocorrelation function for simplicity.

4.6 Physical Meanings of the Autocorrelation Function

Figure 4.5 will be used again to explain the physical meanings of the autocorrelation function. At the zero lag ($\xi = 0$), the black solid line and the red dashed line in Figure 4.5 should overlap each other. That is, when $\xi = 0$ the black solid line is identical to the red dashed line at every location, x. If we plot any pair of the $K(x)$ of the black solid line and the $K(x)$ of the red dashed line on an x-y plot (Figure 4.6), it then should perfectly fall on a straight line with 45° slope, indicating a perfect correlation ($\rho = 1$) between $K(x)$ and $K(x)$. When $\xi \neq 0$ the two lines are offset as shown in Figure 4.5. If we plot all pairs of the K of the black solid line (i.e., $K(x)$) and K of the red dashed line (i.e., $K(x + \xi)$) in the overlapping region in Figure 4.5, these pairs will scatter around the 45° line, implying the correlation coefficient is less than 1. In other words, the two lines within the overlapping region are no longer identical. The scatter of the data points generally grows with the lag, and the correlation coefficient decreases accordingly.

If we now plot these correlation coefficients as a function of lag (ξ), we then obtain the autocorrelation function for this process $K(x)$, as shown in Figure 4.7. This plot is called the autocorrelation function, or autocorrelogram.

All autocorrelation functions have the following properties:

1) $\rho(0) = 1$
2) $|\rho(\xi)| \leq 1$
3) If $K(x)$ is a real value process, then
   $\rho(-\xi) = \rho(\xi)$, for all ξ. That is, $\rho(\xi)$ is always an even function.

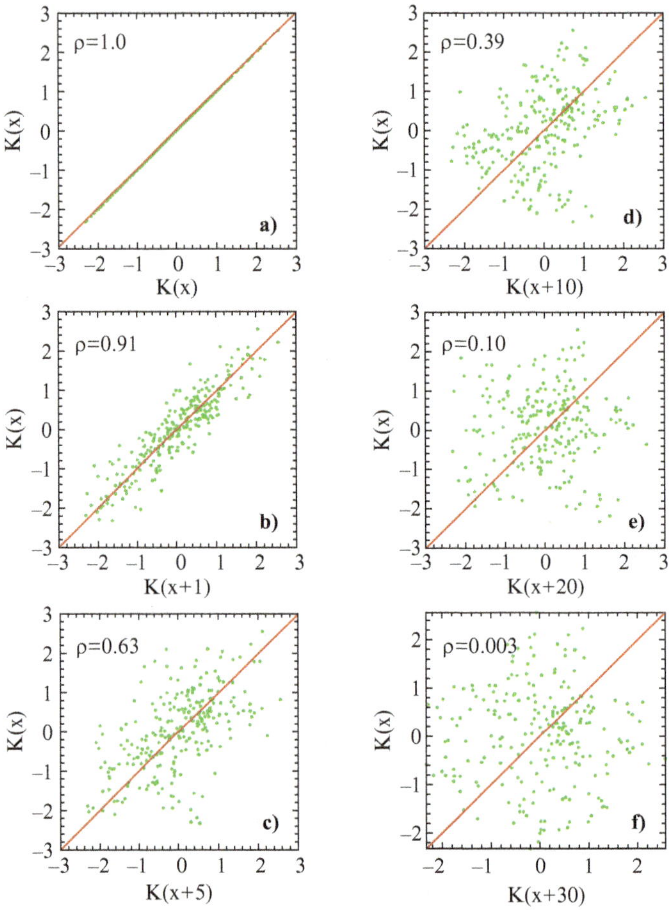

Figure 4.6. Illustration of the concept of autocorrelation analysis: analysis of linear correlation between data series at different lags. Note the lag increases as one moves from the left column to the right column plots (i.e., from (a) to (f)).

As another example for the autocorrelation analysis, Figure 4.8a shows a mean-removed hypothetic hydraulic conductivity distribution, $K'(x)$, along a 20 m transect, which appears as a regular sine wave. Its autocorrelation function as a function of the number of lags (the lag interval is 0.1 m) is plotted in Figure 4.8b. The wave length of the sine wave is 2π (6.28 m). In theory, the autocorrelation function of a sine wave with an infinite record should drop from +1 to 0 and then to −1, as the wave is shifted out of phase with itself, or when peaks are matched with troughs of the sine wave. The autocorrelation increases again until it reaches +1, when the wave is shifted exactly one wavelength. Figure 4.8b, however, shows that the maximum correlation decreases from +1, with an increase in the spatial lag. This is attributed to the fact that a smaller number of pairs of data are used to evaluate the correlation as the lag increases, since this sine wave record used in our example is finite. Nevertheless, the autocorrelation analysis

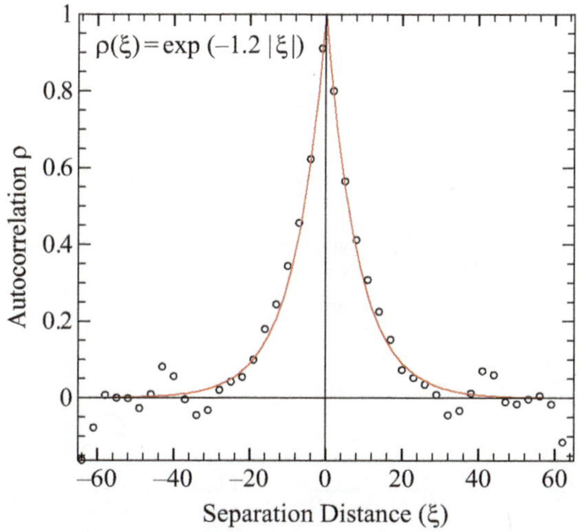

Figure 4.7. Autocorrelation function, or autocorrelogram, for the process $K(x)$. The symbol o represents sample the autocorrelation value and the solid line represents a theoretical autocorrelogram.

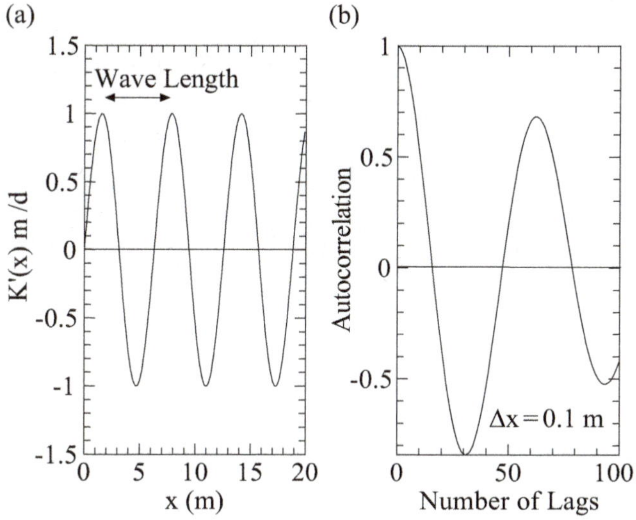

Figure 4.8. a) Spatial distribution of the residuals of a hypothetical hydraulic conductivity field; b) the autocorrelation function of the hydraulic conductivity residuals.

reveals the characteristics of a spatial series: periodicity in this case, which is characterized by the wavelength. In other words, the sine wave repeats itself after one wavelength.

Figures 4.9a and 4.9b illustrate an uncorrelated hydraulic conductivity (white noise) field and its autocorrelation function, respectively. The correlation values of the uncorrelated hydraulic conductivity field drop rapidly to almost 0 just after the first lag, indicating the white noise in the hydraulic conductivity is not spatially

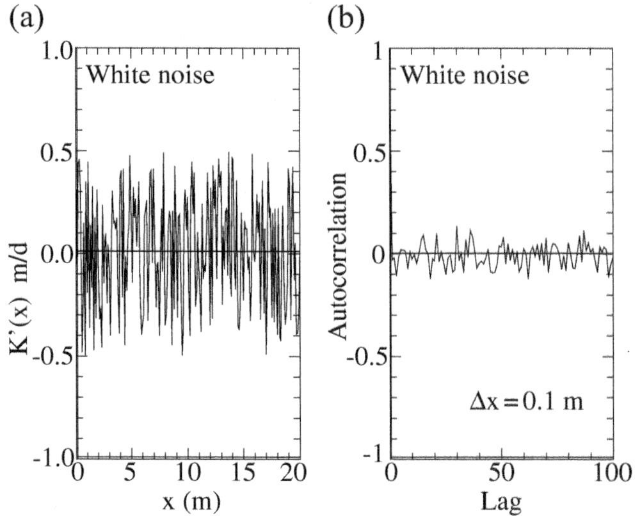

Figure 4.9. a) Spatial distribution of a white noise field b) the autocorrelation function of the white noise field.

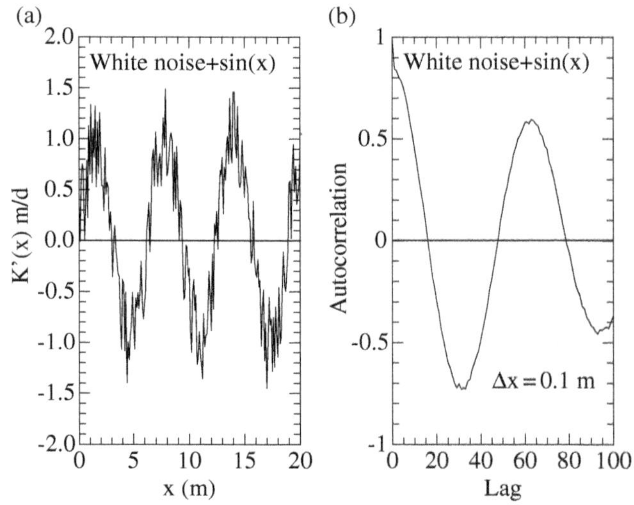

Figure 4.10. a) Spatial distribution of the perturbation of a hydraulic conductivity field; b) autocorrelation function of the hydraulic conductivity field.

correlated. That is, the occurrence of a value at a given location has no relationship to that at the location beside it – a characteristic of white noise.

Figure 4.10a shows a hydraulic conductivity field that is formed by the sum of the white noise (Figure 4.9a) and the sine wave (Figure 4.8a) – heterogeneities at two scales. Its corresponding autocorrelogram is shown in Figure 4.10b. According to this figure, even though the variability of the hydraulic conductivity exists at two scales, the autocorrelogram reveals the underlining periodicity of the major spatial pattern of the hydraulic conductivity field. In comparison with Figure 4.8b, Figure 4.8b shows a rapid decline of the correlation at small lag numbers, indicative of the effects of the random white noise on the autocorrelation function (i.e., the nugget effect in the variogram analysis, which will be discussed in Section 4.8). The overall behavior of the autocorrelogram for this composite signal, however, is dominated by that of the sine wave (i.e., the large-scale trend).

Spatial series of the saturated hydraulic conductivity of geologic formations (strata) generally have been found to exhibit an exponential autocorrelation function: The autocorrelation decreases exponentially as the separation distance increases (e.g., Sudicky, 1986). Such exponential behavior can be described by the mathematical model:

$$\rho(\xi) = \exp\left[-\frac{|\xi|}{\lambda}\right] \tag{4.6.1}$$

where ξ denotes the separation distance and λ is the correlation scale (or integral scale). In the following, we will use a one-dimensional model as an example to define correlation and integral scales and discuss their meanings.

Correlation scale: It is defined as a characteristic length (or the separation distance) beyond which the correlation between data is no longer considered significant or the correlation value becomes 0.

Integral scale, λ: It is defined as the area under an autocorrelation function from $\xi = 0$ to $\xi = \infty$ if the area is positive and non-0. For example, if an exponential autocorrelation function is considered, the integral scale is given as

$$\int_0^\infty \rho(\xi)d\xi = \int_0^\infty e^{-\frac{|\xi|}{\lambda}}d\xi = -\lambda e^{-\frac{|\xi|}{\lambda}}\Big|_0^\infty = \lambda \tag{4.6.2}$$

when $\xi = \lambda$, the autocorrelation is

$$\rho(\lambda) = e^{-1} = 0.3678$$

This means that if the interval between two samples is greater than the integral scale (i.e., $\xi > \lambda$), then the correlation of the two samples will drop below 0.3678 and the two data sets are no longer considered to be correlated (see Figure 4.11). Notice that if the concept of the correlation scale (i.e., the separation distance at which the correlation is 0) is used for the exponential correlation model, the correlation scale for the exponential model will be infinite.

Generally speaking, the correlation scale is a loosely defined term, since any arbitrary value (say, $\rho = 0.1$) can be used to designate the significance of correlation. On the other hand, the integral scale provides a definite cutoff value as long as it exists. However, in some cases the integral scale may not exist (e.g., mathematical

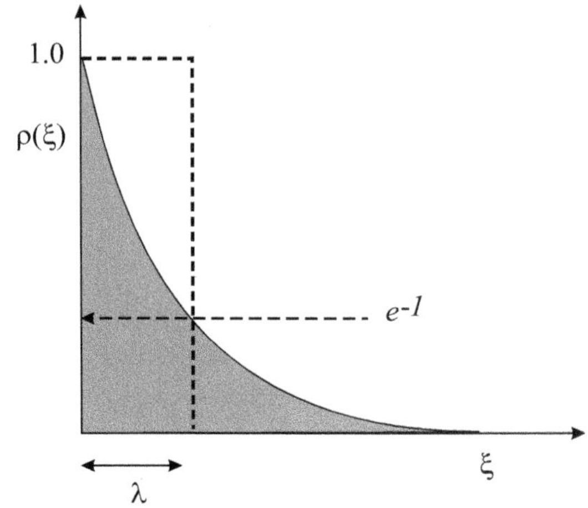

Figure 4.11. The exponential autocorrelation function of a hydraulic conductivity field and the definition of an integral scale for the exponential function.

functions that have some negative correlations). We therefore will use the term "correlation scale" in the rest of the text for consistency, even though the cutoff value for the significance of correlation can be arbitrarily defined.

To illustrate the practical implications of the correlation scale, consider measuring the saturated hydraulic conductivity (K) of a geologic medium at some interval along a transect line. Suppose this geologic medium consists of many clay lenses. Intuitively, we would expect that pairs of hydraulic conductivity values measured at intervals within a clay lens will be highly correlated, and measurements taken at intervals greater than the dimension of the clay lens are uncorrelated. In other words, we may speculate that the correlation function must have some relation to the size of clay lenses.

To substantiate this conjecture, consider an outcrop that consists of alternating layers of sand and clay with a uniform thickness (3 m), representing cyclic or periodic sedimentation processes. The K distribution of the formation along a borehole may be represented by Figure 4.10a, where the K value of sand ranges between 0 and +1, and between 0 and −1 for clay. The K autocorrelogram is illustrated in Figure 4.10b, where the correlation scale is 1.5 m (the distance between +1 and 0 correlation values), which is approximately equal to half of the thickness of each layer. This periodic autocorrelogram and the correlation scale imply that, on average, we expect to find another layer of sand or clay after a distance equal to four times of the correlation scale in this formation. We therefore conclude that the correlation scale is a statistical measure of the spatial structure (or spatial pattern) of hydrologic parameters. We thus can use it to describe the average size of heterogeneity (or geofabric) in the sampled field site.

Unambiguously, geologic heterogeneity is always three-dimensional. It thus should be treated as a 3-D stochastic process or random field, and therefore a 3-D correlation function is necessary to define its 3-D spatial structure. For example, the following 3-D exponential autocorrelation function has been widely adopted:

$$\rho(\xi) = \exp\left\{-\left[\left|\frac{\xi_1}{\lambda_1}\right|^2 + \left|\frac{\xi_2}{\lambda_2}\right|^2 + \left|\frac{\xi_3}{\lambda_3}\right|^2\right]^{\frac{1}{2}}\right\} \tag{4.6.3}$$

where ξ is the separation vector with components ξ_1, ξ_2, and ξ_3 and λ_1, λ_2, and λ_3 are correlation scales in x, y, and z directions, respectively.

4.7 Classification of the Spatial Stochastic Processes

Previously, we classified stochastic processes as statistically homogeneous (stationary) and heterogeneous (nonstationary) processes. Now, on the basis of the correlation scale concept, a three-dimensional spatial stochastic process can be classified as either a statistically isotropic or anisotropic process. This statistical isotropy or anisotropy is different from the hydraulic conductivity isotropy or anisotropy, which is discussed in Chapters 2, 3, and 6. The statistical anisotropy and isotropy define the shape or dimensions of geologic structures in geologic media in a statistical sense, whereas the hydraulic conductivity anisotropy defines the preferred flow directions of fluids in geologic media. Nonetheless, these types of anisotropies are related. For instance, a large-scale effective hydraulic conductivity anisotropy, in effect, can be related to the statistical anisotropy (spatial structure) of small-scale hydraulic conductivity parameters (see Chapter 6). Formal definitions of statistically isotropic or anisotropic processes are presented in the following.

4.7.1 Statistically Isotropic Process

A stochastic process that has the same correlation scales in all directions, that is, $\lambda_1 = \lambda_2 = \lambda_3$, is called a statistically isotropic stochastic process. an aquifer that is said to have a statistically isotropic structure is considered to be made of many spherical inclusions (for instance, a conglomerate formation or a uniformly packed sandbox) with an average diameter equal to the correlation length or λ. Of course, these spherical inclusions would have different shapes and properties. Figure 4.12 illustrates such a conceptualized medium and a numerically generated two-dimensional random field (one realization) using a stochastic field generator (Gutjahr, 1989) with an isotropic exponential correlation function with a correlation scale of 20 units in both x and y directions.

4.7.2 Statistically Anisotropic Process

If the correlation scales of a stochastic process in x, y, and z directions are different (i.e., $\lambda_1 = \lambda_2 = \lambda_3$), then the process is said to be statistically anisotropic. If the

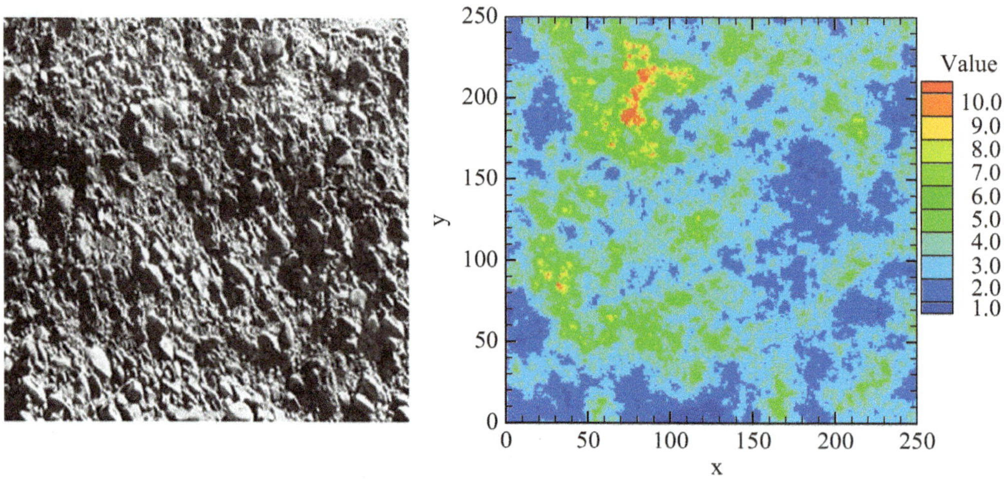

Figure 4.12. A statistically isotropic geologic medium in a field (left) and a numerically generated statistically isotropic hydraulic conductivity field with a correlation of 20 units in both x and y directions (right).

hydraulic property of an aquifer is statistically anisotropic, the aquifer is considered to be made of many inclusions of elliptic shapes – what kind of geologic formation may have this type of heterogeneity? Because of large-scale geologic deposition processes, geologic materials are often spread over larger horizontal than vertical extents. Therefore, a statistical anisotropy is a common and often legitimate spatial structure of geologic heterogeneity. A schematic illustration of a possible pattern of geologic heterogeneity with statistical anisotropy is shown in Figure 4.13, along with a single realization of a generated stochastic random process with an exponential covariance function. The correlation scale of the generated random process in the x direction is 100 units and 20 units in the y direction. This realization was generated with the same random seed number (a required input) for the random field generator and mean and variance as those in Figure 4.12. Comparing Figures 4.12 and 4.13, the two random fields have the same distribution of high and low values with the exception that zones of high and low values are greatly stretched horizontally in Figure 4.13.

The hydraulic property field of a perfectly stratified aquifer represents a special case of a statistically anisotropic process. A perfectly stratified aquifer can be interpreted as a collection of many disc-shaped lenses, which can be represented by a statistically anisotropic random conductivity field with $\lambda_1 \neq \lambda_2 \neq \lambda_3$, where λ_1, λ_2, and λ_3 are correlation scales in the x, y, and z directions, respectively. A schematic illustration of a possible pattern of geologic heterogeneity with statistical anisotropy is shown in Figure 4.14 (left), along with a single realization of the generated stochastic random field (Figure 4.14 right) with an exponential covariance function. The correlation scales are 1000 units in the x direction and 20 units in the y (vertical) direction. This realization was generated with the same seed number for the random

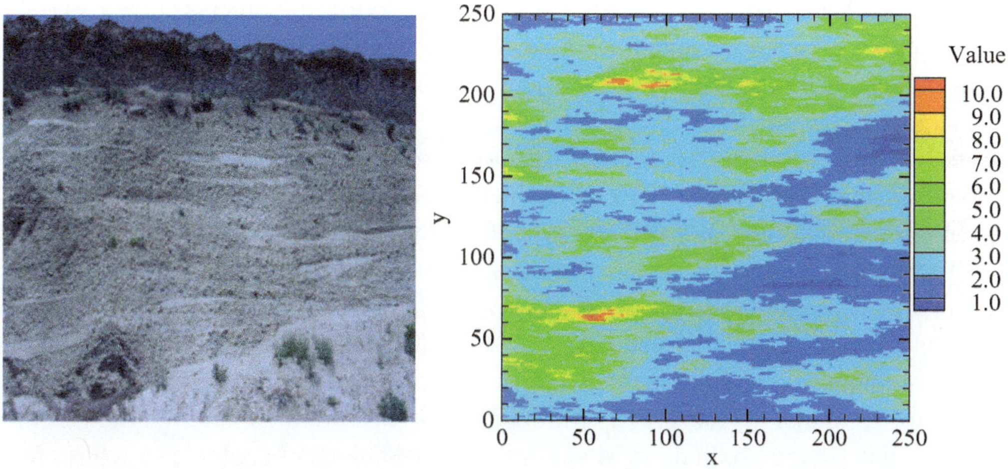

Figure 4.13. A statistically anisotropic geologic medium at a field site (left) and a generated statistically anisotropic hydraulic conductivity field with a correlation length of 100 units in the *x* direction and 20 units in the *y* direction (right).

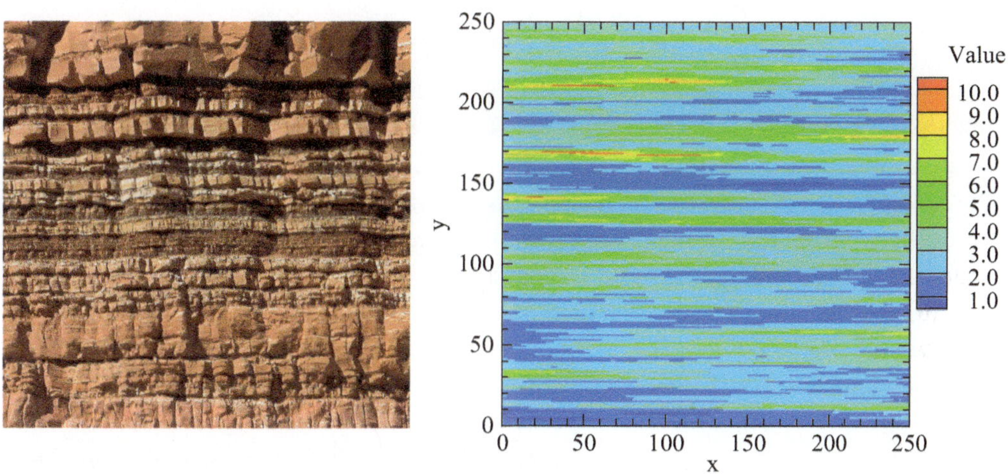

Figure 4.14. Statistically perfectly stratified geologic media (left) and a synthetic, statistically anisotropic hydraulic conductivity field with a correlation of 1000 units in the *x* direction and 20 units in the *y* direction (right).

field generator and mean and variance as in Figures 4.12 and 4.13. These three figures demonstrate how statistical anisotropy affects the distribution of the conceptual geologic heterogeneity and how the statistical anisotropy can be used to describe general geologic fabrics within a medium.

Notice that for the numerically generated stochastic fields, we assumed the joint probability density functions (i.e., log normal distribution, with a given mean and covariance structure) of the stochastic processes are known. Since these generated random fields represent only one realization of the stochastic processes for a particular scenario, estimated moments of these generated fields will be equal to the specified ensemble moments only if (a) a large number of realizations are considered or (b) the generated fields are large enough that they have encompassed all possible heterogeneities, and therefore the ergodicity assumption is applicable. In addition, we must point out that these generated random fields resemble the true property field of a geologic medium only in terms of the statistical properties (i.e., mean and variance).

The statistical anisotropy as discussed earlier assumes that the maximum and minimum correlation scales are in the same directions as our chosen coordinates system (i.e., horizontal and vertical). We will refer to the directions of the maximum and minimum correlation scales as the principal directions of the statistical anisotropy, similar to the principal directions of hydraulic conductivity anisotropy discussed in Chapter 6. In many situations, the orientations of geologic structures may not align with our chosen coordinates. In these cases, a coordinate transformation (see Chapter 6) must be performed to determine the principal directions of the statistical anisotropy.

4.7.3 Statistically Homogeneous Processes

Statistical homogeneity has the same definition as that of stochastic stationarity, discussed in Section 4.4. This concept is also loosely similar to the definition of hydrogeologic homogeneity (see Chapter 2). The stochastic stationarity requires the JPD of a stochastic process to be translationally invariant (see Section 4.4). On the other hand, the hydrogeologic homogeneity demands that any hydraulic property defined over some CV is spatially invariant (or spatially uniform), i.e., existence of REVs.

4.7.4 Statistically Heterogeneous Processes

A stochastic process that does not satisfy the definition of the statistical homogeneity is called a statistically heterogeneous or nonstationary process. Hydraulic properties of most geologic formations are generally nonstationary, although this depends on the scale of observation and the sampling domain. For some cases, a nonstationary stochastic process may possess a nonconstant mean, that is, $E\,[\ln K] = F(x)$ – the mean of the process is a function of location, or the $\ln K$ field has a trend, but with a spatially constant variance. In this case, the trend can be removed and the residuals are then treated as a second-order or strict stationary process.

In theory, stationarity and nonstationarity are defined strictly on the basis of the ensemble statistics, which exist only in a theoretical sense. Therefore, field data cannot be strictly classified as either stationary or nonstationary processes, unless one assumes ergodicity. The assumption of ergodicity, however, is never verifiable in reality. Nevertheless, for practical purposes, a field data set that exhibits a trend and trend-dependent variance is said to be a nonstationary process. In this case, the trend is generally removed first and the residual is then examined. For example, consider a geologic formation that consists of many different geologic units, each unit having its own mean and variance. Of course, we can examine the spatial statistics of each unit separately to describe its heterogeneity and geofabrics (or spatial structures), as discussed in Section 4.10. Alternatively, we can treat the entire formation as one unit, ignoring the trend and considering the variability of the physical property of the entire formation as a stationary process with a constant mean. If we choose to do so, the spatial statistics will include both the trend and the variability around the trend, which is likely dominated by the trend because of its greater variability compared with that of the residual. As a matter of fact, the heterogeneous K field illustrated in Figure 4.10a, where white noise is imposed on a sine wave, is a nonstationary field. The corresponding nested autocovariance function is illustrated in Figure 4.10b, which consists of the effect of the white noise and the structure of the sine wave (the major trend).

4.8 Variogram Analysis

Another way to characterize spatial variability of geologic media statistically is through variogram analysis. Variogram analysis relies on an intrinsic hypothesis, which is less stringent than the second-order stationary assumption. The intrinsic hypothesis assumes that even if the variance of a hydraulic conductivity field along a transet, $K(x)$, is not finite (i.e., the field is a nonstationary process), the variance of the first-order increments of K, $[K(x+\xi)-K(x)]$, where ξ is the separation distance, will be finite and the increments are second-order stationary, satisfying the following requirements:

$$E[K(x+\xi)-K(x)] = m(\xi) \quad (4.8.1)$$

$$\mathrm{var}[K(x+\xi)-K(x)] = 2\gamma(\xi) \quad (4.8.2)$$

In other words, the difference between pairs of Ks, separated by a given distance, ξ, will have a mean, $m(\xi)$, and variance, 2γ (ξ), for a given ξ. This implies that the K field, itself, may not be a second-order stationary process but the increment of the K field is. That is to say, the mean and the variance of the increment of the K field are functions of ξ only, not x. This is simply a way to broaden the second-order stationarity definition.

According to equation (4.8.2), a variogram or semivariogram is then defined as

$$\gamma(\xi) = \frac{1}{2}\mathrm{var}[K(x+\xi)-K(x)] \quad (4.8.3)$$

A variogram should have the following properties:

1) The value of the variogram at the origin is zero, $\gamma(0) = 0$;
2) The values of the variogram are positive, $\gamma(\xi) > 0$ when $\xi > 0$; and
3) The variogram is an even function, $\gamma(\xi) = \gamma(-\xi)$.

Now, if we assume

$$E[K(x+\xi) - K(x)] = m(\xi) = 0 \tag{4.8.4}$$

(i.e., the process K has a constant mean), then

$$\begin{aligned}\gamma(\xi) &= \frac{1}{2}E\{[K(x+\xi) - K(x)]^2\} \\ &= \frac{1}{2}\left[E\left\{[K(x+\xi)]^2\right\} - 2E[K(x+\xi)K(x)] + E\left\{[K(x)]^2\right\}\right]\end{aligned} \tag{4.8.5}$$

Furthermore, if the second moment of the K process is stationary, the variogram can be related to the covariance using the following equation.

$$\gamma(\xi) = C(0) - C(\xi) \tag{4.8.6}$$

Therefore, the variogram of a second-order stationary process is a mirror image of the autocovariance function as illustrated in Figure 4.15. This figure also shows that the variance of a covariance function is equivalent to the sill of its corresponding variogram; the correlation scale in the covariance function is identical to the range in the variogram analysis.

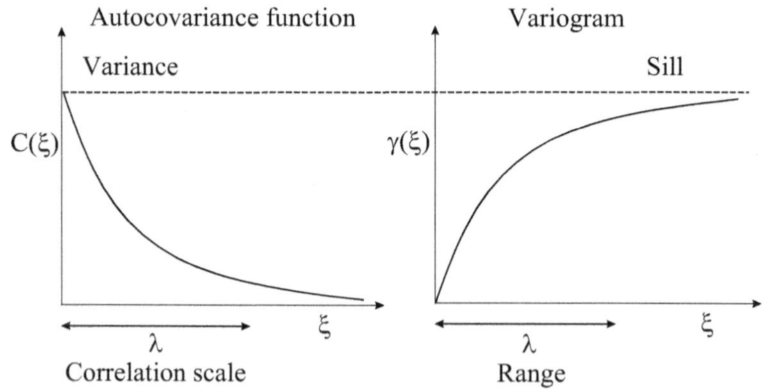

Figure 4.15. An illustration of the relation between the autocovariance function and the variogram for the case where the stochastic process is second-order stationary.

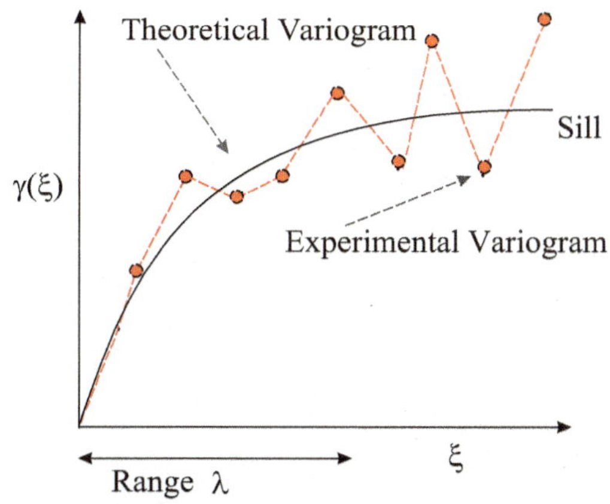

Figure 4.16. An illustration of a sample variogram and theoretical variogram.

4.8.1 Sample Variogram

The variogram defined in equation (4.8.3) is based on the ensemble concept. To derive a variogram for a data set, again, we must invoke the ergodicity assumption. Then, the variogram derived from the data set is called a sample (or experimental) variogram. In general, the sample variograms of a spatial data set are estimated from the observed data using a discrete form of equation (4.8.3). Software packages such as Geo-EAS or GSLIB (Clayton et al., 1992) are readily available for such analysis. As illustrated in Figure 4.16, the sample variogram generally fluctuates, in particular at large separation distances where the number of data pairs becomes small. As a consequence, the sample variograms at large lags are generally of little statistical significance. A sample variogram is often fitted with a continuous theoretical variogram model and the spatial structure of the data set is then described by the name of the theoretical model and its associated model parameters.

4.8.2 Theoretical Variogram

First of all, a theoretical variogram must be positive definite. That is, its counterpart (covariance function) must be nonnegative, or the variance is always positive. This requirement will ensure the existence and uniqueness of the solution to a kriging system of equations (to be discussed later). Many positive definite variograms are available (see de Marsily, 1986; Isaaks and Srivastava, 1989). Some of the commonly used variogram models are discussed below:

1. Power variogram model: This variogram is related to the separation distance, ξ, in a power law relationship:

$$\gamma(\xi) = \sigma^2 |\xi|^n \tag{4.8.7}$$

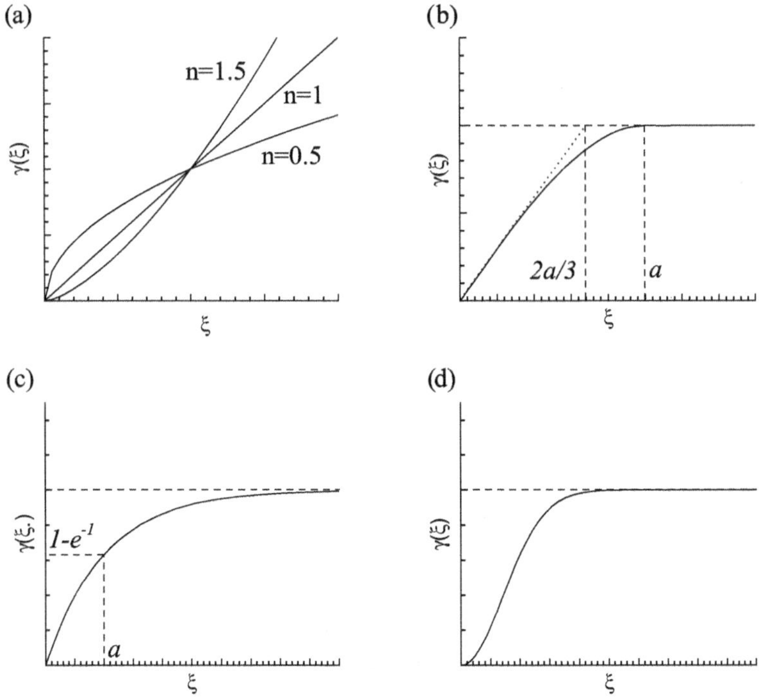

Figure 4.17. Behaviors of four theoretical variogram models: a) power, b) spherical, c) exponential, and d) Gaussian.

where n and σ^2 are the parameters to be fitted. This power model is generally used to represent a nonstationary process; the sill of the process increases as the range increases.

2. Spherical variogram model:

$$\begin{aligned}\gamma(\xi) &= \sigma^2\left[\frac{3}{2}\left(\frac{\xi}{\lambda}\right)-\frac{1}{2}\left(\frac{\xi}{\lambda}\right)^3\right] \qquad |\xi| < \lambda \\ &= \sigma^2 \qquad |\xi| > \lambda\end{aligned} \tag{4.8.8}$$

3. Exponential model:

$$\sigma^2\left[1-\exp\left(\frac{-|\xi|}{\lambda}\right)\right] \tag{4.8.9}$$

4. Gaussian model:

$$\sigma^2\left(1-\exp\left[-\left(\frac{|\xi|}{\lambda}\right)^2\right]\right) \tag{4.8.10}$$

Behaviors of these models are shown in Figure 4.17.

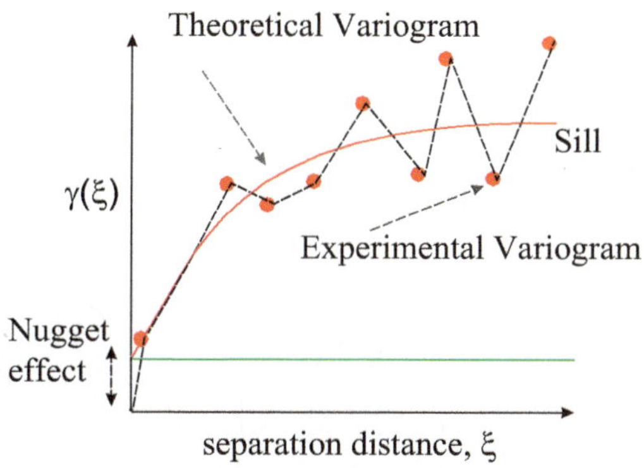

Figure 4.18. A schematic illustration of the nugget effect in a variogram.

Each of the theoretical variogram models, as well as the covariance models, represents some type of heterogeneity pattern, perhaps reflecting a type of geologic depositional environment. This variogram-depositional environment relationship, however, generally has not been defined and is therefore beyond the scope of this book. Overall, only subtle differences exist among spherical, exponential, and Gaussian models. To select an accurate variogram model for a field site, a very densely sampled data set is required. On the other hand, as the number of samples increases, utility of the spatial statistics diminishes for any delineation of the spatial distribution of the data. Therefore, a simple model, such as the exponential model, is highly desirable, and in fact it (the exponential model) is widely used for practical purposes.

Questions:

As discussed previously, there are many theoretical variogram models available. A logical question one should ask is, How accurate must a theoretical model be when it is selected to fit an experimental variogram? In other words, how would an inaccurate variogram model affect our representation of spatial variability and even estimation, such as the kriging estimate to be discussed in Chapter 9?

4.8.3 Nugget Effect

A nugget (a constant variogram – green line), illustrated in Figure 4.18, is often used to force a variogram to be discontinuous (rapid change in the value of a variogram) when the separation distance is close to zero to represent the following: 1) measurement errors or 2) the small-scale variability that is ignored by large sampling scales.

A typical nugget model is

$$\gamma(\xi) = a[1-\delta(\xi)] + \gamma'(\xi) \tag{4.8.11}$$

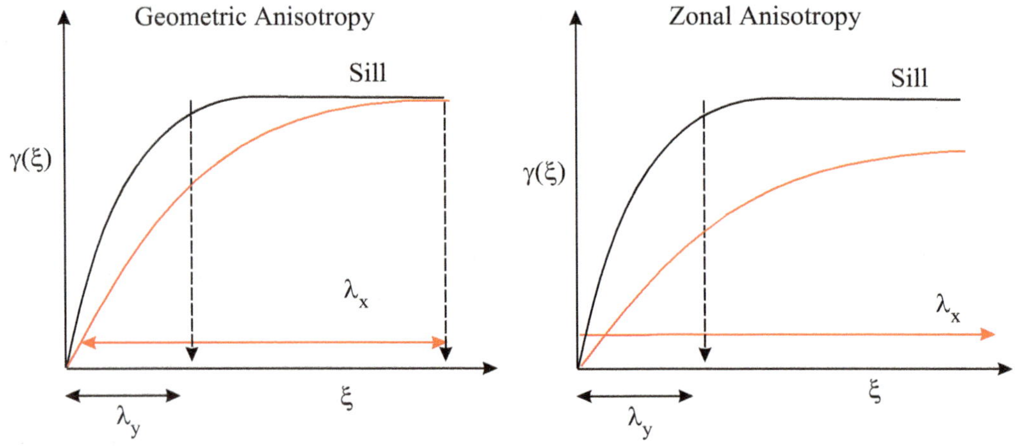

Figure 4.19. Schematic illustration of the two variograms showing the geometric and zonal anisotropies.

where $\delta(\xi) = 1$ if $\xi = 0$ and $\delta = 0$ otherwise (the Kronecker delta), while $\gamma'(\xi)$ is the variogram without the nugget. As a result, the variogram model with the nugget will start with 0 at the origin and become $a + \gamma'(\xi)$ at $\xi > 0$. The a denotes the sill (variance) of noise or small-scale heterogeneity (the green line in Figure 4.18), while $\gamma'(\xi)$ represents the spatial structure and the sill of large-scale heterogeneity (the red line in Figure 4.18). This behavior is identical to the autocorrelation function depicted in Figure 4.10b for the hydraulic conductivity field that is formed by the sum of the sine wave and the white noise. The nugget, in essence, makes the autocorrelation less than 1 at zero separation distance. Estimators such as kriging (to be discussed in Chapter 9) using the variogram with the nugget effect thus will not yield exact observed values at the observation locations, indicative of some measurement errors or small scale variability.

4.8.4 Anisotropic Variogram

Similar to the correlation scales, the variogram of a hydraulic property can have different ranges in different directions. This type of variogram is called an anisotropic variogram. Theoretically, the variograms in different directions should have the same sill but different ranges because the sill or variance is always a scalar.

If variograms of a multidimensional data set have the same sill values but different range values in different directions, the data set has *geometric anisotropy*. However, if variograms of a data set have different sill and range values in different directions, then the data set has *zonal anisotropy*. The geometric and zonal anisotropic variograms are illustrated in Figure 4.19. Zonal anisotropy is a result of the presence of an extensive layering structure of the data set (see Section 4.10). More precisely, the horizontal dimension of the data sampling area is much smaller than the horizontal correlation scale of the data. For example, if a sampling area of the porosity (sampled at 1 m intervals) covers only a 50 m x 50 m cross section of a geologic formation that has layers that extend horizontally over 500 m and vertically over 5 m, the variogram of the porosity data set will exhibit zonal anisotropy. In this example, the ratio of the vertical dimension of the cross section to the vertical correlation scale of the porosity is 10, and the ratio of the horizontal dimension of the cross section to the horizontal correlation scale is 0.1. These ratios suggest that sampling at 1 m intervals over the 50 m in vertical direction will likely include all possible heterogeneity in the vertical direction. On the contrary, the small horizontal aspect ratio suggests that the horizontal sampling extent is not sufficiently long to include all possible heterogeneities. That is, this zonal anisotropy can be avoided if the sampling area is extended over a cross-sectional area that is much larger than the correlation scale in all directions. Using the new sampling area, if that area is extended, the sill of the horizontal variogram of the porosity will reach the same sill as that of the vertical variogram.

One variogram model that is frequently observed but has not been discussed yet is a variogram model with the hole effect: a nonmonotonic variogram. Such variograms often indicate a form of cyclicity, or periodicity, a common and legitimate spatial characteristic of many geologic formations. Two theoretical variogram models of this type are 1) without dampening,

$$\gamma(\xi) = \sigma^2\left[1 - \cos\left(\frac{\xi}{a}\pi\right)\right] \tag{4.8.12}$$

and 2) with dampening (the amplitude of the cyclicity decreases),

$$\gamma(\xi) = \sigma^2\left[1 - \exp\left(\frac{-3|\xi|}{a}\right)\cos\left(\frac{\xi}{a}\pi\right)\right] \tag{4.8.13}$$

Pyrcz and Deutsh (2003) provided an elucidative discussion of the physical meanings of various variograms with hole effects.

4.9 Stochastic Description of Aquifer and Vadose Zone Heterogeneity

To describe the spatial variability of the hydraulic conductivity, $K(x)$, or transmissivity of a geologic formation, we normally use the natural logarithm of K values ($\ln K$) as the variable, not only because $K(x)$ is generally found to have a log normal distribution – which is not universally true for all cases – but also because there are some other advantages. For example, the use of $\ln K$ allows linearization of the governing flow equation (i.e., a mathematical simplicity); it avoids negative K values in generating random fields; it can lead to exact conditional mean estimates if kriging is used; and so on.

After transforming the data set using the natural logarithm and then employing the stochastic representation and invoking the second-order stationarity and ergodicity assumptions, we can quantify the variability in ln$K$ by its mean and autocovariance function without specifying the joint probability density function. We further express the ln$K$ process as

$$\ln K(x) = F + f(x) \tag{4.8.14}$$

where F denotes the mean of ln$K$, which is constant in space, and $f(x)$ is the perturbation (or the random part) about the mean. To describe the spatial variability of the process, we then specify its mean, variance, and autocovariance:

Mean

$$E[\ln K] = F \tag{4.8.15}$$

Variance

$$\sigma_f^2 = E[(\ln K - F)^2] \tag{4.8.16}$$

Autocovariance function (for example, 3-D isotropic stochastic process)

$$C(\xi) = \sigma_f^2 \exp\left(-\frac{|\xi|}{\lambda}\right) \tag{4.8.17}$$

Of course, variograms can be used to replace the covariance function. Notice that the variance of ln$K$ (equation (4.8.16)) is independent of its units for hydraulic conductivity. As a consequence, the variance of ln$K$ becomes a universal index for the variability of the hydraulic conductivity of a geologic medium.

On the basis of the preceding stochastic representation of hydraulic conductivities, if an aquifer is said to be homogeneous in terms of its hydraulic conductivity, we mean that the correlation scale for the hydraulic conductivity approaches infinity, $\lambda = \infty$ (i.e., the conductivity values are correlated everywhere). Its variance, however, does not have to vanish, since it can represent the uncertainty about the hydraulic conductivity value, even though it is uniform or homogeneous.

On the other hand, if we say the correlation scale of a hydraulic conductivity field is zero, $\lambda = 0$, we mean that the hydraulic conductivity field is a purely random process (white noise; see Figures 4.9a and 4.9b). In this case, knowledge of hydraulic conductivity at a location tells us nothing about the hydraulic conductivity at any adjacent locations.

Certainly, different shapes of the autocovariance (e.g., a Gaussian and a spherical with similar correlation scales) may indeed represent different structures of geologic heterogeneity or reflect different depositional environments. But the covariance function in itself represents the structural characteristics of a geologic medium only in an averaged sense. Therefore, the use of a mathematically simple model with similar correlation

Table 4.1. *Aquifer material log-hydraulic conductivity variability and correlation scales for some field sites*

Site	Variance of lnK	Horizontal Correlation Scale (m)	Vertical Correlation Scale (m)	Comments	Sources
Borden site					
	0.38	2.80	0.12		Sudicky (1986)
	0.24	5.10	0.21	Transect A-A	Woodbury and Sudicky (1991)
	0.37	8.30	0.34	Transect B-B	Woodbury and Sudicky (1991)
	0.37	1–7.5	0.1–0.3		Robin et al. (1991)
Cape Cod site					
	0.24	5.10	0.26	Best estimates	Hess (1989)
	0.20–0.32	3.1–8.8	0.16–0.91	plus/minus 2 sigma intervals	
Columbus site					
	3.50	12.00		15 wells	Rehfeldt et al. (1989)
	4.50	12.00	1.50	48 wells	Boggs et al. (1990)

Source: After Gelhar, 1993.

scales to those of a more complex model is highly justifiable. As a consequence, the exponential model, which offers considerable mathematical advantages, has been a popular autocovariance or variogram model for hydraulic conductivity as well as other properties. Table 4.1 shows the mean, variance, and correlation scales of ln*K* of various types of geologic formations based on the exponential autocovariance functions.

It is often useful to recognize the relation between the mean of ln*K* (or *F*) and *K*:

$$\mu_k = \exp\left(F + \frac{\sigma_f^2}{2}\right) \tag{4.8.18}$$

where F is the mean of ln*K* and σ_f^2 is the variance of ln*K*. μ_k and σ_k^2 are the mean and variance of *K*, respectively. Similarly, the variance of *K* can be derived if the mean and variance of ln*K* are known.

$$\sigma_k^2 = \left(\exp\left(\sigma_f^2\right) - 1\right)\exp\left(2F + \sigma_f^2\right) \tag{4.8.19}$$

4.9.1 Advantages of Stochastic Conceptualization of Heterogeneity

Generally, characterizing the spatial distribution of hydraulic properties at small scales over a groundwater basin requires numerous measurements, considerable time, and great expense. Such a sampling effort is practically impossible. To alleviate these difficulties, the stochastic representation of aquifer heterogeneity offers several advantages, such as the following:

1) It allows us to use the spatial statistics from a densely sampled field site to infer the spatial variability of the entire basin, if the property is spatially stationary.
2) It allows us to use documented spatial statistics to infer the spatial variability of a given basin if the deposition environment of the basin is known, even though the measurements of the hydrologic properties are not available.
3) It also allows us to describe heterogeneity of geologic formations quantitatively. This information is useful for the parameter estimation as prior information (see Chapter 9).

Finally, one must recognize that stochastic characterization of spatial variability is merely a means to quantify the spatial variability of a property of geologic formations. Besides, it depicts the general spatial relationship between the property at one location and that at other locations over the sampling domain (i.e., the average geologic structure). We must emphasize that it does not necessarily provide us with detailed and exact distribution of any hydrologic property in space. Also, one should notice that while our discussions centered on the stochastic conceptualization of spatial variability of hydrologic properties, the same stochastic tools can be applied to describing spatial variability of any other hydrologic processes (such as hydraulic head, concentration, and moisture content distributions), as illustrated in case studies presented later in this chapter (Section 4.10).

4.9.2 Field Examples for Aquifer and Vadose Zone Heterogeneity

We next review some of the intensively studied research sites, where the spatial variability in hydraulic properties has been characterized for either aquifer material or vadose zone soils.

The Borden site, which comprises fine to medium sands of glaciolacustrine origin (Sudicky, 1986), is the first field site at which extensive 3-D hydraulic conductivity characterization was undertaken. Core samples were collected along two perpendicular transects, with maximum horizontal separation of about 20 m. These samples were repacked in the laboratory permeameters to estimate the in situ hydraulic conductivity. The estimated horizontal and vertical correlation scales and variance of $\ln K$ are listed in Table 4.1. Later analyses of Borden data by Woodbury and Sudicky (1991) and Robin et al. (1991) yielded somewhat different statistical parameter estimates, as shown in Table 4.1, although they are of the same order of magnitude. Such differences in the estimated statistical parameters illustrate the uncertainty that can be encountered in characterizing heterogeneity. Generally, the Borden site aquifer is considered relatively homogeneous. Two other sites where detailed characterization of hydraulic conductivity (Table 4.1) has been conducted are the Cape Cod site (Hess, 1989), and the Columbus site (Rehfeldt et al., 1989; Boggs et al., 1990). The Cape Cod site correlation scales and variance of ln$K$ values are similar to those of the Borden site, whereas the Columbus site has significantly larger variance of ln$K$ values.

Next, we report vadose zone heterogeneity, as observed in a number of field sites (Table 4.2). Russo, in a series of papers in the 1980s and 1990s, provided extensive description of heterogeneity characterization of Bet Dagan soils in Israel. Russo and Bouton (1992) analyzed core-scale estimates of soil parameters of Gardner and Mualem-van Genuchten models (see Chapter 3) for 417 undisturbed soil cores taken from a trench wall. The estimated covariance functions of various soil parameters exhibited statistical anisotropy. Components of the estimated correlation scales of various parameters in the horizontal direction were about four times larger than their respective components in the vertical direction, as shown in Table 4.2.

4.10 Field Applications

In this section, we present an application of the stochastic characterization approach by Ye et al. (2005) to observed moisture distributions at a field site, which are often considered to be nonstationary (or statistically heterogeneous). The site of investigation is located within the U.S. Department of Energy Hanford Site in southeastern Washington. Ward et al. (2000) and Gee and Ward (2001) conducted water and tracer injection experiments at the site, which was originally envisioned and designed by Sisson and Lu (1984). The injection experiments were conducted over a period of two months during the summer of 2000. Using neutron probes, Gee and Ward collected the initial and postinjection moisture contents (θ) for thirty-two wells at forty-three depth intervals over an area of 15 m $\times$ 15 m.

Figure 4.20 shows a lithostratigraphic cross section near a field experiment site at Hanford, Washington (Last and Caldwell, 2001; Last et al., 2001). The site mainly consists of sandy deposits, with some presence of a distinctly larger fraction of fine sediments in the middle two layers. In addition, well-developed laminations exist within layers.

All the moisture content (θ) values collected on May 5, 2000 (initial condition before injection), and July 31, 2000 (during redistribution), are plotted as a function of depth in Figures 4.21a and 4.21b, respectively, showing five distinct zones of low and high θ values. The five-zone pattern is similar to that found in the 1980 θ measurements (not shown here). The consistency of these θ patterns over the twenty-year interval suggests that the May 5 θ distribution is under a state of natural equilibrium, with larger θ values associated with fine-textured materials, and smaller values with coarse-textured materials.

Since this interpretation of the θ pattern is in general agreement with the geologic cross section (Figure 4.20), the site geology was classified into five sediment layers. A bottom layer of coarse-textured sediments (~ 3 m thick) is overlain by a layer of fine-textured sediments (~ 2 m thick). Another layer of coarse-textured sediments

Table 4 2. *Variability and correlation scales of unsaturated hydraulic properties for some field sites*

Parameter	Mean	Variance	Correlation Scale, Horizontal	Correlation Scale, Vertical	Range, Horizontal	Range, Vertical	Comments	Sources
θ_s	0.397	0.01					Panoche soil, 150 ha, 20 plots, 6 depths, data for 30 cm depth shown	Nielsen et al. (1973)
K_s, cm/h	0.85	0.76						
lnK_s, cm/h	−3.5616	0.90611	0.8 m	0.2 m			Hamra Red Mediterranean soil, Bet Dagan, Israel, trench (20 m long, 1.7 m wide, and 2.5 m deep), 417 soil cores	Russo and Bouton (1992)
ln, cm^{-1}	−1.9946	0.77201	0.3 m	0.1 m				
lnK_s, cm/h	−0.636	1.242	0.87 m	0.80 m	2.5 m	0.6 m	40 m long x 6 m wide x 2 m deep plot; N = 130; in situ measure-ments of discharge and head are used to estimate *Ks* and *α by Guelph permeameter method*	Russo et al. (1997)
Lnα, cm^{-1}	−3.933	0.425	0.49 m	0.45 m	1.4 m	0.4 m		
lnK_s, cm/h	0.301	0.203	0.5 m		2 m		In situ measure-ments using disc permeameters, N = 40	White and Sully (1992)
lnα, cm^{-1}	0.0461	0.256						
lnK_s, cm/day	−0.3383	0.391	5.68 m to 8.25 m				Panoche soil, 3 m x 75 m plot instrumented with neutron probes and tensiometers, 75 cm depth data show.	Unlu et al. (1990)
α,cm^{-1}	0.0225	2.32×10^{-5}	19.98 m					
$C(d\theta/dh)$, cm^{-1}	3.23×10^{-4}	2.09×10^{-9}	5.96 m to 7.50 m					

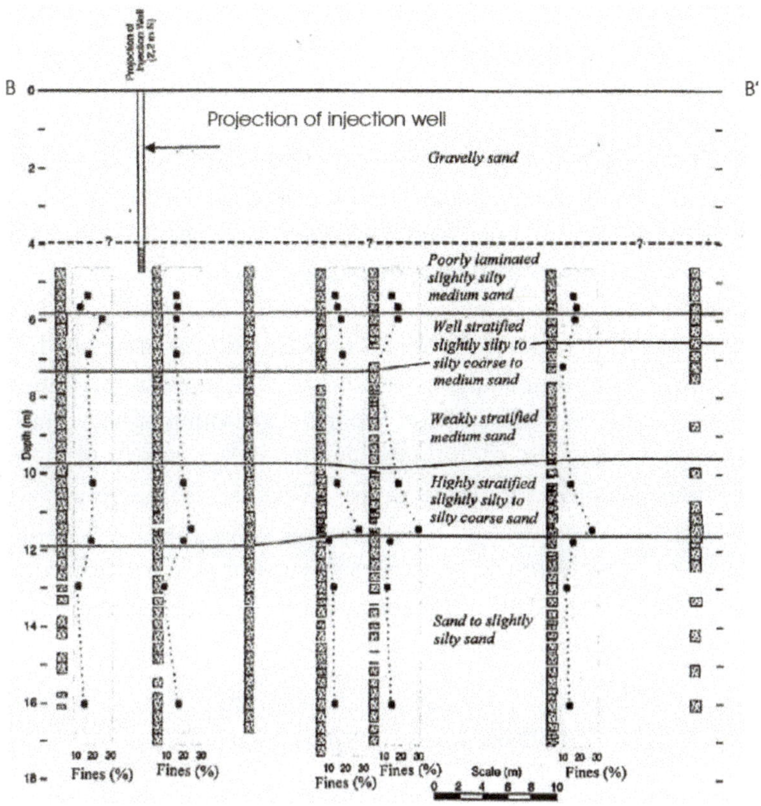

Figure 4.20. A lithostratigraphic cross section near a water injection experiment site at the vadose zone at Hanford, Washington. (Ye et al., 2005)

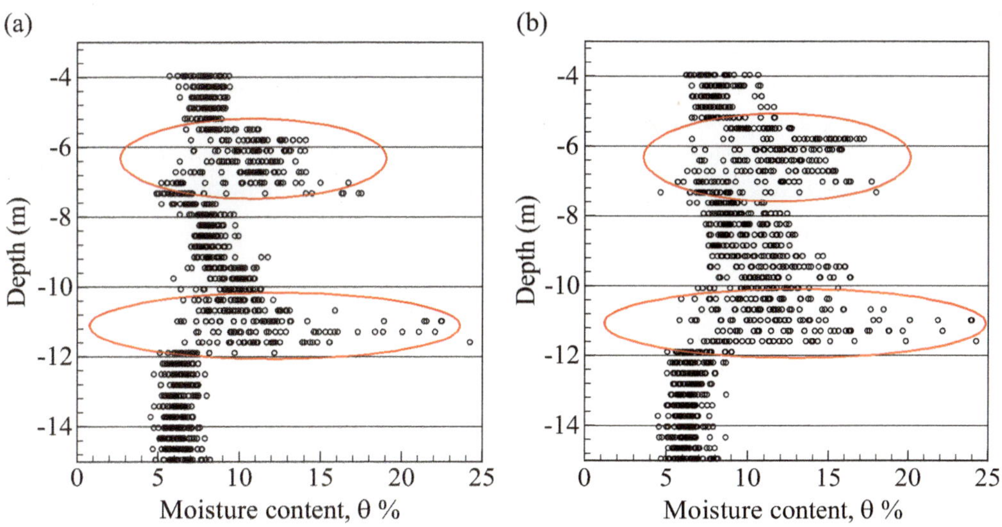

Figure 4.21. Observed moisture distribution over depth from 32 boreholes on a) May 5 (initial conditions) and b) July 31, 2000 (during redistribution of the soil moisture). (Modified from Ye et al., 2005)

(~ 3 m thick) appears at a depth of about 10 m and is overlain by a layer of fine-textured sediments (~2 m thick). Finally, a layer of coarse-textured sediments appears at a depth of about 5 m. These layers are assumed to be perfectly stratified. Table 4.3 summarizes the layer boundaries and statistics (mean, standard deviation, and coefficient of skewness) of θ for each observed layer during the experimental period. According to Table 4.3, the initial moisture content within each layer has different statistical properties and thus the θ distribution over the entire domain can be considered as a nonstationary process if ergodicity is assumed (Priestley, 1981).

The apparent correspondence between the initial θ distribution and the geologic cross section prompted the use of the initial θ distribution as a means for describing geology at the site. Variogram analysis was therefore applied to the initial θ distribution to yield a quantitative description of the three-dimensional geologic fabric (structure) at the field site. In addition, a hierarchical approach (e.g., Barrash and Clemo, 2002) was adopted for this purpose. That is, the sampling domain-level geologic structure was first assessed by evaluating variograms of θ in the horizontal and the vertical direction. Then, separating θ measurements according to the identified layers, the layer-level structure was investigated by examining the horizontal and vertical variograms of θ for each layer.

4.10.1 Domain-Level Spatial Variability

The vertical experimental variogram (Figure 4.22) reveals a hole effect with a double-hump feature. The double hump of the vertical variogram implies a cyclicity in the θ field and, in turn, the presence of a cyclic sequence of coarse and fine sediments or irregular layers (strata) as revealed in the θ profiles (Figure 4.21) and the geologic cross section (Figure 4.20). The cyclicity occurs about every 3.5 m, as is corroborated by the average thickness of the geologic layers at the field site. The horizontal variogram increases with lag without reaching a sill within the maximum lag distance of the sampling domain, implying a long correlation scale in the horizontal direction, which is greater than the domain's horizontal dimension (16 m). Such a correlation scale agrees with the relatively continuous bedding observed in the geologic cross section (see Figure 4.20).

The sill of the vertical variogram (Figure 4.22) is larger than that of the horizontal variogram, implying zonal anisotropy. Similar behavior was also observed by Barrash and Clemo (2002) for porosity data at the Boise Hydrogeophysical Site, Idaho. As discussed previously, the zonal anisotropy can be attributed to the way the variograms are computed and the presence of nearly perfectly stratified bedding at the site. Specifically, calculation of the vertical variogram considers only data pairs along the vertical direction, while the horizontal variogram analysis considers only horizontal data. For the vertical variogram, data pairs most likely cross different geologic boundaries. As the θ values in the fine-textured layers are greater than those in the

Table 4.3. *Statistics of the moisture content (volumetric) for each layer and for the entire sampling domain*

	Layer 1	Layer 2	Layer 3	Layer 4	Layer 5	All Layers
Depth (m)	15~11.7	9.9~11.7	6.9~9.9	5.4~6.9	4.0~5.4	4.0~11.7
Number of measurements	352	192	320	160	160	1184
			May 5, 2000			
Mean (percentage)	6.29	11.12	8.49	10.41	7.93	8.46
Standard Deviation (percentage)	0.72	3.28	1.92	1.86	0.73	2.55
Skewness	−0.09	1.70	1.71	−0.13	−0.46	1.77
June 2, 2000						
Mean (percentage)	6.29	11.01	8.83	12.54	8.51	8.89
Standard Deviation (percentage)	0.71	3.27	2.07	3.74	2.00	3.16
Skewness	1.14	1.48	1.71	0.43	2.85	1.61
			June 9, 2000			
Mean (percentage)	6.26	11.00	9.55	13.11	8.57	9.16
Standard Deviation (percentage)	0.67	3.26	3.06	3.92	2.11	3.50
Skewness	0.21	1.35	1.62	0.16	2.69	1.43
			June 16, 2000			
Mean (percentage)	6.33	11.16	10.34	13.23	8.61	9.44
Standard Deviation (percentage)	0.77	3.27	3.60	3.99	2.25	3.72
Skewness	2.25	1.45	1.21	0.16	2.68	1.31
June 23, 2000						
Mean (percentage)	6.34	11.42	10.66	13.53	8.69	9.62
Standard Deviation (percentage)	0.71	3.30	3.65	3.90	2.35	3.80
Skewness	0.39	1.05	1.03	−0.048	2.59	1.16
			July 7, 2000			
Mean (percentage)	6.29	11.83	10.11	12.66	8.42	9.49
Standard Deviation (percentage)	0.70	3.68	2.66	2.97	1.51	3.45
Skewness	0.20	1.00	1.12	−0.16	1.73	1.02
July 17, 2000						
Mean (percentage)	6.34	11.82	10.23	12.33	8.34	9.36
Standard Deviation (percentage)	0.70	3.57	2.58	2.72	1.33	3.23
Skewness	0.51	1.05	0.80	−0.10	1.05	1.06
			July 31, 2000			
Mean (percentage)	6.37	11.78	10.01	12.05	8.21	8.21
Standard Deviation (percentage)	0.78	3.50	2.35	2.58	1.14	1.14
Skewness	0.63	1.17	0.88	−0.12	0.91	0.91

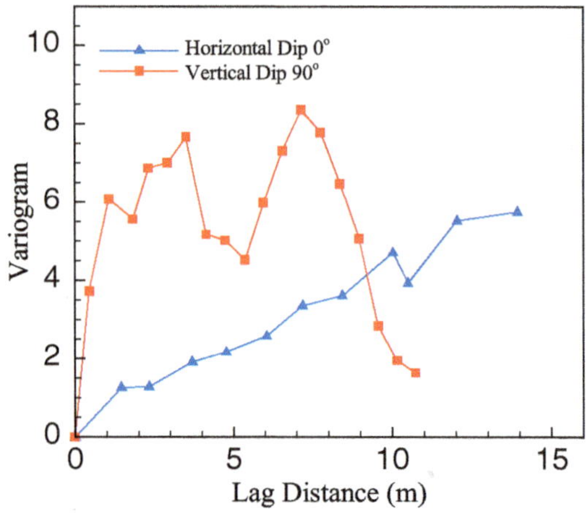

Figure 4.22. Variograms of the measured initial moisture content at the vadose zone injection field experiment site, Hanford, Washington. (Modified from Ye et al., 2005)

coarse-textured layers, θ differences for the data pairs crossing the boundary are expected to be large. In contrast, horizontal variogram analysis only considers data pairs within either the coarse- or fine-textured layer, and θ differences for the data pair within the same layer are likely to be small. Hence, a greater sill value of the vertical variogram than the horizontal is expected.

4.10.2 Layer-Level Spatial Variability

Directional sample variograms for layer-level spatial variability of the initial θ measurements evaluated along horizontal directions (Figure 4.23a) appear to increase with lag without reaching any sill within the maximum lag distance of the sampling domain. This suggests a long correlation scale in the horizontal direction, which may be greater than the horizontal dimension (16 m) of the sampling domain. The vertical variograms (Figure 4.23b), on the other hand, appear to stabilize at lag distances that are of the same order and are either equal to or slightly greater than the sampling interval (0.305 m). The vertical correlation scale of the θ field apparently is smaller than the average layer thickness (3.5 m). This conclusion is consistent with the presence of fine-scale laminations in different layers (Last et al., 2001). In general, although the θ measurements within different layers have similar horizontal and vertical spatial correlation lengths, they have different means and variances (Table 4.3). Note that the uniform thickness assumption embedded in our classification of the five layers may have inevitably included materials from coarse-textured layers into fine-textured layers. This assumption may thus have had some effects on the results (such as the increase in the sill of the variogram for layers 3 and 5 in Figure 4.23).

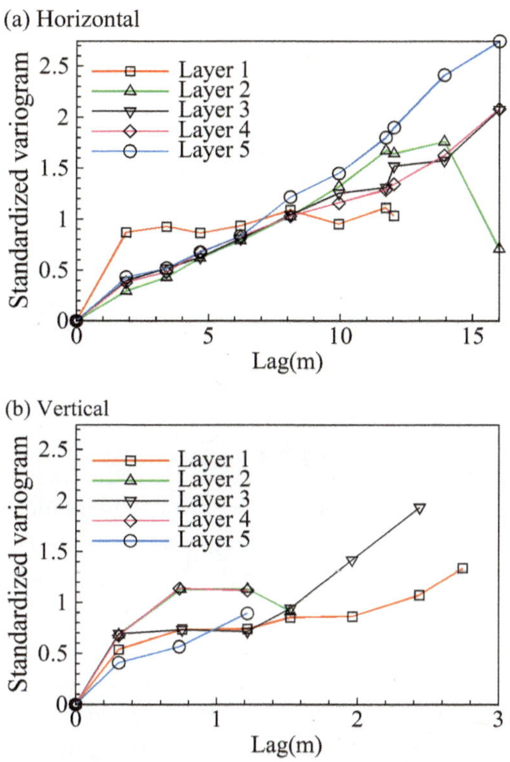

Figure 4.23. a) Horizontal and b) vertical variograms of the moisture content within each layer normalized by the variance of the moisture content of all samples at the field site. (Yeh et al., 2005)

4.11 Summary

- Any deterministic property or event can always be conceptualized as a random variable or event.
- A stochastic process or random field consists of an infinite number of random variables, which are characterized by the JPD. The distribution is characterized by its spatial moments (mean, variance, and correlation structures or scales).
- Mean specifies the most likely value of a property; variance represents the most likely contrast between values of the property; the correlation scales indicate the relationship between data points at different distances, that is, average dimensions of heterogeneity (the clusters of properties of the same kind).
- Statistics of a nonstationary field without removing the trend mainly reflect variability of dominant large-scale heterogeneity.
- Statistics of perturbations resulting from trend removal represent variability of smaller-scale heterogeneity.

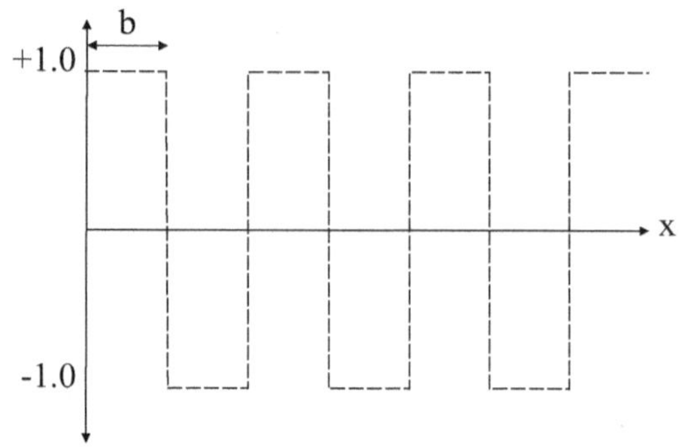

Figure 4.24. Illustration for problem 1.

4.12 Exercise

1. Figure 4.24 illustrates a spatial series of a hypothetical hydraulic conductivity perturbation field. It has a regular pattern: It is +1.0 m/d for a distance of b and it then becomes −1.0 m/d for another distance of b. Such a pattern is repeated continuously. Generate a 200 meter long series with $\Delta x = 0.1$ m for each of the five cases: 1) $b = 1$ m, 2) $b = 5$ m, 3) $b = 10$ m, 4) $b =$ 15m, and d) b = 20 m. Calculate the autocorrelation function of each space series and determine its correlation scale. Study the relationship between correlation scale and the width of the pattern, b, and draw a conclusion. Since this series has a regular pattern, which can be expressed in a mathematical form, you may derive the relationship analytically.
2. Can you develop an equation for the exponential covariance function that considers the nugget effect (or noise)?
3. Discuss how to extend the 1-D autocorrelation analysis, that is, equation (4.4.7), to 2-D or 3-D random fields. Hint: First, consider the random field is stratified and the direction of stratification is known. Then discuss the case if the direction is unknown.
4. Discuss the limitation of stochastic representation of fracture conductivity in fractured geologic media.
5. As illustrated in Figure 4.22, the vertical variogram for the moisture distribution at the Hanford experiment site has a higher sill than the horizontal variogram. Speculate and discuss the values of the sills and the shapes of variograms, on the basis of pairs of moisture contents in directions other than the vertical and horizontal directions.

5

Governing Flow Equations for Heterogeneous Conceptual Models

5.1 Introduction

In this chapter, we first define the meaning of a conceptual model for analyzing flow through any geologic medium. We then classify three-dimensional conceptual models into 1) heterogeneous and 2) equivalent homogeneous conceptual models, on the basis of the control volume (CV) and representative elementary volume (REV) concepts developed in Chapter 2. Subsequently, we develop mathematical models for the heterogeneous conceptual models. Note that mathematical models for the equivalent homogeneous conceptual models are discussed in Chapter 6. To develop these mathematical models, we start with the fundamental mass balance equation for flow. Three-dimensional governing partial differential equations for flow through fully saturated media and for flow through unsaturated media are then presented. For unsaturated media, mixed-form, moisture content–based, and pressure head–based governing flow equations are introduced and their advantages and disadvantages are subsequently discussed. Combining the governing equations for flow through both saturated and unsaturated media, we then formulate the governing equations for flow through variably saturated media. Then, equations for flow through fractured rocks are presented and discussed. The auxiliary conditions necessary for deriving particular solutions to these governing equations are presented afterward.

After presenting the governing equations for various 3-D conceptual models, we devote the rest of this chapter to the development of plan-view, depth-averaged, two-dimensional groundwater flow equations. We first define depth-averaged properties, including transmissivity, storage coefficient, and depth-averaged hydraulic head. We then present a depth-averaged approach that vertically averages the three-dimensional governing equation for flow through saturated media to derive the governing flow equations for two-dimensional confined and unconfined aquifers. The effective hydraulic properties arising from the depth-averaged procedure are then introduced and are followed by a discussion of the Dupuit assumption.

5.2 Conceptual Models

A conceptual model in this book is used to refer to the way we think about how the flow through geologic media should be described qualitatively. Conceptual modeling is a necessary step before any development or selection of a quantitative mathematical model. Many different conceptual models can be developed for flow through geologic media. However, we will classify conceptual models used in this book in two broad categories according to their predictive resolution: (1) **a heterogeneous conceptual model** and (2) **an equivalent homogeneous model**. The heterogeneous approach can generally be further subdivided into 1) a ***highly parameterized conceptual model*** and 2) *a* ***geologic conceptual model*** **(or** ***zonation conceptual model*****)**. All these conceptual models assume that Darcy's law established for laboratory-scale problems is valid for describing groundwater flow in these conceptual models.

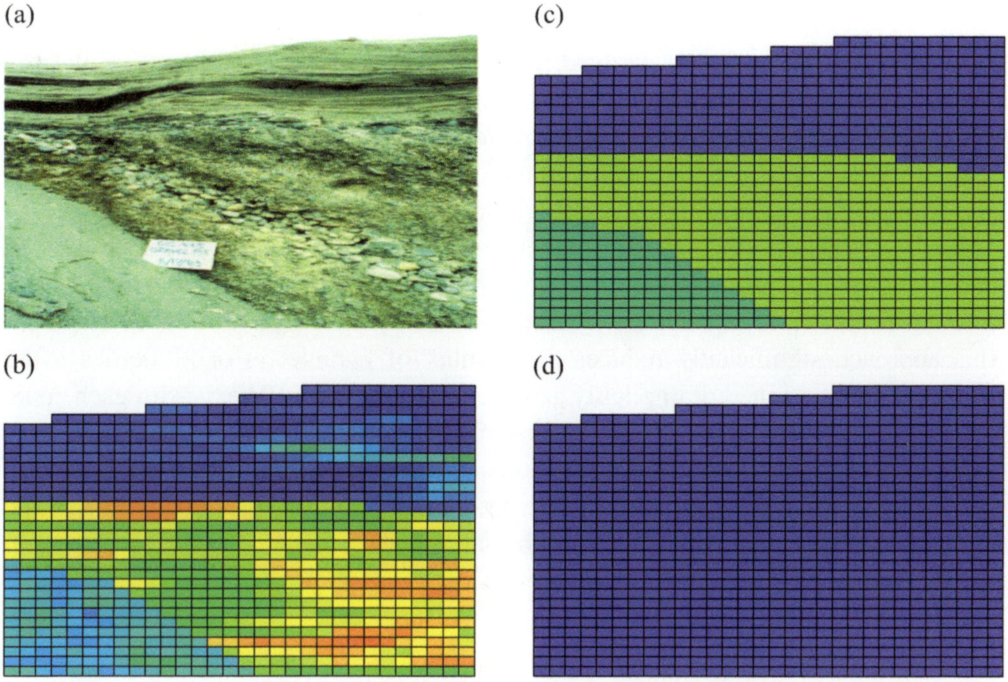

Figure 5.1. a) an angular unconformity in a heterogeneous geologic outcrop; b) a highly parameterized conceptual model, which has a different property (e.g., hydraulic conductivity) at each grid block to represent the heterogeneous outcrop; c) a geologic conceptual model that consists of three different zones and in which each layer has its own uniform properties; d) an equivalent homogeneous conceptual model that has uniform properties.

5.2.1 Highly Parameterized Conceptual Model

The highly parameterized conceptual model aims at investigating behaviors of water flow or transport of contaminants at the highest possible resolution. It conceptualizes a heterogeneous geologic medium as a collection of many small geologic blocks with different hydraulic properties (or parameters), mimicking the detailed spatial distribution of geologic deposits. For example, we consider the outcrop shown in Figure 5.1a as the geologic medium to be modeled; it is discretized into 854 blocks for numerical modeling purposes. On the basis of this conceptual model, hydraulic properties of each block may vary from block to block (e.g., Figure 5.1b). That is, it treats hydraulic properties of the domain as spatial variables. Implementation of this approach requires specification of hydraulic parameters defined over a small CV (e.g., each block whose size can be chosen arbitrarily as long as the Darcian continuum holds) and a highly parameterized mathematical model. As a result, many technical difficulties and practical issues arise. For example, it is difficult, if not formidable, to solve this mathematical problem analytically, and, more importantly, it is practically impossible to measure hydraulic properties for each block of the entire outcrop.

5.2.2 Geologic or Zonation Conceptual Model

An alternative approach has evolved, namely, the **geologic conceptual model (or zonation conceptual model)**, which parameterizes a field in terms of hydrogeologic information. For instance, the outcrop shown in Figure 5.1a may be mapped into three large-scale geologic formations based on textures, grain sizes, or facies. Each formation is then assumed to have a representative value for each property (see Figure 5.1c). Specifically, the 854 blocks are grouped into three zones; blocks in each zone have the same hydraulic property and blocks within different zones have different properties. Therefore, only three sets of hydraulic properties are required. This approach significantly reduces the number of parameters or properties to be measured or determined. It implicitly assumes that a spatial REV exists for each zone. If not, each zone is assumed to be an ensemble REV (see Chapters 2 and 4).

This approach encounters the same issues as the equivalent homogeneous conceptual model, which is to be discussed next: that is, how to measure or determine the "effective hydraulic properties" of such a large-scale, zonal REV, or any relationship between the effective hydraulic properties of the zonal REV and those of small-scale cores or CVs (i.e., upscaling rule). These are the subjects of discussion in Chapter 6.

5.2.3 Equivalent Homogeneous Conceptual Model

The equivalent homogeneous conceptual model approach expands the laboratory-scale CV to a field-scale CV, which is sufficiently large to include heterogeneities at scales much greater than pores (such as layers, zones, and formations, which can be characterized by the correlation scales or ranges discussed in Chapter 4). As in the geologic or zonation conceptual model, if values of a hydraulic property defined over such a field-scale CV at various spatial locations in a geologic medium are the same, the medium is then considered as a spatially homogeneous medium and a field-scale spatial REV exists. If this REV does not exist, the entire medium can still be conceptualized as a homogeneous geologic medium. In this situation, homogenization of the geologic medium is tantamount to saying that the REV exists only in an ensemble sense (ensemble REV). For example, while the field-scale spatial REV does not exist for the outcrop shown in Figure 5.1a, the outcrop nevertheless can be conceptualized as an equivalent homogeneous medium such that all of the 854 blocks have the same hydraulic properties (Figure 5.1d). Again, such a homogeneity assumption implies that a REV exists in an ensemble space (our imagination) rather than in a spatial sense. As a consequence, the predicted behavior based on the equivalent homogeneous model may capture major features for the bulk fluid flow but likely will be quite different from the one observed in an observation well.

Despite its low resolution predictions, such an equivalent homogeneous conceptual modeling approach has been widely adopted over the past decades. Its popularity is largely due to our limited ability to derive analytical solutions for flow and solute transport in complex heterogeneous media, in addition to our inability to characterize heterogeneities at different scales in detail. While the homogeneous conceptual model may mitigate the mathematical problem, many other issues arise with the use of this model. For instance, how can the hydraulic properties (effective hydraulic properties) of this homogeneous conceptual model be defined or measured? Furthermore, how different is the predicted aquifer behavior based on this approach from the observed aquifer behavior in the field? A detailed discussion of the equivalent homogeneous conceptual model for geologic media is provided in Chapter 6.

5.2.4 Other Conceptual Models

Before presenting mathematical models for heterogeneous conceptual models, we will define a few terms that are commonly used to describe and conceptualize large-scale geohydrological features in groundwater basins.

Aquifers are typically saturated regions of geologic media, which can produce an economically feasible quantity of water (e.g., saturated media with high hydraulic conductivity such as sand and gravel or fractured bedrock or karstic systems). An **aquitard**, **aquiclude,** or **aquifuge** is a geologic unit that has a very low hydraulic conductivity and restricts the flow of groundwater from one aquifer to another.

Generally, two types of aquifers are defined: *unconfined and confined.* **Unconfined** aquifers are sometimes also called *water table* or *phreatic* aquifers, because their upper boundary is the water table or phreatic surface. This boundary may vary in time and space. Furthermore, the **water table** is defined as a pressure surface where the pressure head is zero in a gauge pressure sense (i.e., atmospheric pressure). Geologic media below the water table are generally considered fully saturated and the water below is under positive pressure. Additionally, because of capillary forces in the vadose zone above the water table, the portion of the media slightly above the water table is almost fully saturated and under slightly negative pressure. This portion of the media is generally called the capillary fringe. **Confined** aquifers, on the other hand, are water-bearing formations with an aquitard or aquiclude as their upper and lower boundaries. Confined aquifers are pressurized (i.e., the pressure head within the aquifer is always higher than the upper boundary of the aquifer). Therefore, confined aquifers sometimes are also called artesian aquifers, if water level in the aquifers rises above the ground surface. Along with these two types of major aquifer types (i.e., unconfined and confined aquifers), the literature on groundwater hydrology refers to a third type of aquifer – the perched water table aquifer. The term "perched" refers to a condition in which groundwater accumulates above a small patch of a low-permeability unit in the vadose zone (explained later).

The **vadose zone**, also termed the **unsaturated zone**, is the portion of earth between the land surface and the water table (*vadose* is a Latin word for "shallow"). Pore spaces in the vadose zone are generally partly filled with water and partly filled with air. Water in the vadose zone has a negative pressure.

It is important to recognize that these confined and unconfined aquifers, as well as vadose zones, are merely simplified conceptual models for heterogeneous geologic media under variably saturated conditions.

5.3 Continuity Equation

Consider mass balance for a stationary CV, $\Delta x \Delta y \Delta z$ (Figure 5.2). The rate of mass accumulation within the volume equals the rate of total mass flowing into the element minus the rate of total mass flowing out of the element. That is,

$$\left\{ \begin{array}{c} \text{rate of mass} \\ \text{accumulation} \end{array} \right\} = \left\{ \begin{array}{c} \text{rate of} \\ \text{mass} \\ \text{in} \end{array} \right\} - \left\{ \begin{array}{c} \text{rate of} \\ \text{mass} \\ \text{out} \end{array} \right\} \tag{5.3.1}$$

Translating this mass balance principle into a mathematical equation for the control volume, we have

$$\begin{aligned} \Delta x \Delta y \Delta z \frac{\partial \rho n}{\partial t} = & \Delta y \Delta z [\rho q_x|_x - \rho q_x|_{x+\Delta x}] + \\ & \Delta x \Delta z \left[\rho q_y|_y - \rho q_y|_{y+\Delta y}\right] + \\ & \Delta x \Delta y \left[\rho q_z|_z - \rho q_z|_{z+\Delta z}\right] \end{aligned} \tag{5.3.2}$$

where ρ is water density, [M/L^3]; n is the porosity; [ ], q_x, q_y, and q_z are the specific discharge in x, y, and z directions, respectively, [L/T]. The term $\rho q_x|_x$ represents the mass flux per unit area in the x direction perpendicular to the $\Delta y \Delta z$ plane at location x. Note that a similar notation applies to the mass flux per unit area normal to other planes in other directions as well.

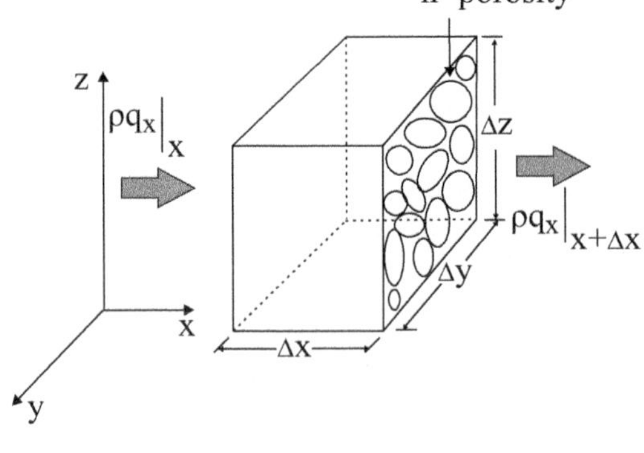

Figure 5.2. A control volume embedded in a porous medium for mass balance consideration.

Dividing equation (5.3.2) by $\Delta x \Delta y \Delta z$ on both sides, one obtains the following mass balance equation per unit volume of the porous medium:

$$\frac{\partial \rho n}{\partial t} = -\left[\frac{\Delta \rho q_x}{\Delta x} + \frac{\Delta \rho q_y}{\Delta y} + \frac{\Delta \rho q_z}{\Delta z}\right] \tag{5.3.3}$$

Next, let the element volume approach an infinitesimally small value (i.e., taking the limit as $\Delta x \Delta y \Delta z$ approaches zero), we have a mass balance equation in terms of the partial differential equation format:

$$\frac{\partial \rho n}{\partial t} = -\left[\frac{\partial \rho q_x}{\partial x} + \frac{\partial \rho q_y}{\partial y} + \frac{\partial \rho q_z}{\partial z}\right] \tag{5.3.4}$$

Using vector notation, the mass balance equation (also known as the continuity equation) takes the following form:

$$\frac{\partial \rho n}{\partial t} = -(\nabla \cdot \rho \mathbf{q}) \tag{5.3.5}$$

where $\nabla \cdot \rho \mathbf{q}$, also written as div($\rho \mathbf{q}$), is called divergence of the fluid mass flux, representing the net fluid influx/efflux over the CV. The continuity equation states that the rate of the increase in the mass density of fluid within a small volume element fixed in space is equal to the net rate of mass influx to the CV divided by its volume.

If the fluid is incompressible (i.e., constant ρ) and the porous medium is rigid or constant n in time, equation (5.3.5) becomes

$$(\nabla \cdot \mathbf{q}) = 0 \quad or \quad \frac{\partial n \rho}{\partial t} = 0 \tag{5.3.6}$$

Note that equation (5.3.6) does not imply that $\mathbf{q}$ is independent of spatial coordinates (i.e., uniform flow). It merely states that the flow is steady and there is no change in storage.

5.4 Equations for Flow through Saturated Porous Media

The equations developed in Section 5.3 are merely mass balance equations over a CV, which does not describe how the water flows into or out of the volume. In order to do that, we will include Darcy's law in equation (5.3.5).

Recall Darcy's law for flow through a fully saturated porous medium:

$$\mathbf{q} = -\mathbf{K}\nabla H \tag{5.4.1}$$

In equation (5.4.1), $\mathbf{K}$ is the saturated hydraulic conductivity tensor [L/T] (to be discussed in Chapter 6) at the CV scale, and H denotes the hydraulic head, [L].

Substituting equation (5.4.1) for equation (5.3.5), one can derive the following equation:

$$\nabla \cdot (\mathbf{K} \nabla H) = S_s \frac{\partial H}{\partial t} \tag{5.4.2}$$

Here, the rate of change in mass in the CV (i.e., the term on the right-hand side of equation (5.3.5)) is now related to the rate of change of hydraulic head (or total head). In equation (5.4.2), S_s is the specific storage of the porous medium, [1/L], which is the volume of water, $\Delta V_{w,}$ that a unit volume of aquifer, ΔV_T, releases from storage, under a unit decline in hydraulic head, Δh. Mathematically, it can be written as

$$S_s = \frac{\Delta V_w}{\Delta V_T \Delta h} \left[\frac{1}{\mathrm{L}}\right] \tag{5.4.3}$$

If we assume that compaction of the medium occurs in the vertical direction only, the specific storage term can also be expressed as

$$S_s = \rho g(\alpha + n\beta) \tag{5.4.4}$$

where α is the compressibility of the porous medium, $[\mathrm{M/LT^2}]^{-1}$; n is the porosity [dimensionless]; and β is the compressibility of water, $[\mathrm{M/LT^2}]^{-1}$ (Bear, 1972). This relationship [i.e., equation (5.4.4)] implies that the release of water from a porous medium is contributed from 1) the release of water in storage due to its expansion per unit aquifer volume and per unit decline of head, denoted by the term $\rho g n\beta$; 2) the amount of water in storage released because of the compression of the intergranular skeleton per unit aquifer volume and per unit decline of head, represented by the term $\rho g\alpha$.

Using a Cartesian coordinate system, the general equation for flow in saturated, heterogeneous, and anisotropic porous media is expressed as

$$\frac{\partial}{\partial x}\left[K_x \frac{\partial h}{\partial x}\right] + \frac{\partial}{\partial y}\left[K_y \frac{\partial h}{\partial y}\right] + \frac{\partial}{\partial z}\left[K_z \left(\frac{\partial h}{\partial z} + 1\right)\right] = S_s \frac{\partial h}{\partial t} \tag{5.4.5}$$

Notice that the preceding equation assumes that the coordinate system is aligned with that of the principal hydraulic conductivity tensor, and it is written in terms of pressure head, h, instead of the hydraulic head. Additionally, notice that the last term on the left-hand side of equation (5.4.5) includes the gravity term, which is unity. Here it is worthwhile to notice that some textbooks on groundwater hydrology use ψ for pressure head and h for the total head.

Hereafter, *heterogeneity* implies that the hydraulic property (K or S_s), defined over a laboratory-scale CV, varies from location to location, that is, $K(x, y, z)$ in a geologic medium. *Anisotropy* means that the laboratory-scale hydraulic conductivity at every location has a directional preference, for example, $K_x(x, y, z)$, $K_y(x, y, z)$, $K_z(x, y, z)$, where the subscript denotes the direction of the hydraulic conductivity. Generally speaking, hydraulic conductivity in the direction parallel to bedding is greater than that in the direction perpendicular to bedding. Again, hydraulic conductivity anisotropy is a result of the omission of substructures (e.g., microscale layers or laminations) and variations in the hydraulic properties of these substructures in the control volume that define the hydraulic conductivity. Thus, the anisotropy varies with the size and degree of heterogeneity within the CV. In general, as one chooses a highly parameterized conceptual model, one implicitly focuses on the importance of spatial variability of the hydraulic conductivity defined over the CV and ignores its laboratory-scale anisotropy. The specific storage is a scalar and is independent of direction, and it can vary in space and/or with the size of the control volume.

In terms of the hydraulic head, equation (5.4.5) can be written in the following form:

$$\frac{\partial}{\partial x}\left[K_x\frac{\partial H}{\partial x}\right]+\frac{\partial}{\partial y}\left[K_y\frac{\partial H}{\partial y}\right]+\frac{\partial}{\partial z}\left[K_z\left(\frac{\partial H}{\partial z}\right)\right]=S_s\frac{\partial H}{\partial t} \tag{5.4.6}$$

If we assume that the medium is homogeneous and isotropic, then equation (5.4.6) becomes

$$D\frac{\partial^2 H}{\partial x^2}+D\frac{\partial^2 H}{\partial y^2}+D\frac{\partial^2 H}{\partial z^2}=\frac{\partial H}{\partial t} \tag{5.4.7}$$

where $D = K/S_s$ and is often called hydraulic diffusivity [L^2/T]. This equation is similar to the diffusion equation that is used to describe molecular diffusion. In other words, as an excitation is imposed on the aquifer, the hydraulic head (energy/weight of fluid) propagates outward from the location of the excitation as a diffusion process. A diffusivity length is often defined as $\ell_D=\sqrt{4Dt}$. If the diffusivity length is much greater than the system size, the profiles of the hydraulic head will be essentially linear; they will be nonlinear, otherwise. The time for 32 percent of head perturbations to travel a distance L where a constant head is specified then becomes

$$t^*=\frac{L^2}{4D}=\frac{L^2 S_s}{4K} \tag{5.4.8}$$

In essence, the time for the system to reach a steady-state flow condition is approximately three times the t^*. Since the magnitude of S_s is extremely small (10^{-4}), the time to reach the steady state condition would be rapid in a confined aquifer or any rigid geologic medium with constant head boundaries (see Section 5.8).

5.5 Equations for Flow through Unsaturated Porous Media

Assuming Darcy's law is valid for flow in unsaturated porous media, we have

$$q_x = -K(h)\frac{\partial H}{\partial x},\ q_y = -K(h)\frac{\partial H}{\partial y},\ q_z = -K(h)\frac{\partial H}{\partial z} \tag{5.5.1}$$

The mass balance equation for flow through unsaturated porous media can be written as

$$\frac{\partial \rho\theta}{\partial t} = -\left[\frac{\partial \rho q_x}{\partial x} + \frac{\partial \rho q_y}{\partial y} + \frac{\partial \rho q_z}{\partial z}\right] \tag{5.5.2}$$

Notice that moisture content replaces the porosity term for the saturated porous media in the preceding mass balance equation. If we assume ρ is constant and apply Darcy's law, the governing equation for flow through unsaturated porous media takes the following form

$$\frac{\partial}{\partial x}\left[K(h)\frac{\partial H}{\partial x}\right] + \frac{\partial}{\partial y}\left[K(h)\frac{\partial H}{\partial y}\right] + \frac{\partial}{\partial z}\left[K(h)\frac{\partial H}{\partial z}\right] = \frac{\partial \theta}{\partial t} \tag{5.5.3}$$

Recognizing that $H = h + z$, one can rewrite the equation as

$$\frac{\partial}{\partial x}\left[K(h)\frac{\partial h}{\partial x}\right] + \frac{\partial}{\partial y}\left[K(h)\frac{\partial h}{\partial y}\right] + \frac{\partial}{\partial z}\left[K(h)\frac{\partial h}{\partial z} + 1\right] = \frac{\partial \theta}{\partial t} \tag{5.5.4}$$

Equation (5.5.4) is called the mixed form of Richards' equation because it contains both pressure head and moisture content terms in the formulation. Richards' equation only considers water flow by assuming that the gas phase is immobile or equilibrates very quickly. Notice that the first two terms involve pressure gradients $\partial h/\partial y$ and $\partial h/\partial y$, but the third includes both the pressure gradient $\partial h/\partial z$ and the gravity term (i.e., 1). In addition, the equation assumes hydraulic conductivity isotropy. Equation (5.5.4) is applicable to both homogeneous and heterogeneous media. If the medium is homogeneous (i.e., has spatially invariant constitutive relationships), the unsaturated hydraulic conductivity can vary in time and space as a result of its dependence on the pressure head and moisture content. On the other hand, if the medium is heterogeneous, the parameters (e.g., K_s, α, m, n, θ_s, and θ_r – see Chapter 3) in the moisture release and conductivity curves are functions of spatial locations.

Now, applying chain rule to the term on the right-hand side of equation (5.5.4), we have the following relationship

$$\frac{\partial \theta}{\partial t} = \frac{d\theta}{dh}\frac{\partial h}{\partial t} = C(h)\frac{\partial h}{\partial t} \tag{5.5.5}$$

in which *C(h)*, equal to $d\theta/dh$, is the moisture capacity term. It carries the same physical meaning as specific storage: the amount of change in water storage per unit volume of the porous medium and per unit change of head (equation 5.4.3). However, the process causing the change in water storage is actually dewatering or filling the voids of the porous medium, rather than expansion or compaction of the water or porous medium.

A pressure-based Richards' equation for either a heterogeneous or a homogeneous medium takes the form

$$\frac{\partial}{\partial x}\left[K(h)\frac{\partial h}{\partial x}\right]+\frac{\partial}{\partial y}\left[K(h)\frac{\partial h}{\partial y}\right]+\frac{\partial}{\partial z}\left[K(h)\left(\frac{\partial h}{\partial z}+1\right)\right]=C(h)\frac{\partial h}{\partial t} \tag{5.5.6}$$

If the moisture content is solely a function of pressure head, the following relationship is valid:

$$K(h)\frac{\partial h}{\partial x}=K(h)\frac{dh}{d\theta}\frac{\partial\theta}{\partial x}=\frac{K(h)\partial\theta}{C(h)\partial x} \tag{5.5.7}$$

The preceding transformation (equation (5.5.7)) is valid only for homogeneous media with spatially invariant constitutive relationships (see the Exercise). If we define $K(h)/C(h)$ as the moisture diffusivity, $D(h)$ or $D(\theta)$, $[L^2/T]$, a moisture-based Richards' equation is formed:

$$\frac{\partial}{\partial x}\left[D(\theta)\frac{\partial\theta}{\partial x}\right]+\frac{\partial}{\partial y}\left[D(\theta)\frac{\partial\theta}{\partial y}\right]+\frac{\partial}{\partial z}\left[D(\theta)\frac{\partial\theta}{\partial z}\right]+\frac{\partial K(\theta)}{\partial z}=\frac{\partial\theta}{\partial t} \tag{5.5.8}$$

or

$$\frac{\partial}{\partial x}\left[D(\theta)\frac{\partial\theta}{\partial x}\right]+\frac{\partial}{\partial y}\left[D(\theta)\frac{\partial\theta}{\partial y}\right]+\frac{\partial}{\partial z}\left[D(\theta)\frac{\partial\theta}{\partial z}\right]+\frac{dK(\theta)}{d\theta}\frac{\partial\theta}{\partial z}=\frac{\partial\theta}{\partial t} \tag{5.5.9}$$

Equation (5.5.9) has a form similar to the advection-diffusion equation for solute transport. The first three terms on the left-hand side of equation (5.5.9) represent a diffusive, fluid-flow spreading process of moisture content due to the moisture content gradient. The advection in the vertical direction is denoted by the fourth term, where $dK(\theta)/d\theta$ is analogous to the advective velocity. More appropriately, we can refer to equation (5.5.9) as a nonlinear advection-diffusion equation.

Several important points about the moisture-based Richards' equation should be recognized and they are discussed in the following.

1) The moisture diffusivity term does not exist near saturation (i.e., as $D(\theta)$ approaches ∞) since $C(h)$ approaches zero.

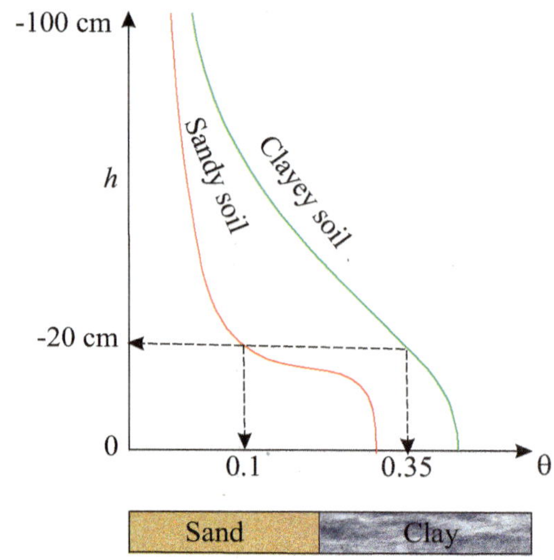

Figure 5.3. A schematic explanation of a no flow condition in a soil block consisting of two types of soils of different moisture content values.

2) The moisture content term in equation (5.5.9) becomes porosity if the medium is fully saturated. Thus, this equation is only suitable for describing flow in unsaturated porous media.
3) More importantly, the moisture content term in the equation is merely a surrogate of the pressure head (i.e., energy/weight of fluid). Hydraulic head gradients drive fluid flow. Moisture content is assumed (in this form of the equation) to be an indirect measure of the head or pressure gradient. This assumption is valid only in homogeneous media. In other words, the application of the chain rule in equation (5.5.7) assumes that the θ-h relationship is invariant in space.

Hence, the moisture-based equation is inappropriate for heterogeneous media. To illustrate this fact, consider a horizontal, 1-D soil sample consisting of two soils (clay and sand) in Figure 5.3. Suppose that the clay has a higher moisture content (say, 0.35) than the sand (0.10). Thus, the moisture gradient is toward the sand. Such a moisture gradient, however, does not necessarily imply that water flows from clay to sand. According to Darcy's law, water flows according to the hydraulic head (not moisture) gradient: from high energy regions to low energy regions. As shown in the moisture release curves in the figure, the sand and clay have different moisture contents, but they have the same pressure head and, in turn, the same energies. The total head gradient between the two soils is thus zero and there is no flow between the two soils. In conclusion, the moisture-based equation is not suitable for describing flow in heterogeneous unsaturated media.

Lastly, while the hydraulic diffusivity for the saturated media and moisture diffusivity for the unsaturated media have the same dimensions (i.e., L^2/T), the former is a measure of the rate of spreading of an energy front and the latter denotes the rate of spreading of a moisture front. Spread of energy is quite different from the spread of moisture or solutes. To show the difference, we use the following analogy. Consider a tube of toothpaste. If it is squeezed at the end of the tube, the pressure travels almost instantaneously to the rest of the tube and forces toothpaste at the outlet to emerge. This is the propagation of energy. On the other hand, the toothpaste at the end of the tube moving toward the outlet is analogous to the transport of moisture or a tracer. Therefore, hydraulic diffusivity is much larger than the moisture diffusivity.

5.6 Equations for Flow through Variably Saturated Porous Media

For heterogeneous and isotropic porous media under variably saturated conditions (i.e., part of the media is saturated and part is unsaturated.), the governing flow equation takes the following form:

$$\frac{\partial}{\partial x}\left[K(h)\frac{\partial h}{\partial x}\right]+\frac{\partial}{\partial y}\left[K(h)\frac{\partial h}{\partial y}\right]+\frac{\partial}{\partial z}\left[K(h)\left(\frac{\partial h}{\partial z}+1\right)\right]=(C(h)+\beta S_s)\frac{\partial h}{\partial t} \tag{5.6.1}$$

in which h is the pressure head. It is positive if the medium is fully saturated, and it is negative if the medium is unsaturated. A saturation index, β, is used to control the storage property of the medium under fully saturated conditions: β is set to 1, and otherwise its value is 0. The governing equation for variably saturated flow in heterogeneous anisotropic media is

$$\frac{\partial}{\partial x_i}\left[K_{ij}(h,x)\left(\frac{\partial h}{\partial x_j}+\delta_j\right)\right]=(C(h,x)+\beta S_s(x))\frac{\partial h}{\partial t} \tag{5.6.2}$$

where i and $j = 1, 2, 3$, corresponding to x, y, and z; the position vector is denoted by $\mathbf{x}$; $\delta_j = 1$ where $j = 3$, representing the gravity term and its value is 0 otherwise.

There are many advantages for adopting equation (5.6.2) for describing flow through geologic media. First, a geologic formation is treated realistically as a continuous medium without explicitly defining the location of the water table, which is often used to divide a geologic medium into the vadose zone and the aquifer. The free-surface boundary condition for the water table (Bear, 1972) is implicitly included in equation (5.6.2). Further, this equation permits perched water table aquifers in the thick vadose zone, and it automatically includes unsaturated flow processes when a fully saturated medium is dewatered. Generally speaking, this equation is far more realistic and flexible than classical approaches that artificially partition a geologic medium into unsaturated and saturated zones and employ different equations for the processes in different zones.

5.7 Equations for Flow through Fractured Media

Currently available conceptual models of flow through fractured rocks fall into four broad categories: 1) an equivalent porous media approach based on the representative elementary volume (REV) concept (Bear, 1972) that treats the rock as a single homogeneous porous continuum representing the fracture network (see Chapter 6); 2) a dual continuum approach (discussed in Chapter 6) that treats the rock as two overlapping continua of matrix and fracture (Barenblatt et al., 1960; Duguid and Lee, 1977; Pruess and Narasimhan, 1985; Dykhuizen, 1990; Gerke and van Genuchten, 1993a, b); 3) a stochastic continuum approach (i.e., the highly parameterized conceptual model discussed in Section 5.2) that treats the rock properties as a random variable (Neuman, 1987; Tsang et al., 1996); and 4) a discrete fracture network (DFN) model that idealizes fractures as parallel plates (Long et al., 1982; Dershowitz and Einstein, 1988; Cacas et al., 1990; Dverstörp et al., 1992, National Research Council, 1996). Notice that all these conceptual models rely on the continuum and laminar flow assumption.

A fractured geologic medium is merely a mixture of solids and two distinct populations of voids (i.e., fractures and pores of porous medium). Fractures belong to a void population that has voids large in size and planar in shape. On the other hand, the pores of the matrix belong to a void population that consists of voids with small sizes and granular shapes. In the past, inabilities to characterize these two void populations in geologic media and considerations of practicality contributed to the creation of the equivalent homogeneous porous media concept. Effects of the contrast of the two void populations and distinctly different responses of the fractures and the matrix then led to the development of dual porosity/mass transfer models. These models are, however, suitable only for describing or predicting bulk flow and transport behaviors averaged over a large volume of fractured media, which often do not satisfy the resolution of our interest. The demand for high-resolution predictions thus promoted the development of DFN models. These models, however, dictate detailed mappings of fracture distributions. Uncertainty in characterizing their distributions due to our limited characterization technologies then becomes the logic behind the stochastic, heterogeneous continuum concept based on the highly parameterized conceptual model described earlier.

All of these conceptual models are built on the same assumption: laminar flow. Specifically, all of these models rely on the linear relationship between specific discharge and hydraulic gradient, analogous to Darcy's law for flow through porous media. The only difference between these fracture models and porous media models is that the fracture models relate the coefficient of the linear relationship to fracture characteristics, such as fracture aperture opening, length, or width (i.e., the cubic law). The discrete fracture model describes flow in fractures only. The dual porosity model further considers the interaction between flow in fractures and the surrounding matrix using a mass transfer function. The dual porosity model is in fact a simplified mathematical model for flow through highly heterogeneous porous media. In other words, flow through a fractured rock is analogous to flow through heterogeneous geologic media with strong contrast in hydraulic characteristics. Therefore, the equation derived for flow through porous media is unequivocally appropriate for flow through fractured rock. The only concern about these models is that the Reynolds number of flow in fractured rock may exceed that for flow through porous media under some conditions. As a consequence, the linear relationship between the specific discharge and the hydraulic gradient for laminar flow may be in question. For modeling laminar flow in fractured rocks, the real hurdle is not whether these mathematical models are valid or not. Rather, it is whether or not a technology exists that can characterize the complex spatial pattern of fractures to the extent needed for prediction.

5.8 Initial and Boundary Conditions

In order to obtain a particular solution to the aforementioned governing equations for transient flow in variably saturated media or confined or unconfined aquifers, an initial condition and boundary conditions must be specified. For steady-state flow, only boundary conditions are required. Otherwise, only general solutions (i.e., many possible solutions) can be obtained.

The **initial condition** represents the pressure head, h, or moisture content, θ, distribution in the flow domain at the time before a simulation or prediction of a perturbation to the geologic media takes place. That is,

$$h(x, y, z, 0) = h^*(x, y, z) \quad \text{and} \quad \theta(x, y, z, 0) = \theta^*(x, y, z) \tag{5.8.1}$$

A boundary condition specifies the flow behavior at the boundaries of a geologic medium under consideration. Several types of boundary conditions exist.

1. **Prescribed head boundary condition (or Dirichlet boundary condition in mathematics)**: Pressure or total head is specified at the boundary of a simulation domain. For example, one can specify the river stage as the prescribed head boundary condition at the interface between the river and the geologic medium; this boundary condition is illustrated in Figure 5.4 along segment B-C, and mathematically it can be expressed as

$$h(x, y, z, t)|_{x,y,z} = h^*(x, y, z, t) \tag{5.8.2}$$

2. **Prescribed flux boundary condition (or Neumann boundary condition in mathematics)**: Fluxes are specified at the boundary locations. For example, the flux at the bottom boundary of the aquifer is specified as zero, representing an impermeable boundary (segment C-D) in Figure 5.4.

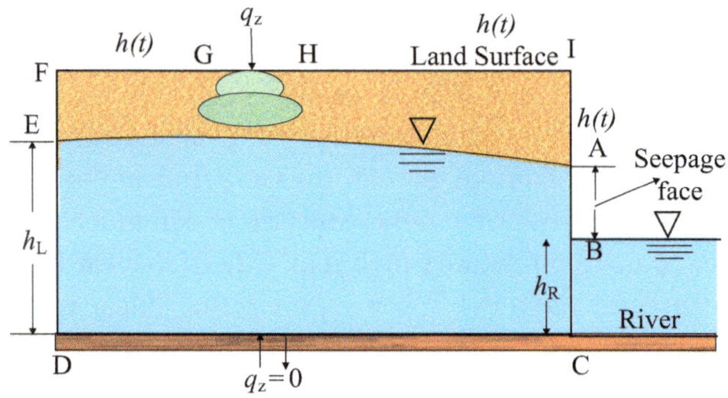

Figure 5.4. Conceptual diagram of an unconfined aquifer adjacent to a river illustrating several commonly used boundary conditions.

$$q_z(x, y, z, t)|_{x,y,z} = 0 \tag{5.8.3}$$

3. **Seepage face boundary condition**: When the water table approaches the downstream-side boundary a flow domain (the right-hand side boundary of Figure 5.4), it terminates on the boundary at a point (point A in the figure) that is above the water surface elevation of the body of open water outside the flow domain. Then, segment A-B of the boundary above the river stage and below the water table is called the seepage face. This seepage face is the boundary where the pressure head is equal to the atmospheric pressure head. That is to say, the pressure head along this seepage face is zero, and the elevation head is the total head. Water seeps out from the face freely. That is,

$$h(x, y, z, t) = 0 \tag{5.8.4}$$

The location of the seepage face varies with the location of the water table. In order to determine the location of the seepage face, the governing equation must be solved for the water table position. However, to solve the equation for the water table position, the location of the seepage face must be given. Thus, determination of the seepage face adds nonlinearity to the governing flow equation.

4. **Unit gradient boundary condition**: At this boundary, the pressure head gradient in the vertical direction is specified as zero, and flow crossing this boundary is driven by gravity alone. This boundary condition is useful for modeling a semi-infinitely thick vadose zone.

$$\left.\frac{\partial h}{\partial z}\right|_{x,y,z} = 0 \tag{5.8.5}$$

5.9 Equations for Flow in Two-Dimensional Horizontal Aquifers

Because of the presence of the multiscale heterogeneities in geologic media, groundwater flow in geologic formations always is a three-dimensional phenomenon.

Geologic formations through which groundwater flows can be fully saturated at some locations and/or partially saturated at other locations. Therefore, the 3-D equation for flow in variably saturated media is most appropriate for such field applications. A two-dimensional, depth-averaged horizontal flow equation (or conceptual model) to be introduced later, however, often is found to be useful for many practical purposes (for example, analysis of groundwater flow at regional scales where our interest is horizontal patterns of flows only). The 2-D horizontal flow equation also avoids many mathematical complexities – this is especially true when analytical solutions are the only available tools.

5.9.1 Depth-Averaged Properties

Prior to introducing the equation for 2-D depth-averaged flow, we will define some new terms. These terms are used in the formulation of the 2-D governing flow equations.

Aquifer transmissivity, *T*, is defined as the integral of all conductivity values over the depth of an aquifer. That is,

$$T(x,y) = \int_{z_b(x,y)}^{z_t(x,y)} K(x,y,z)dz \tag{5.9.1}$$

where $z_b(x, y)$ and $z_t(x, y)$ are the elevation of the bottom and top boundaries of the aquifer, respectively. In other words, $z_t(x, y) - z_b(x, y) = b(x, y)$, which is the thickness of the confined aquifer at location (x, y). The transmissivity also can be expressed as

$$T(x,y) = b(x,y)\overline{K}(x,y) \text{ where } \overline{K}(x,y) = \frac{1}{b(x,y)} \int_{0}^{b(x,y)} K(x,y,z)dz \tag{5.9.2}$$

which is a depth-averaged hydraulic conductivity. As a result, the transmissivity, $[L^2/T]$, is defined as the volume of water transmitted by a unit thickness of an aquifer under a unit gradient condition. It is often recognized as a depth-averaged property of an aquifer, which is invariant in the vertical direction but varies in the horizontal plane. If discrete horizontal strata of different hydraulic conductivity values are considered, the transmissivity can be defined as

$$T(x,y) = \sum_{i=1}^{n} b_i(x,y)K_i(x,y) \tag{5.9.3}$$

in which $b_i(x, y)$ is the thickness of each individual stratum, $K_i(x, y)$ is the hydraulic conductivity value associated with the ith stratum, and n is the total number of strata at location (x, y).

Aquifer storage coefficient (storativity) *S* [dimensionless]. Similar to transmissivity, the storage coefficient is defined as

$$S(x,y) = \int_0^b S_s(x,y,z)dz = \int_0^{b(x,y)} \rho g(\alpha(x,yz) + n(x,y,z)\beta)dz \tag{5.9.4}$$
$$= b(x,y)\overline{S}_s(x,y) = b\Delta V_w / V\Delta H$$

where S_s is the specific storage. Therefore, the storage coefficient (a dimensionless parameter) represents the volume of water, V_w, released from (or increased in) storage per unit horizontal area, A, of a confined aquifer under per unit decline (or increase) of hydraulic head, Δh. That is,

$$S = \Delta V_w / A\Delta H \quad \text{or} \quad S = \overline{S}_s b \tag{5.9.5}$$

Depth-averaged hydraulic head, $\overline{H}$. Similarly, the depth-averaged head is defined as an integrated head value over the depth of the aquifer.

$$\overline{H}(x,y,t) = \frac{1}{b(x,y)} \int_0^{b(x,y)} H(x,y,z,t)dz \tag{5.9.6}$$

Generally speaking, the water level registered in a piezometer represents the pressure head at the point, where the piezometer is located in the aquifer (Figure 5.5). A piezometer is a well that is open only at its bottom tip. On the other hand, a screened well is a well that is perforated over a section of the casing, and it measures an integrated pressure head over the length of the perforated interval (i.e., a depth-averaged head). Water levels measured by these two devices can be identical if the flow field is purely horizontal. Piezometers are specifically designed for monitoring

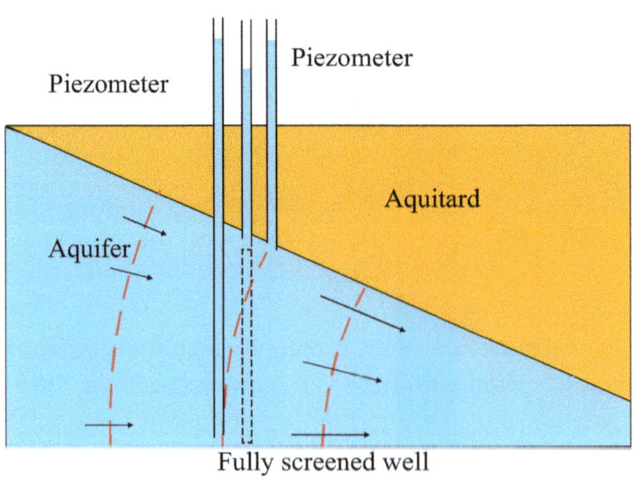

Figure 5.5. Water levels in a fully screened well and piezometers.

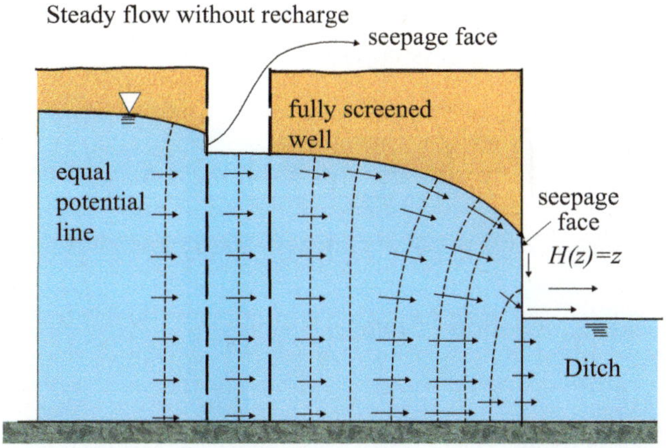

Figure 5.6. A cross-sectional view of a steady-state flow field through a fully screened well to a ditch in an unconfined aquifer. Within the fully screened well, the flow is horizontal, as is the flow entering the ditch.

hydraulic head. On the other hand, wells are constructed for both water production and monitoring purposes. Wells used for water supply generally are constructed with long well screens to maximize the amount and longevity of water production.

The measurements of head in a piezometer and a fully screened well, however, can be significantly different in aquifers where vertical flow components exist. As shown in Figure 5.5, the two piezometers are located at the same equal potential (hydraulic head) line and, thus, the water levels in the piezometers are at the same height. The water level in the fully screened well in the figure is, however, lower than that in the piezometer, because this water level is an integrated hydraulic head over many equal potential lines crossing the screened depth.

Figure 5.6 explains the reason why the water level in a fully screened observation well is a depth-averaged hydraulic head. In a fully screened well, as shown in Figure 5.6, vertical circulation of water does not exist and the equal-potential line in the well must be vertical. The water level in the well is thus a depth-averaged hydraulic head.

5.9.2 Depth-Averaged Equation for Flow in Confined Aquifer

There are many ways to derive the governing flow equation for areal, two-dimensional, groundwater flow. Vertically averaging a three-dimensional groundwater flow equation in fully saturated media is the most rigorous approach and it is discussed below.

Consider the three-dimensional flow equation in fully saturated anisotropic and heterogeneous porous media with the coordinate system in alignment with the principal directions of hydraulic conductivity:

$$\frac{\partial}{\partial x}\left(K_x\frac{\partial H}{\partial x}\right)+\frac{\partial}{\partial y}\left(K_y\frac{\partial H}{\partial y}\right)+\frac{\partial}{\partial z}\left(K_z\frac{\partial H}{\partial z}\right)=S_s\frac{\partial H}{\partial t} \tag{5.9.7}$$

Now, if we integrate this equation over the depth from the elevation of the bottom of the aquifer, $z_b(x, y)$, to the top of the aquifer, $z_t(x, y)$, we then have

$$\int_{z_b(x,y)}^{z_t(x,y)} \frac{\partial}{\partial x}\left(K_x \frac{\partial H}{\partial x}\right)dz + \int_{z_b(x,y)}^{z_t(x,y)} \frac{\partial}{\partial y}\left(K_y \frac{\partial H}{\partial y}\right)dz + \int_{z_b(x,y)}^{z_t(x,y)} \frac{\partial}{\partial z}\left(K_z \frac{\partial H}{\partial z}\right)dz = \int_{z_b(x,y)}^{z_t(x,y)} S_s \frac{\partial H}{\partial t} dz \tag{5.9.8}$$

Leibnitz's rule for differentiation of integrals states

$$\frac{\partial}{\partial x}\int_A^B f(x,t)dt = \int_A^B \frac{\partial f(x,t)}{\partial x} dt + f(x,B)\frac{dB}{dx} - f(x,A)\frac{dA}{dx} \tag{5.9.9}$$

In order to apply the rule to equation (5.9.8), we will let

$$f = K_x \frac{\partial H}{\partial x} \tag{5.9.10}$$

and A = $z_b(x, y)$ and B = $z_t(x, y)$. According to the rule, we have

$$\frac{\partial}{\partial x}\int_{z_b(x,y)}^{z_t(x,y)} K_x \frac{\partial H}{\partial x} dz = \int_{z_b(x,y)}^{z_t(x,y)} \frac{\partial}{\partial x}\left(K_x \frac{\partial H}{\partial x}\right)dz + K_x \frac{\partial H}{\partial x}\frac{dz_t(x,y)}{dx} - K_x \frac{\partial H}{\partial x}\frac{dz_b(x,y)}{dx} \tag{5.9.11}$$

If we further assume that a) the thickness of the aquifer is uniform and b) both $z_t(x, y)$ and $z_b(x, y)$ are invariant in x and y, the resultant equation can be simplified to

$$\frac{\partial}{\partial x}\int_{z_b}^{z_t} K_x \frac{\partial H}{\partial x} dz = \int_{z_b}^{z_t} \frac{\partial}{\partial x}\left(K_x \frac{\partial H}{\partial x}\right)dz \tag{5.9.12}$$

Next, we consider that $K_x(x, y, z)$, $S_s(x, y, z)$ and $H(x, y, z, t)$ can be expressed as the sum of depth-averaged quantities and perturbations:

$$\begin{aligned} K_x(x,y,z) &= \overline{K}_x(x,y) + K'_x(x,y,z) \\ S_s(x,y,z) &= \overline{S}_s(x,y) + S'_s(x,y,z) \\ H(x,y,z,t) &= \overline{H}(x,y,t) + H'(x,y,z,t) \end{aligned} \tag{5.9.13}$$

where the quantities with an overhead bar denote their depth-averaged values. Substituting equation (5.9.13) into equation (5.9.12), and assuming $z_b = 0$ and $z_t = b$ (i.e., thickness of the aquifer is constant), we have

$$\frac{\partial}{\partial x}\left[b\overline{K}_x\frac{\partial \overline{H}}{\partial x}\right]+\frac{\partial}{\partial x}\int_0^b K'_x\frac{\partial \overline{H}}{\partial x}dz+\frac{\partial}{\partial x}\int_0^b \overline{K}_x\frac{\partial H'}{\partial x}dz+\frac{\partial}{\partial x}\int_0^b K'_x\frac{\partial H'}{\partial x}dz=\int_0^b\frac{\partial}{\partial x}\left(K_x\frac{\partial H}{\partial x}\right)dz \tag{5.9.14}$$

The second and third terms on the left-hand side of equation (5.9.14) are zero because the integration of the perturbation equals zero. We thus have

$$\frac{\partial}{\partial x}\left[b\overline{K}_x\frac{\partial \overline{H}}{\partial x}\right]+\frac{\partial}{\partial x}\int_0^b K'_x\frac{\partial H'}{\partial x}dz=\int_0^b\frac{\partial}{\partial x}\left(K_x\frac{\partial H}{\partial x}\right)dz \tag{5.9.15}$$

Similarly, we can derive the following,

$$\frac{\partial}{\partial y}\left[b\overline{K}_y\frac{\partial \overline{H}}{\partial y}\right]+\frac{\partial}{\partial y}\int_0^b K'_y\frac{\partial H'}{\partial y}dz=\int_0^b\frac{\partial}{\partial y}\left(K_y\frac{\partial H}{\partial y}\right)dz \tag{5.9.16}$$

for the second term on the left-hand side of equation (5.9.8). Note that the last term on the left-hand side of equation (5.9.8), after integration over z, becomes

$$\int_0^b\frac{\partial}{\partial z}\left(K_z\frac{\partial H}{\partial z}\right)dz=K_z\frac{\partial H}{\partial z}\bigg|_b-K_z\frac{\partial H}{\partial z}\bigg|_0=q_z|_b-q_z|_0=0 \tag{5.9.17}$$

The result of the integration is zero as a result of the assumption that the upper and lower boundaries of the confined aquifer are impermeable.

Next, we consider the integration of the storage term. We have

$$\int_0^b S_s\frac{\partial H}{\partial t}dz=\int_0^b S'_s\frac{\partial H'}{\partial t}dz+b\overline{S}_s\frac{\partial \overline{H}}{\partial t}=\int_0^b S'_s\frac{\partial H'}{\partial t}dz+S\frac{\partial \overline{H}}{\partial t} \tag{5.9.18}$$

Substituting equations (5.9.16) through (5.9.18) into equation (5.9.8), the final depth-averaged equation takes the form

$$\frac{\partial}{\partial x}\left[T_x\frac{\partial \overline{H}}{\partial x}\right]+\frac{\partial}{\partial x}\int_0^b K'_x\frac{\partial H'}{\partial x}dz+\frac{\partial}{\partial y}\left[T_y\frac{\partial \overline{H}}{\partial y}\right]+\frac{\partial}{\partial y}\int_0^b K'_y\frac{\partial H'}{\partial y}dz=\int_0^b S'_s\frac{\partial H'}{\partial t}dz+S\frac{\partial \overline{H}}{\partial t} \tag{5.9.19}$$

where $T=b\overline{K}$ is the transmissivity, as defined previously. Notice that equation (5.9.19) is composed of terms that involve the depth-averaged properties as well as the depth-averaged product of perturbations. Suppose we assume that the terms $\int_0^b K'_x \frac{\partial H'}{\partial x} dz$ and $\int_0^b K'_y \frac{\partial H'}{\partial y} dz$ are linearly proportional to $\frac{\partial \overline{H}}{\partial x}$ and $\frac{\partial \overline{H}}{\partial y}$, respectively, and that the term $\int_0^b (S'_s \partial H'/\partial t) dz$ can be related to $\partial \overline{H}/\partial t$; the final form of the depth-averaged flow equation becomes

$$\frac{\partial}{\partial x}\left[\hat{T}_x \frac{\partial \overline{H}}{\partial x}\right] + \frac{\partial}{\partial y}\left[\hat{T}_y \frac{\partial \overline{H}}{\partial y}\right] = \hat{S}\frac{\partial \overline{H}}{\partial t} \tag{5.9.20}$$

where $\hat{T}_x$, $\hat{T}_y$, and $\hat{S}$ are the effective transmissivity in x and y directions and effective storage coefficient. Notice again that $\overline{H}$ is the depth-averaged head and that the effective properties are not equal to the depth-averaged transmissivity and storage coefficient that we defined previously. Specifically, the **effective properties** are

$$\hat{T}_x = T_x + \left[\int_0^b K'_x \frac{\partial H'}{\partial x} dz\right] \Big/ \left[\frac{\partial \overline{H}}{\partial x}\right], \quad \hat{T}_y = T_y + \left[\int_0^b K'_y \frac{\partial H'}{\partial y} dz\right] \Big/ \left[\frac{\partial \overline{H}}{\partial y}\right], \text{ and}$$

$$\hat{S} = S + \left[\int_0^b S'_s \frac{\partial H'}{\partial t} dz \Big/ \frac{\partial \overline{H}}{\partial t}\right] \tag{5.9.21}$$

According to equation (5.9.21), the effective properties will be the same as the depth-averaged properties only if the flow is purely horizontal (i.e., no vertical variation in either hydraulic conductivity or head). Otherwise, the depth-averaged properties are merely approximations of the effective properties, which will vary with the flow field. Further, the accuracy of these approximations depends on the magnitude of vertical components of the flow.

Figure 5.7 shows steady flow field (head and streamline distributions) in a vertical cross section of a heterogeneous aquifer. Suppose we assume the aquifer is homogeneous. Can you imagine what the hydraulic head and streamline would look like? Then, how does heterogeneity affect head and velocity in the aquifer? If the depth-averaged approach is applied to this flow field, how will the spatially varying head and velocity fields affect the result?

In the case where the aquifer thickness varies spatially, the depth-averaged equation also reflects this variation. Hence, the "depth averaging" procedure may introduce many artifacts into the conceptual model, which may vary from one flow

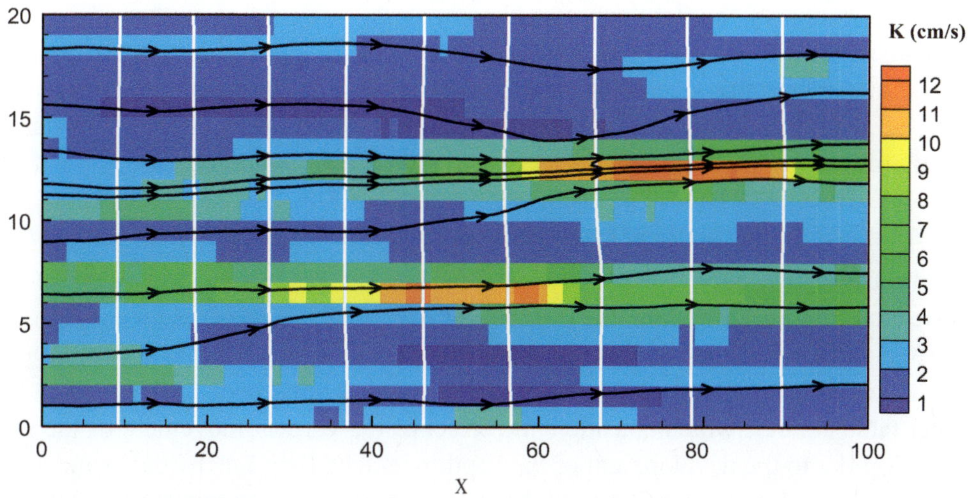

Figure 5.7. The hydraulic head and streamline patterns in a heterogeneous aquifer with a constant thickness.

situation to another. Therefore, it is generally most desirable to use a fully three-dimensional model to prevent simplifications that can lead to unnecessary artifacts.

5.9.3 Depth-Averaged Equation for Flow in Phreatic Aquifers

The general governing equation for three-dimensional flow through fully saturated porous media (i.e., equation (5.4.5)) is valid for flow in the saturated region of either water table or confined aquifers. The primary difference in modeling water flow in water table and confined aquifers using equation (5.4.5) lies in the specification of boundary conditions. Generally speaking, the location of the upper boundary condition for a confined aquifer is temporally invariant and can be a prescribed head, a no-flux, or a flux boundary. On the other hand, the upper boundary for a water table aquifer is not necessarily stationary in time and space because of the unsaturated zone above. If the flow process in the unsaturated zone, including the capillary fringe, is neglected and if we focus on the flow in the saturated region between the water table and bottom impermeable boundary of the aquifer, the upper boundary of water table aquifers will be the water table, whose position can vary in time and space depending on flow. This dynamic upper boundary for the water table aquifer is often called the "free-surface boundary." Notice that the free-surface boundary condition is a by-product of our ignorance of the variably saturated zone. In other words, had we used the governing equation for flow in variably saturated media (equation (5.6.1)), the free-surface boundary condition would not be necessary, since the evolution of the water table (the zero pressure surface) is part of the evolution of the flow process. Nevertheless, modeling flow through variably saturated media is by far more complex and involved, as a result of the nonlinear nature of the governing equation, than modeling flow in fully saturated media. The free-surface boundary approach therefore is considered practical, especially when our ability to analyze flow through variably saturated media is limited.

Depth-Averaged Approach

As discussed previously, for many field problems, we may be interested in only horizontal flow. Additionally, the difficulty in solving the three-dimensional flow and the free-surface boundary equations forces us to take a shortcut. As a consequence, we will introduce a depth-averaged two-dimensional horizontal flow equation for a water table aquifer, which considers the effect of the vertical movement of the water table. Similar to the development of the depth-averaged two-dimensional equation for a confined aquifer, this equation will be derived by integrating the three-dimensional equation over the vertical direction.

Suppose we average the hydraulic head from the bottom of an aquifer to the water table, where the pressure head is zero. That is, we neglect the vadose zone and the effects of its capillary pressure. We have the depth-averaged head of a water table aquifer. That is,

$$\overline{H}(x, y, t) = \frac{1}{b(x, y)} \int_0^{b(x,y)} H(x, y, z, t)dz \tag{5.9.22}$$

where *b(x, y)* is the elevation of the water table at location (*x*, *y*), and it is the saturated thickness of the aquifer if the elevation of the bottom of the aquifer is set to zero (see equation (5.9.22)). As explained previously, the depth-averaged hydraulic head can be different from the height of the water table (or the saturated thickness; see Figure 5.8). For mathematical simplicity, we will assume that the difference is negligible. That is,

$$\overline{H}(x, y, t) \approx b(x, y, t) \tag{5.9.23}$$

Integration of equation (5.3.4) or (5.4.5) over the depth results in

$$-\int_0^{\overline{H}} \left[\frac{\partial q_x}{\partial x} + \frac{\partial q_y}{\partial y} + \frac{\partial q_z}{\partial z}\right] dz = S_s \int_0^{\overline{H}} \frac{\partial H}{\partial t} dz \tag{5.9.24}$$

Unlike in the previous depth-averaged analysis for confined aquifers, we will assume the aquifer is homogeneous to simplify the mathematics. Following the same integration procedure as used in the confined aquifer, we obtain

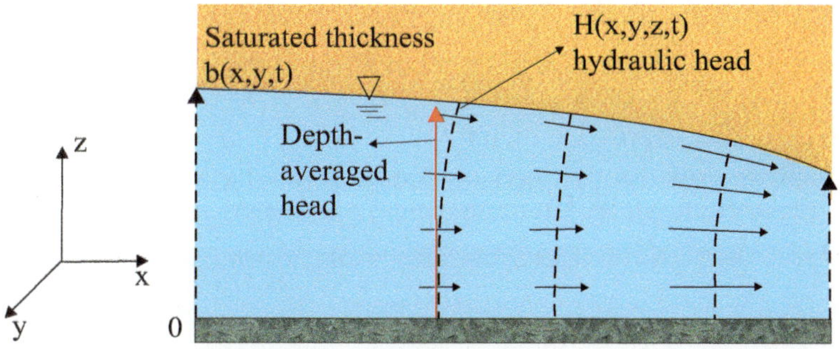

Figure 5.8. A schematic illustration of the differences in hydraulic head, depth-averaged head, and saturated thickness in a water table aquifer.

$$\frac{\partial}{\partial x}\left(K_x\overline{H}\frac{\partial\overline{H}}{\partial x}\right)+\frac{\partial}{\partial y}\left(K_y\overline{H}\frac{\partial\overline{H}}{\partial y}\right)-q_x\Big|_b\frac{\partial\overline{H}}{\partial x}-q_y\Big|_b\frac{\partial\overline{H}}{\partial y}+q_z|_b=S_s\overline{H}\frac{\partial\overline{H}}{\partial t} \qquad (5.9.25)$$

Notice that there is no leakage at the bottom boundary of the aquifer. Also, the last three terms on the left-hand side of equation (5.9.25) are associated with the boundary specific discharges at the water table (at $z = b$). These terms do not exist in the analysis of confined aquifers because of the assumption of the uniform thickness and impermeable upper boundary. An application of Darcy's law to the last three terms yields

$$-q_x\Big|_b\frac{\partial\overline{H}}{\partial x}-q_y\Big|_b\frac{\partial\overline{H}}{\partial y}+q_z\Big|_b=K_x\left(\frac{\partial\overline{H}}{\partial x}\right)^2+K_y\left(\frac{\partial\overline{H}}{\partial y}\right)^2-K_z\frac{\partial\overline{H}}{\partial z}\Big|_b \qquad (5.9.26)$$

Equation (5.9.26) is essentially a part of the free-surface boundary condition in the case where accretion (recharge from the vadose zone) is zero. The full equation for the free-surface boundary condition is

$$\frac{\partial\overline{H}}{\partial t}-\frac{1}{n}\left[K_x\left(\frac{\partial\overline{H}}{\partial x}\right)^2+K_y\left(\frac{\partial\overline{H}}{\partial y}\right)^2-K_z\frac{\partial\overline{H}}{\partial z}\right]=0 \qquad (5.9.27)$$

where n is the porosity (see Bear, 1979). According to equations (5.9.26) and (5.9.27), we then have the following relationship

$$-q_x|_b\frac{\partial\overline{H}}{\partial x}-q_y|_b\frac{\partial\overline{H}}{\partial y}+q_z|_b=K_x\left(\frac{\partial\overline{H}}{\partial x}\right)^2+K_y\left(\frac{\partial\overline{H}}{\partial y}\right)^2-K_z\frac{\partial\overline{H}}{\partial z}=n\frac{\partial\overline{H}}{\partial t} \qquad (5.9.28)$$

This equation describes the rate of change in the water table elevation (the thickness of the saturated zone), caused by non-horizontal flow in the aquifer – not due to recharge from the vadose zone. Substituting equation (5.9.28) back to equation (5.9.25), equation (5.9.25) becomes

$$\frac{\partial}{\partial x}\left(K_x\overline{H}\frac{\partial\overline{H}}{\partial x}\right)+\frac{\partial}{\partial y}\left(K_y\overline{H}\frac{\partial\overline{H}}{\partial y}\right)-n\frac{\partial\overline{H}}{\partial t}-S_s\overline{H}\frac{\partial\overline{H}}{\partial t}=0 \qquad (5.9.29)$$

or

$$\frac{\partial}{\partial x}\left(K_x\overline{H}\frac{\partial\overline{H}}{\partial x}\right)+\frac{\partial}{\partial y}\left(K_y\overline{H}\frac{\partial\overline{H}}{\partial y}\right)=\left(n+S_s\overline{H}\right)\frac{\partial\overline{H}}{\partial t}=S_y\frac{\partial\overline{H}}{\partial t} \qquad (5.9.30)$$

where S_y is the specific yield, a dimensionless parameter, and is defined as

$$S_y=\frac{\Delta V_w}{A\Delta\overline{H}} \qquad (5.9.31)$$

It is the depth-averaged unconfined aquifer storage term, which represents the amount of water released from the storage of the water table aquifer, per unit area, per unit decline in the water table. This release of water results from drainage of pores as well as compressibility of water and aquifer materials below the water table. Since the specific storage is generally very small, and the saturated thickness of the water table aquifer is generally thin, the specific yield is often considered to be equal to the porosity. Notice that specific yield is a product of the depth-averaging procedure, and it is suitable for the two-dimensional depth-averaged equation only. We should also remember that the unsaturated flow process related to falling or rising of the water table is neglected in the preceding formulation. That is, the specific yield concept assumes instantaneous and complete release or storage of water in pores as the water table drops or rises. This assumption is certainly not true and has led to estimates of specific yield values too small and often time dependent in real-world problems (see Chapter 8).

Equation (5.9.30), called the **Boussinesq equation,** is also nonlinear. However, this equation is much easier to cope with mathematically than the three-dimensional flow equation and associated free-surface boundary conditions. For example, if the medium is assumed to be homogeneous, equation (5.9.30) for steady flow can be written as

$$K_x\frac{\partial^2\overline{H}^2}{\partial x^2}+K_y\frac{\partial^2\overline{H}^2}{\partial y^2}=0 \qquad (5.9.32)$$

Equation (5.9.32) is linear in terms of $\overline{H}^2$ and analytical solutions for this type of equation are readily available for different types of boundary conditions.

In spite of its popularity, the Boussinesq equation suffers from even greater limitations than does the depth-averaged confined aquifer equation. A three-dimensional variably saturated flow equation for heterogeneous media is more appropriate for describing the temporal and spatial evolution of the water table, since it considers effects of the unsaturated zone and the capillary fringe and the unsaturated flow process associated with the rising and falling of the water table. If the variably saturated flow equation is adopted, the specific yield concept is virtually unnecessary.

5.9.4 Dupuit Assumptions

The **Dupuit assumption** assumes that groundwater moves horizontally in an unconfined aquifer, and that the groundwater discharge is proportional to the saturated aquifer thickness.

$$Q_x = -KH\frac{\partial H}{\partial x} \quad \text{and} \quad Q_y = -KH\frac{\partial H}{\partial y} \tag{5.9.33}$$

This idea was first proposed by Jules Dupuit in 1863 to simplify the groundwater flow equation for obtaining analytical solutions. The Dupuit assumption requires that the water table is relatively flat, and that the equipotential lines are vertical as well as the nonexistence of the seepage face. In fact, one can invoke the Dupuit assumption and then directly derive the Boussinesq equation without involving mathematical complexity, and without considering the free-surface boundary condition as we did previously. Our previous development, however, provides an in-depth understanding of the physics behind the formulation of the Boussinesq equation.

5.10 Control Volume and Observation Scale

We reiterate the fact that the aforementioned governing flow equations are built upon the CV concept. This CV (i.e., each block in Figure 5.1b), which is the volume over which mass balance is considered – see equation (5.3.5) – and a representative hydraulic property is defined, must be sufficiently large to contain many pores and/or other larger-scale heterogeneity. As a result, the head or the specific discharge at a point in the space described by these equations represents the average of all possible values (ensemble average) at that point in the block. Because the size of the CV or the block is much smaller than the entire domain, the degree of heterogeneity within the CV is likely mild. Therefore, as long as the property of each block of the heterogeneous conceptual model is known and specified, the simulated ensemble-averaged process at a point can be in close agreement with the corresponding observed process, which is a spatially averaged value over a small volume in the block. Therefore, a highly parameterized, heterogeneous conceptual/mathematical model generally can lead to predictions close to reality (see discussion in Section 6.7 of Chapter 6).

5.11 Summary

- The 3-D governing equation for flow through variably saturated heterogeneous media is the most general equation if effects of air and other factors can be ignored. There is no need to define saturated zone or unsaturated zone; there is no need to consider the water table spatial and temporal fluctuations explicitly.

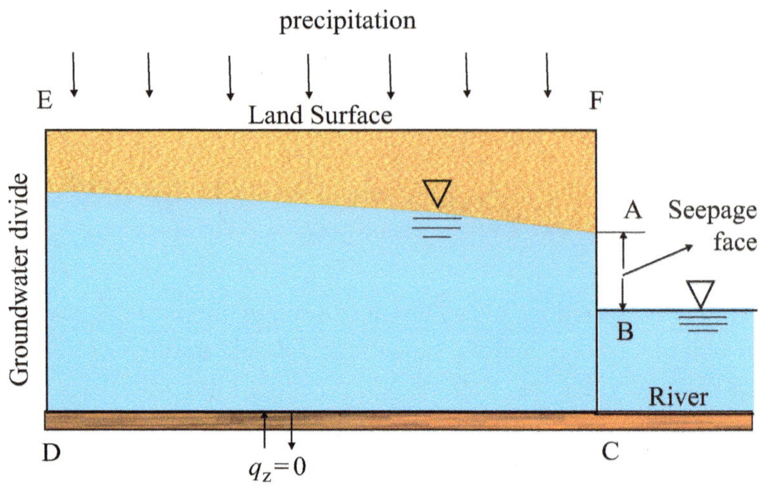

Figure 5.9

- Likewise, using the 3-D (or 2-D vertical section) governing equations, based on the heterogeneous conceptual model, avoids defining confined or unconfined aquifers explicitly.
- Moisture content in the moisture-based Richards equation is a surrogate of fluid energy, which is accurate only when the medium is homogeneous. In other words, the moisture-based Richards equation is valid for homogeneous media only.
- Simplification of a 3-D heterogeneous model to a 2-D depth-averaged model introduces fictitious transmissivity and storage coefficient (or specific yield), which can vary with flow and time. Therefore, transmissivity and specific yield are not invariant properties of the aquifer.

5.12 Exercise

1. A cross-sectional view of a water table aquifer, which consists of the vadose zone and saturated aquifer and is connected to a stream, is shown in Figure 5.9. You are asked to formulate a mathematical model that describes the flow in this two-dimensional cross section (i.e., governing partial differential equation, and boundary and initial conditions that describe the transient flow field in the aquifer). Define all variables used in the equations.
2. Now, derive a depth-averaged equation for the preceding problem and again define all variables.
3. Using the Dupuit assumption, derive a plan view 2-D Boussinesq equation.
4. If a confined aquifer is conceptualized as a homogeneous aquifer with anisotropic hydraulic conductivity, write a 3-D governing flow equation for this aquifer, assuming the principal directions are unknown.
5. If the parameters of the moisture retention curve of the vadose zone are spatially variable, that is, $h(x, \theta, \alpha, n, \theta_r, \theta_s)$, derive a formula that converts the pressure head gradient to the moisture gradient.

6

Equivalent Homogeneous Media Conceptual Models

6.1 Introduction

This chapter discusses mathematical models of the equivalent homogeneous media conceptual model. As stated in Chapter 5, this conceptual model expands the laboratory-scale control volume (CV) (see Chapter 2) to a field-scale CV. This field-scale CV is sufficiently large to include heterogeneities at scales much greater than pores (such as layers, zones, and formations). If the hydraulic properties of a medium, defined over this CV, are spatially invariant at a field site, this medium is then considered to be spatially homogeneous, and a spatial REV exists. The hydraulic property field of this medium is thus considered a realization of a stationary process (see Section 4.7 of Chapter 4). On the other hand, if the spatial REV does not exist, the property field is a realization of a nonstationary process. Nevertheless, this medium can still be treated as homogeneous over the entire field site. This homogenization of this geologic medium is tantamount to saying that an REV exists only in the ensemble sense (ensemble REV or ensemble stationarity), rather than in the spatial sense. That is, the spatial statistics of the medium property over this site are identical to the spatial statistics of all other possible realizations at the same site. As a consequence, the predicted behavior at any location based on this equivalent homogeneous model represents the ensemble-averaged behavior at that location, resulting from all possible heterogeneities over the entire field. Since the heterogeneity is nonstationary, its variance must be large, and the ensemble-averaged behavior likely deviates greatly from the observed behavior. This model thus yields low resolution predictions (see Section 6.7 and Chapter 9).

Despite its low resolution predictions, such an equivalent homogeneous conceptual modeling approach has been widely adopted over the past decades. Its popularity is largely due to our limited ability to derive analytical solutions for governing equations for flow and solute transport in complex heterogeneous media, in addition to our inability to characterize heterogeneities at different scales in detail. The homogeneous conceptual model undoubtedly offers practical solutions to complex problems, but it confronts many other issues. For instance, how can the hydraulic properties (effective hydraulic properties) of this equivalent homogeneous conceptual model be defined? How different is the predicted aquifer behavior based on the approach from the observed behavior in the field? These topics are discussed in this chapter.

6.2 Effective Hydraulic Conductivity for Saturated Media

In this section, examples are used to explain the concept of effective hydraulic conductivity for geologic media. They are also used to study the relationship between field-scale effective hydraulic conductivities and laboratory-scale hydraulic conductivities. Consequently, this relationship sheds light on some upscaling principles for effective hydraulic conductivity.

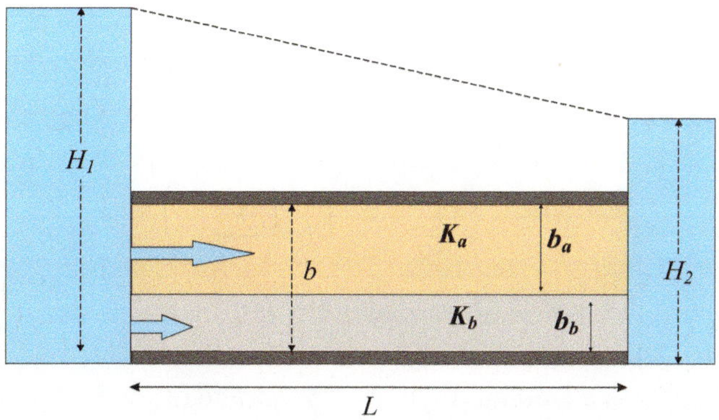

Figure 6.1. Schematic illustration of flow parallel to bedding.

6.2.1 Flow Parallel to Stratification (Layering or Bedding)

Consider the case where an aquifer consists of two layers (homogeneous within each layer) of different geologic media (Figure 6.1), and the direction of flow is parallel to the orientation of the layering. The upper layer has hydraulic conductivity K_a with a uniform thickness of b_a; the hydraulic conductivity of the lower layer is K_b and its thickness is b_b. Suppose that a hydraulic head, H_1, is imposed on one side of the aquifer and H_2 is on the other side. If $H_1 > H_2$ and the flow induced by this hydraulic-head gradient has reached a steady condition, the total discharge (per unit width normal to the section) from the aquifer will be the sum of the discharges through each layer:

$$Q = -b_a K_a \frac{(H_2 - H_1)}{L} - b_b K_b \frac{(H_2 - H_1)}{L} \tag{6.2.1}$$

If this aquifer is treated as an equivalent homogeneous aquifer, and Darcy's law is valid for this fictitious aquifer, then the total discharge from this homogeneous aquifer under the same boundary condition is

$$Q = -b\overline{K}\frac{(H_2 - H_1)}{L} \quad \text{where } b = b_a + b_b \tag{6.2.2}$$

where $\overline{K}$ is the effective hydraulic conductivity of this fictitious aquifer. To determine the $\overline{K}$ that can yield the same total discharge Q of the layered aquifer under the same hydraulic head gradient, we will equate the quantities on the right sides of equations (6.2.1) and (6.2.2) and solve for $\overline{K}$. The expression for $\overline{K}$ after a little algebraic manipulation is

$$\overline{K} = \frac{K_a b_a + K_b b_b}{b} \quad \text{where } b = b_a + b_b \tag{6.2.3}$$

Equation (6.2.3) thus is the hydraulic conductivity of the fictitious aquifer, and we will call it the effective hydraulic conductivity for flow parallel to bedding. This effective hydraulic conductivity formula can be generalized for aquifers consisting of n layers:

$$\overline{K} = \left(\sum_{i=1}^{n} b_i K_i\right) \Big/ \left(\sum_{i=1}^{n} b_i\right) \tag{6.2.4}$$

where K_i and b_i are the hydraulic conductivity and the thickness of layer i, respectively. Equation (6.2.4) states that the effective hydraulic conductivity of an equivalent homogeneous aquifer is a weighted arithmetic average of the hydraulic conductivities of all of the layers in the layered aquifer. If the thickness is the same for each stratum, equation (6.2.4) can be simplified as

$$\overline{K} = \sum_{i=1}^{n} K_i / n \tag{6.2.5}$$

which is simply the arithmetic mean of all of the hydraulic conductivity values of the aquifer.

Notice that if the number of layers in this domain increases, a spatial REV smaller than the entire domain can then be defined such that the medium can be visualized as homogeneous. Alternatively, a CV that is as large as the entire domain can be used to define the effective hydraulic conductivity. For such a case, an ensemble REV is assumed.

6.2.2 Flow Normal to Stratification

Figure 6.2 shows an aquifer with flow perpendicular to bedding. The boundary conditions are the same as those in Figure 6.1 and the flow direction is the same, except that the orientation of stratification is vertical. The length of layers (i.e.,

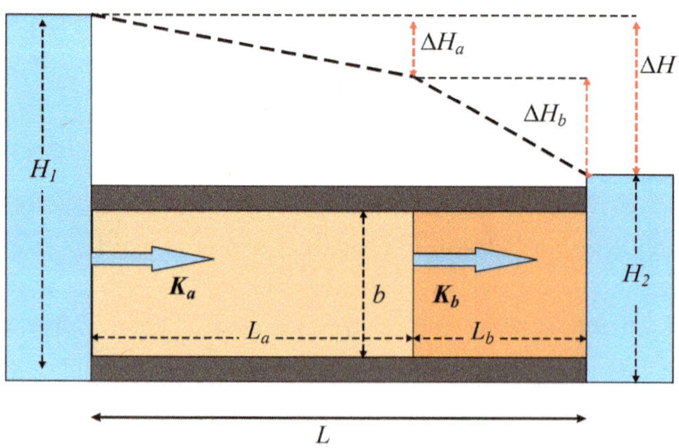

Figure 6.2. Schematic illustration of the case where flow is perpendicular to bedding.

blocks) a and b is denoted by L_a and L_b, respectively. Suppose block a is more conductive than block b and the flow is steady. The steady hydraulic head distribution along the aquifer is depicted by the heavy dashed line, indicating a smaller head gradient (head loss) over block a than block b. According to the Bernoulli equation, the total head loss in the aquifer must be equal to the sum of the head loss in block a and the head loss in block b. That is,

$$\Delta H = \Delta H_a + \Delta H_b. \tag{6.2.6}$$

Now, we seek the effective hydraulic conductivity of an equivalent homogeneous aquifer that yields the same Q as that in the layered aquifer with the same total head loss. Using Darcy's law, the head losses in equation (6.2.6) can be expressed as

$$\frac{QL}{b\overline{K}} = \frac{QL_a}{bK_a} + \frac{QL_b}{bK_b} \tag{6.2.7}$$

where the left-hand side of equation (6.2.7) is the total head loss expressed using Darcy's law for the equivalent homogeneous aquifer, and $L = L_a + L_b$. Further, $\overline{K}$ is its effective hydraulic conductivity. Solving equation (6.2.7) for $\overline{K}$, we have

$$\overline{K} = \frac{L}{\frac{L_a}{K_a} + \frac{L_b}{K_b}} \tag{6.2.8}$$

That is, $\overline{K}$ is a length-weighted harmonic mean of the hydraulic conductivities of the two blocks. In the case where an aquifer consists of n blocks, the $\overline{K}$ becomes

$$\overline{K} = \frac{L_1 + L_2 + \ldots + L_n}{\frac{L_1}{K_1} + \frac{L_2}{K_2} + \ldots + \frac{L_n}{K_n}} = \frac{\sum_{i=1}^{n} L_i}{\sum_{i=1}^{n} \frac{L_i}{K_i}} \tag{6.2.9}$$

Equation (6.2.9) manifests that under flow perpendicular to bedding, the effective hydraulic conductivity is a weighted harmonic mean of all of the hydraulic conductivity values in the aquifer. If $L_1 = L_2 = \ldots = L_n$, then equation (6.2.9) can be simplified as

$$\overline{K} = \frac{n}{\sum_{i=1}^{n} \frac{1}{K_1}} \tag{6.2.10}$$

In this case, $\overline{K}$ is a harmonic mean of the hydraulic conductivity values of the individual layers (subscale heterogeneity). Again, a spatial REV for this case can be defined only if the number of blocks is large. Otherwise, we assume the REV exists in the ensemble sense.

According to the preceding analyses, effective hydraulic conductivity is the hydraulic conductivity of a fictitious homogeneous medium that produces the same discharge as the layered one under the same boundary conditions. The analyses show that for flow parallel to the layers, the effective hydraulic conductivity is a weighted arithmetic mean of the individual hydraulic conductivity values and is a weighted harmonic mean for flow perpendicular to the layers. An arithmetic mean is heavily weighted toward the highest hydraulic conductivity in the aquifer, while the harmonic mean is strongly influenced by the lowest hydraulic conductivity in the aquifer. As a result, the arithmetic mean is generally greater than the harmonic mean. The effective hydraulic conductivity is therefore flow-direction dependent and has a directional preference (i.e., anisotropic).

The preceding analyses essentially establish the principle for upscaling small-scale hydraulic properties to large-scale effective hydraulic properties. They also suggest that the effective hydraulic conductivity of a heterogeneous medium depends on the following:

- Contrast in hydraulic conductivity of layers or blocks (i.e., K_1, K_2, ... K_n), or the variation of the hydraulic conductivity among blocks, which can be quantified by its statistical variance, σ_K^2 (see Chapter 4)
- Dimensions or the shape of heterogeneity (i.e., thickness of layers and the length of blocks), which can be defined by the statistical correlation scales of the hydraulic conductivity of the layers or blocks (discussed in Chapter 4)
- Flow direction: parallel or perpendicular to bedding

Because of its dependence on the preceding factors, effective hydraulic conductivity is not the same as the statistical mean of hydraulic conductivities of small-scale samples. It should be apparent that the variation in local-scale hydraulic conductivity (assuming local isotropy, Chapter 2) and shape of the layers or blocks introduce the large-scale anisotropy in the effective hydraulic conductivity of the equivalent homogeneous media.

6.2.3 Predicted Heads Based on the Effective Hydraulic Conductivity

While the effective hydraulic conductivity of the fictitious homogeneous aquifer reproduces the discharge in a heterogeneous aquifer for any given boundary condition, does it also reproduce the hydraulic head distribution in the heterogeneous medium? If not, what does it represent? In order to answer this question, we will revisit the preceding two flow scenarios.

For flow parallel to stratifications of uniform thickness, the hydraulic head and hydraulic gradients, along a vertical line at any distance *L*, will be the same, in spite of the difference in the hydraulic conductivity of each layer. However, the specific discharges of the strata will be different. In other words, the larger the hydraulic

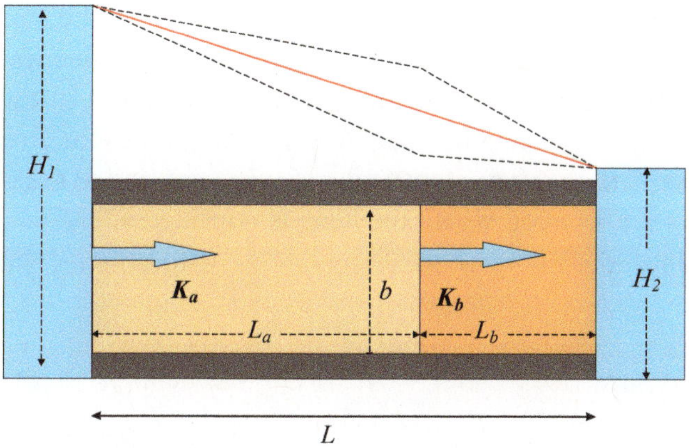

Figure 6.3. Possible hydraulic head distributions along an aquifer with two blocks of hydraulic conductivities.

conductivity of a layer, the greater will be the specific discharge through the layer, and vice versa. Thus, in this particular case, the effective hydraulic conductivity can also reproduce the head everywhere in this layered aquifer.

Now, consider the case where the flow direction is perpendicular to the bedding. The straight solid line between the two boundaries in Figure 6.3 represents the hydraulic head distribution calculated for the equivalent homogeneous aquifer using the effective hydraulic conductivity (the harmonic weighted hydraulic conductivity). This head distribution is quite different from that of the heterogeneous aquifer, which depends on the magnitude of K_a and K_b. For example, if $K_a > K_b$, the hydraulic head distribution in the heterogeneous aquifer is then depicted by the upper dashed line. The head distribution shows that the hydraulic gradient (or head loss) in block a is smaller than that in block b. The sum of the head losses over the two blocks must be equal to the total head loss between the two boundary heads. If $K_a < K_b$, the hydraulic head distribution is indicated by the lower dashed line, and the head loss in block a is greater than that in block b.

Based on the analysis of these two examples, we may conclude that the solid line in Figure 6.3 (the hydraulic head distribution of the equivalent homogeneous aquifer) is the average of all possible head distributions resulting from two scenarios (i.e., $K_a > K_b$ and $K_a < K_b$), if the lengths of the block are fixed, and even though the two K values are known. This type of average is called an **ensemble average**. This **ensemble-averaged hydraulic head** is different from the actual hydraulic head distribution in either block. Notice that the specific discharge values are the same regardless of the permutations. The preceding analysis demonstrates that the effective hydraulic conductivity reproduces the specific discharge of the heterogeneous aquifer but yields a fictitious head distribution, which is a statistically most likely one, without a knowledge of the K values of the two blocks.

The analysis of the head distribution for the two cases (i.e., flow parallel and perpendicular to the bedding) also informs us of factors that influence the deviation of the ensemble-averaged head distribution from the true heads in the heterogeneous aquifers. They are 1) the direction of the flow with respect to the stratification, 2) the contrast of the two K values (i.e., statistical variance, σ_K^2), and 3) the distance between the boundary and the location where the head is considered.

Question:

Can you extend the previous analysis to the aquifers consisting of many blocks with unknown block sizes?

6.2.4 *Effective Hydraulic Conductivity for Complex Heterogeneity*

The effective hydraulic conductivity for general heterogeneous media can be derived via stochastic methods. These methods developed mathematical formulae that relate the effective hydraulic conductivity to the mean of the natural logarithm of the local scale (or the block) hydraulic conductivities (or geometric mean of K, K_g, or the mean of ln$K$, F); its variance, σ_f^2; and the correlation scales, λ_x, λ_y, λ_z, in x, y, and z directions (Chapter 4), respectively; and direction of the mean gradient in x, y, and z (i.e., J_x, J_y, J_z) (see Gelhar, 1993). That is,

$$K_{eff} = G(F, \sigma_f^2, \lambda_x, \lambda_y, \lambda_z, J_x, J_y, J_z) \tag{6.2.11}$$

where G is a mathematical function. Normalized effective hydraulic conductivity values are shown in Figure 6.4 for 3-D mean flow parallel or perpendicular to bedding in heterogeneous media with different aspect ratios of correlation scales (λ_h/λ_v) and variance (σf^2). The aspect ratio is defined as the ratio of the horizontal correlation scale to the vertical correlation scale, λ_h/λ_v. These effective hydraulic conductivities in the figure are normalized with respect to the geometric mean of the hydraulic conductivities, or K_g = exp(F), where F is the arithmetic mean of ln$K$ values of small-scale samples.

As the aspect ratio approaches infinity (i.e., perfectly stratified aquifers), the effective hydraulic conductivity for flow parallel to bedding approaches the arithmetic mean: $\exp\left(\sigma_f^2/2\right)$. The same plot also shows the normalized effective hydraulic conductivities for flow perpendicular to bedding. They approach the harmonic mean, $\exp\left(-\sigma_f^2/2\right)$, as the aspect ratio increases to infinity. As the aspect ratio becomes 1, the effective hydraulic conductivity is isotropic and is slightly greater than the geometric mean, suggesting that the effective hydraulic conductivity is not the statistical mean of all small-scale samples. Figure 6.4 provides a convenient way to estimate effective hydraulic conductivities if the spatial statistics (the mean, variance, and correlation scales) of the small-scale (or block) hydraulic conductivity values are known in advance.

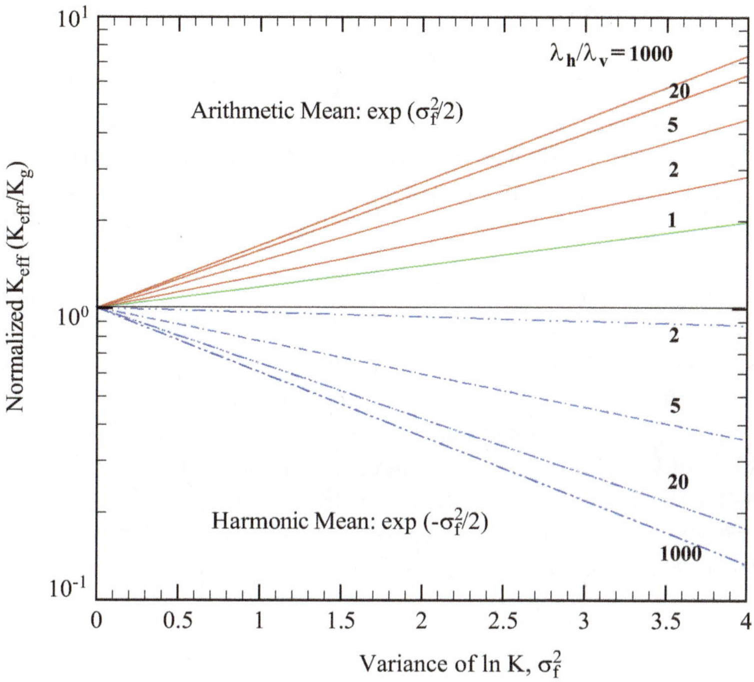

Figure 6.4. Normalized effective hydraulic conductivities for flow parallel to bedding (red lines) and perpendicular to bedding (blue dashed lines) as a function of variance of ln*K* and the aspect ratios of correlation scales. The green line is the effective hydraulic conductivity of the statistically isotropic media. (Modified from Gelhar and Axness, 1983)

In addition, the stochastic methods also provide estimates of deviation (uncertainty) of the model predicted head from the true head (see Gelhar, 1993). For example, for steady-state uniform flow, with mean gradient J, the deviation of the mean head predicted using the effective hydraulic conductivity, in statistically isotropic media (Chapter 4) with the correlation scale, λ, is given as

$$\sigma_h^2 = A\sigma_f^2 J^2 \lambda^2 \tag{6.2.12}$$

In equation (6.2.12), σ_h^2 stands for the variance of heads in a heterogeneous medium around its mean value, and A is a constant, which is 1 for one-dimensional flow, and 0.333 for three-dimensional flow.

Equation (6.2.12) yields several important results: 1) The uncertainty in the predicted head, based on the effective hydraulic conductivity of a 3-D equivalent homogeneous model, is smaller than that of a 1-D model, indicating that a 3-D model is more realistic. 2) Effects of hydraulic heterogeneity on head variation in groundwater systems are not significant, since the mean hydraulic gradient is very small in general. 3) Geologic structures (correlation scales) play an important role on head variation. Lastly, the small impacts of heterogeneity on head variance (uncertainty) explain why the traditional homogeneous media approach has been widely accepted for practical water resources applications, where small deviations in head predictions are often acceptable. This is also the reason that propagation of the head is referred to as a **highly diffusive process:** it homogenizes heterogeneity. To the contrary, solution transport is dominated by the advection process, which is dictated by heterogeneity.

Questions.

1. Does the effective K remain the same if the flow direction is reversed? If it does, will it produce the same distribution for different flow directions?
2. Do you expect that the K determined from a soil column experiment predicts head distribution in the column or head measured in a pore? Why is it critical to use many manometers to determine the hydraulic conductivity in a soil column?
3. Note that the expressions for the effective hydraulic conductivities are derived under steady flow conditions. How would transient or unsteady flow affect the effective hydraulic conductivity estimate (see Chapter 7)?

6.2.5 Flow Inclined to Stratification

Figure 6.5 shows a situation where steady groundwater flow crosses an interface of a stratified aquifer at an angle. The aquifer consists of two layers: Layer A with hydraulic conductivity, K_A, overlays the other layer that has hydraulic conductivity, K_B, and $K_A < K_B$. The two parallel dashed lines represent two equal head potential lines (H_1 and H_2; $H_1 > H_2$), and perpendicular to them, there are two parallel solid lines, denoting two streamlines, which form a flow tube. The incident angle, θ_A, is

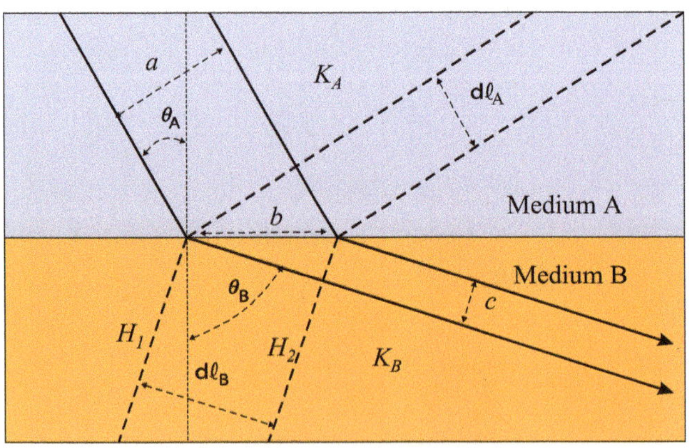

Figure 6.5. Plots of streamlines and equal potential lines when steady flow crosses the interface of two media at an angle.

defined as the angle between a streamline (also referred as the flow line) and the line normal to the interface in layer A. The deflection angle, θ_B, is defined as the angle between the flow line and the line normal to the interface in layer B. According to this figure, the flow tube deflects toward the stratification of stratum B after it crosses the interface. Likewise, the two equal potential lines bend and the distance between the potential lines becomes wider as the flow tube enters layer B.

Since the flow is steady, the total inflow into the tube must equal the outflow from the tube: $Q_A = Q_B$. or

$$K_A a \frac{dH}{d\ell_A} = K_B c \frac{dH}{d\ell_B} \tag{6.2.13}$$

where a and c are the cross-sectional areas of the tube in strata A and B, respectively, and $dH = H_1 - H_2$. Recognizing the following geometric relationships

$$a = b\cos\theta_A \text{ and } c = b\cos\theta_B, \text{ and } \frac{b}{d\ell_A} = \frac{1}{\sin\theta_A} \quad \text{and} \quad \frac{b}{d\ell_B} = \frac{1}{\sin\theta_B} \tag{6.2.14}$$

the following relationship between K_A and K_B is thus derived:

$$\frac{K_A}{K_B} = \frac{\tan\theta_A}{\tan\theta_B} \tag{6.2.15}$$

Equation (6.2.15) is called the tangent law, which defines the angle of deflection of the flow line θ_B after the flow enters a medium at a given angle θ_A from a medium of a different hydraulic conductivity under steady-state flow conditions (see Figure 6.5). Specifically,

$$\theta_B = \tan^{-1}\left(\frac{K_B \tan\theta_A}{K_A}\right) \tag{6.2.16}$$

For a given incident angle, θ_A not equal to zero, the magnitude of the ratio of K_B to K_A dictates the deflection angle, θ_B. If $K_A = K_B$, no deflection will occur. If $K_A > K_B$, the deflection angle, θ_B, will be smaller than the incident angle, θ_A. This result reflects the fact that when flow moving from a high conductivity layer encounters a low conductivity layer, it inclines to move through this less permeable layer rapidly using the shortest path. On the other hand, if $K_A < K_B$, the deflection angle, θ_B, will be greater than the incident angle, θ_A. It implies that the flow tends to use this high conductivity layer as a conduit to move its water through. This phenomenon has, in effect, been used to justify the modeling approach that uses a one-dimensional vertical flow model for flow in leakage aquitards (low conductivity) and a two-dimensional horizontal flow model for flow in aquifers (high conductivity).

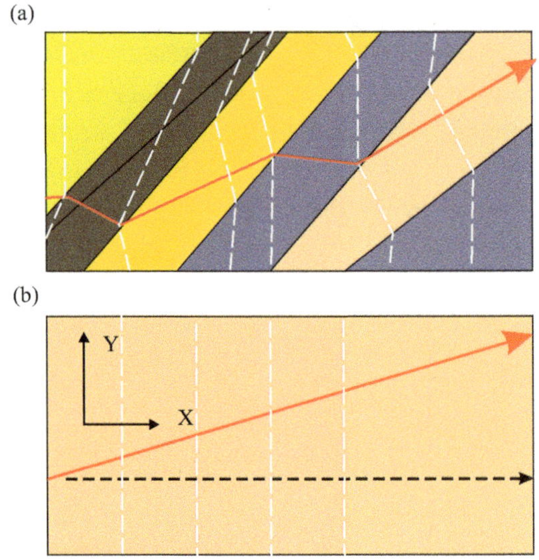

Figure 6.6. Flow lines (red solid) and equal potential lines (white dotted) in a) a stratified aquifer (top) and b) an equivalent homogeneous and anisotropic aquifer (bottom).

We now expand the preceding analysis to flow through a geologic formation comprising a series of slant stratifications (Figure 6.6). The left- and right-hand sides of the geologic medium in Figure 6.6 are bounded by two constant head reservoirs and the other two sides of the medium are unbounded. Further, water level of the reservoir on the left-hand site is assumed to be higher than that of the right-hand side. Suppose the flow in the geologic formation is steady and we use the tangent law to analyze the flow path through the medium. The result of this analysis is shown in Figure 6.6a, where the white dotted lines denote the equal potential lines, and the solid line depicts a flow path. As indicated by this solid line, flow zigzags through the medium (or through a torturous path), controlled by the contrast in the hydraulic conductivities according to the tangent law. Similarly, equal potential lines bend as they encounter layers of different hydraulic conductivity values. Nevertheless, the bulk (averaged) flow moves diagonally as depicted by the solid flow line in Figure 6.6b.

Suppose the layered structure of the geologic medium is unknown, and the medium is assumed to be homogeneous and isotropic for analysis. If the head difference of the two reservoirs is the driving force for flow, then the apparent bulk hydraulic gradient (dH/dx) will be from left to right, as reflected in the ensemble mean head distribution (the equally spaced, vertical dotted white lines in Figure 6.6b). Accordingly, the flow path (i.e., specific discharge vector) predicted by the homogeneous conceptual model is a straight line from left to right (the dotted black line in Figure 6.6b), parallel to the apparent hydraulic gradient. This predicted flow path differs greatly from the bulk flow path in the heterogeneous medium in Figure 6.6a, which is not mathematically **collinear** with (or parallel to) the apparent hydraulic gradient vector. That is, the preceding equivalent homogeneous model fails to reproduce the direction of the bulk flow in such a stratified medium. In order to correct this problem, anisotropy of the effective hydraulic conductivity for the equivalent porous media model must be introduced. This is the topic of our next section.

6.2.6 Anisotropy of Effective Hydraulic Conductivity for Saturated Media

On the basis of the previous analyses, the effective hydraulic conductivity for flow that is parallel to bedding is a weighted arithmetic mean (K_A), while it is a weighted harmonic mean (K_H) for flow perpendicular to bedding. Since the arithmetic mean is always greater than the harmonic mean, the anisotropy ratio (K_A/K_H) of the effective hydraulic conductivity (the ratio of the effective hydraulic conductivity for flow parallel to bedding to that for flow perpendicular to bedding) is always greater than 1. That is:

$$\frac{K_A}{K_H} > 1$$

As discussed earlier, the effective hydraulic conductivities in different flow directions are controlled by the contrast in hydraulic conductivity of each stratum (i.e., variance of $\ln K$) and the shape or dimensions (length, width, thickness, correlation scales) of the stratum. The anisotropy of the effective hydraulic conductivity depends on these factors as well. The anisotropy for general heterogeneity as a function of variance of $\ln K$ and the aspect ratios of correlation scales is depicted in Figure 6.7. This figure offers a practical tool for estimating effective hydraulic conductivity anisotropy of geologic media.

At last, we emphasize again that the effective hydraulic conductivity anisotropy is an artifact resulting from the volume averaging over a CV of heterogeneities at a multiplicity of scales, or a by-product of our upscaling attempt (i.e., our simplification of a complex system). In fact, the concept of such a large-scale anisotropy is unnecessary, if the multiscale heterogeneities of a geologic formation can be depicted in detail and used in our analysis. However, when specification of detailed heterogeneity is not possible, the effective hydraulic conductivity and anisotropy approach is deemed to be useful and practical for investigating the bulk behavior of the flow system through a complex porous medium.

Questions:

1. What is the anisotropy in the effective hydraulic conductivity if layers within the geologic formation are not perfectly stratified (i.e., continuous) or folded? How would

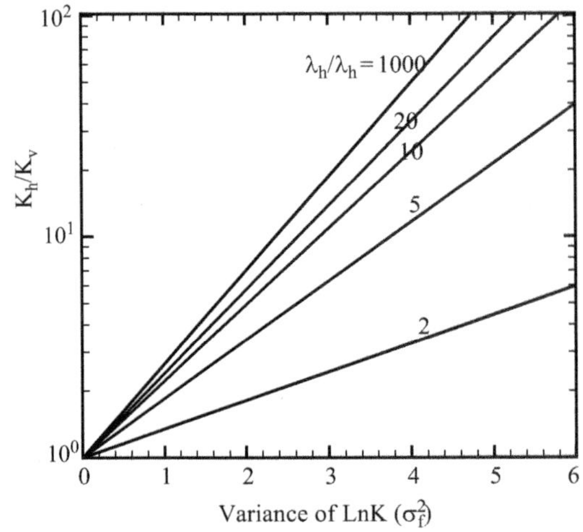

Figure 6.7. Anisotropy of the effective hydraulic conductivities as a function of variance of ln$K$ and the aspect ratio of correlation scales (λ_h/λ_v). (Modified from Gelhar and Axness, 1983)

this anisotropy be related to the average thickness and length of the layers and the harmonic and arithmetic means?

2. If the layer thickness varies with distance, are the harmonic and arithmetic averages valid?
3. How would the anisotropy of the effective hydraulic conductivity be affected by transient flow?

6.3 Effective Hydraulic Conductivity and Anisotropy for Unsaturated Media

Derivation of the effective hydraulic conductivities of unsaturated heterogeneous porous media is not as straightforward as that of the saturated medium. The nonlinear relationship between the hydraulic conductivity and pressure head makes the analysis difficult, and therefore no explicit formulas will be given here. Rigorous analyses are available from Yeh et al. (1985a, b, c), Yeh (1989), Khaleel et al. (2002), and Yeh et al. (2005). Highly simplified approximations based on the principles used in the fully saturated flow cases follow. While they are first-cut approximations, they provide insights into general behaviors of unsaturated effective hydraulic conductivity functions of heterogeneous porous media.

6.3.1 Unsaturated Flow Parallel to Bedding

Consider vertical infiltration into soils with vertically parallel layers. If flow within each layer is driven by gravity, and we follow the approach that derives the effective hydraulic conductivity for saturated flow in stratified geologic media, the effective unsaturated hydraulic conductivity for flow parallel to bedding, $K_H(h)$, at a given pressure head, h, will be the arithmetic average of unsaturated hydraulic conductivity values of all layers at the given pressure head. That is,

$$K_H(h) = \left(\sum_{i=1}^{n} b_i K_i(h)\right) \Big/ \left(\sum_{i=1}^{n} b_i\right) \tag{6.3.1}$$

In equation (6.3.1), b_i and $K_i(h)$ are the thickness and the local-scale hydraulic conductivity of the ith layer, respectively. The number of layers is denoted by n.

6.3.2 Unsaturated Flow Perpendicular to Bedding

In the case where infiltration is perpendicular to bedding, the total head loss of the layered medium is the sum of the head loss within each layer, as is in the saturated flow case. Unlike the pressure head in saturated media, the pressure head variation within each unsaturated layer is nonlinear, and the hydraulic conductivity within each layer varies with pressure head. Consequently, analytical expressions for the effective unsaturated hydraulic conductivity cannot be easily derived. To make the problem tractable, we will assume that flow within each layer is gravity driven (i.e., pressure head is uniform within each layer even at the boundary between layers). Accordingly, the field-scale effective unsaturated hydraulic conductivity in the direction perpendicular to bedding, K_V, at a given pressure head is given by

$$K_V(h) = \sum_{i=1}^{n} b_i \Big/ \sum_{i=1}^{n} [b_i / K_i(h)] \tag{6.3.2}$$

That is, unsaturated hydraulic conductivity values at a given pressure head of all layers are harmonically weighted to represent the effective unsaturated hydraulic conductivity for flow perpendicular to bedding at that pressure head. A collection of the averaged hydraulic conductivity values at different pressure heads thus constitutes the effective unsaturated hydraulic conductivity and pressure head relationship for unsaturated flow perpendicular to bedding.

6.3.3 Anisotropy of Effective Hydraulic Conductivity for an Unsaturated Medium

Notice that most studies have treated anisotropy of unsaturated hydraulic conductivity as an intrinsic property of porous media, the same as the anisotropy in saturated hydraulic conductivity. As a result, the unsaturated hydraulic conductivity anisotropy has often been modeled by scaling the unsaturated hydraulic conductivity and head relationship in different directions according to the anisotropy under fully saturated conditions. The anisotropy thus remains constant over the full range of saturation or pressure head – a constant anisotropy concept. According to equations (6.3.1) and (6.3.2), the anisotropy of the unsaturated effective hydraulic conductivity of an equivalent homogeneous medium may, however, vary with pressure head. This phenomenon is referred to as moisture- or pressure-dependent anisotropy (Yeh et al., 1985a, b, c; Yeh et al., 2005).

Yeh et al. (1985a, b, c) employed a stochastic approach to derive the effective unsaturated hydraulic conductivity of vadose zones with statistical heterogeneity (see Chapters 4 and 9). They assumed that the unsaturated hydraulic conductivity function and the moisture release function at the local scale (i.e., each block in Figure 5.1b) can be represented by Gardner's exponential model (equation 3.5.1 in Chapter 3). They showed that the effective unsaturated hydraulic conductivity, $K_{eff}(H)$, depends on the spatial statistics of the parameters of the exponential model for the local-scale unsaturated properties:

$$K_{eff}(H) = G\left(F, \sigma_f^2, B, \sigma_\beta^2, H, \lambda_x, \lambda_y, \lambda_z, J_x, J_y, J_z\right) \quad (6.3.3)$$

In equation (6.3.3), H is the mean pressure head, G is a mathematical function, F denotes the mean of $\ln K_s$ (K_s represents the saturated hydraulic conductivity), and σ_f^2 is the variance of $\ln K_s$. The parameters, B and σ_β^2, are the mean and variance of $\ln\beta$, respectively. Correlation scales of K_s and β are assumed to be the same, and λ_x, λ_y and λ_z are their values in the x, y, and z directions. The mean gradients in the x, y, and z directions are J_x, J_y, and J_z, respectively. While development of equation (6.3.3) is mathematically rigorous, the effective hydraulic conductivity functions for flow parallel to and perpendicular to bedding approximated by equations 6.3.1 and 6.3.2, respectively, resemble those derived from equation (6.3.3).

An interesting finding from the analysis by Yeh et al. (1985a, b, c) is that anisotropy of the effective unsaturated hydraulic conductivity varies with mean pressure head or moisture content (i.e., moisture-dependent anisotropy). That is, the ratio of the effective unsaturated hydraulic conductivity in the direction parallel to the bedding to that in the direction perpendicular to bedding increases as the medium becomes less saturated. Such a unique behavior provides explanations for commonly observed large lateral spreading of contaminants or moisture plumes in the stratified vadose zone (e.g., Ye et al., 2005; Glass et al., 2005). Yeh et al. (1985a, b, c) also discovered that the head variance (deviation of the true head from the head calculated from the effective parameters) increases as the medium becomes less saturated. We will discuss these findings below.

An illustrative sketch of the pressure head distribution during an infiltration event in a stratified vadose zone is shown in Figure 6.8. Near the infiltration source where the degree of saturation is high, the pressure head contour is generally smooth and

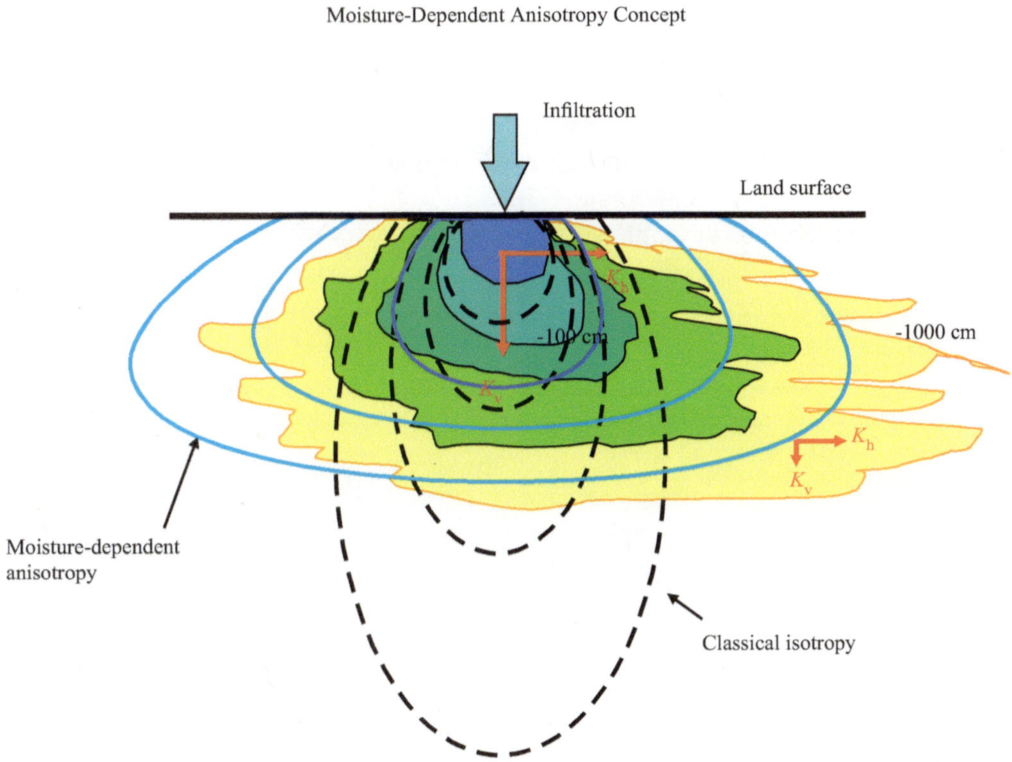

Figure 6.8. Schematic pressure head distributions during an infiltration experiment to explain moisture-dependent hydraulic conductivity anisotropy and moisture-dependent head variability. The lengths of the red arrows denote the magnitudes of unsaturated hydraulic conductivity in the horizontal direction (K_h) and in the vertical direction (K_v) at the pressure head of the given location.

symmetrical (dark blue contour). Away from the source, pressure head contours (light green and yellow contours) become more irregular and a symmetric (i.e., large variability). Overall, the pressure head distributions spread out to greater distances horizontally due to stratifications.

In Figure 6.8, the pressure head distributions calculated from an equivalent homogeneous and isotropic conceptual model are depicted by the black dashed lines. They are smooth and symmetrical, and are elongated in the vertical direction, reflecting effects of gravity and hydraulic conductivity isotropy. This homogeneous and isotropic conceptual model apparently overestimates the vertical migration and underestimates the horizontal spreading of the actual moisture plume.

The light blue solid lines (Figure 6.8) are the pressure head distributions derived from an equivalent homogeneous model with moisture-dependent anisotropy. That is, the hydraulic conductivity values in the horizontal and the vertical direction are almost the same near the infiltration source where the soil is wet. In the region where the soil is dry, both the horizontal and vertical hydraulic conductivities are

Figure 6.9. The moisture and dye distributions at 32.7 hours after an infiltration transport experiment undertaken in a heterogeneous sandy fluvial deposit. (Glass et al., 2005)

smaller than those in the wet region, but the ratio of the horizontal hydraulic conductivity to the vertical is much greater than this ratio in the wet region. Such a moisture-dependent anisotropy in the effective unsaturated hydraulic conductivity allows the equivalent homogeneous model to yield significant lateral spreading as in the stratified vadose zone.

Notice that the pressure head distribution contours in the wet region of the stratified vadose zone are smooth and in general agreement with the simulated based on the moisture-dependent anisotropy. This agreement, however, deteriorates at dry regions. Such behaviors imply that the head is not as diffusive in the dry region as in the wet region, and the prediction of head in the dry region involves large uncertainty (i.e., moisture-dependent variability).

Results of an infiltration transport experiment in a sandy fluvial deposit by Glass et al. (2005) corroborate the findings by Yeh et al. (1985a, b, c). Figure 6.9 shows the moisture and dye distributions in a 2 m wide and 1.2 m high vertical cross section at the experimental site, 32.7 hours after the infiltration transport experiment started. During the experiment, water containing a red dye followed by a blue dye was ponded on the surface using a square infiltrometer. The advancement of the dye and the wetting front on the vertical face were photographed several times throughout the infiltration event. After infiltration was stopped, vertical slices were excavated at regular intervals to the midpoint of the infiltrometer, providing the ability to observe the tracer in a quasi-3-D manner. While initial infiltration of the red dye enhanced the ability to observe wetting front patterns, sequential infiltration of the two dyes resulted in color contrast that allows us to observe flow field behavior. According to the photo, the red dye front significantly lagged behind the wetting front. Why? (Hint: The difference in the void space available for water and dye). Dye fronts are also more irregular than are wetting fronts, suggesting a less diffusive process. Lateral spreading of both wetting and dye fronts was twice that of the vertical wetting and dye fronts, indicating a pronounced effect of horizontal layering (or anisotropy). Such lateral spreading further suggests the need for including moisture-dependent anisotropy for an equivalent homogeneous conceptual model is used.

6.4 Darcy's Law for Field-Scale Geologic Media

Applications of field-scale CV and REV to field-scale geologic media in previous sections have led to the concept of effective hydraulic properties and field-scale anisotropy, and ensemble-averaged head or mean head. These properties are now ready to be included in Darcy's law for field-scale geologic media. Again, Darcy's law is assumed to be valid for these field-scale REVs.

6.4.1 Hydraulic Conductivity Tensor

For one-dimensional flow in saturated or unsaturated media, Darcy's law can be written as

$$q = -K\frac{dH}{dx} \tag{6.4.1}$$

where K is the effective hydraulic conductivity, and H is the ensemble-averaged head, or ensemble mean head. Notice that specific discharge, q; effective hydraulic conductivity, K; and gradient, dH/dx, are scalar for the 1-D problem. If the medium is unsaturated, the hydraulic conductivity will depend on the pressure head and moisture content. Darcy's law in 2-D or 3-D saturated or unsaturated media is generally rewritten in a vector notation as

$$\mathbf{q} = -\mathbf{K}\cdot\nabla H \tag{6.4.2}$$

where the symbol **K** denotes the hydraulic conductivity tensor. Another common notation is the expression based on Einstein's summation convention:

$$q_i = K_{ij}J_j \tag{6.4.3}$$

where i and j = 1, 2, or 3 or x, y, z, and J_1, J_2, and $J_3 = -\partial H/\partial x$, $-\partial H/\partial y$ and $-\partial H/\partial z$, respectively. According to Einstein's summation convention for any product terms, a suffix (subscript or superscript) repeated twice (and only twice) is held to be summed over its range of values. For instance, Darcy's law in three dimensions is

$$\begin{aligned} q_1 &= K_{11}J_1 + K_{12}J_2 + K_{13}J_3 \\ q_2 &= K_{21}J_1 + K_{22}J_2 + K_{23}J_3 \\ q_3 &= K_{31}J_1 + K_{32}J_2 + K_{33}J_3 \end{aligned} \quad \text{or in 2-D,} \quad \begin{aligned} q_1 &= K_{11}J_1 + K_{12}J_2 \\ q_2 &= K_{21}J_1 + K_{22}J_2 \end{aligned} \tag{6.4.4}$$

Matrix forms of Darcy's law in 3-D and 2-D are

$$\begin{bmatrix} q_1 \\ q_2 \\ q_3 \end{bmatrix} = \begin{bmatrix} K_{11} & K_{12} & K_{13} \\ K_{21} & K_{22} & K_{23} \\ K_{31} & K_{32} & K_{33} \end{bmatrix} \begin{bmatrix} J_1 \\ J_2 \\ J_3 \end{bmatrix} \text{ and } \begin{bmatrix} q_1 \\ q_2 \end{bmatrix} = \begin{bmatrix} K_{11} & K_{12} \\ K_{21} & K_{22} \end{bmatrix} \begin{bmatrix} J_1 \\ J_2 \end{bmatrix} \quad (6.4.5)$$

Here, K_{ij} or $[K]$ defines the hydraulic conductivity tensor of the porous medium. Note that a tensor is a quantity (in general, a matrix) that satisfies a specific coordinate transformation rule.

6.4.2 Properties of a Hydraulic Conductivity Tensor

A hydraulic conductivity tensor should have the following properties:

1) It is a symmetric tensor of rank 2. Rank 2 implies two indices, *i*, and *j*. That is, K_{ij} should have nine components in a three-dimensional tensor, and four components in a two-dimensional tensor.
2) The tensor is always symmetric. That is, $K_{ij} = K_{ji}$ when $i \neq j$. For example, $K_{12} = K_{21}$, $K_{13} = K_{31}$, $K_{32} = K_{23}$ for a three-dimensional tensor. As a result, the nine components in the tensor can be reduced to six components.

$$\underbrace{\begin{bmatrix} K_{11} & K_{12} & K_{13} \\ K_{21} & K_{22} & K_{23} \\ K_{31} & K_{32} & K_{33} \end{bmatrix}}_{\text{9 components}} \overset{\text{reduced}}{\Rightarrow} \underbrace{\begin{bmatrix} K_{11} & K_{12} & K_{13} \\ & K_{22} & K_{23} \\ & & K_{33} \end{bmatrix}}_{\text{6 components}} \quad (6.4.6)$$

For a two-dimensional tensor, it requires the information about three components only.

$$\underbrace{\begin{bmatrix} K_{11} & K_{12} \\ K_{21} & K_{22} \end{bmatrix}}_{\text{4 components}} \text{ reduced to } \underbrace{\begin{bmatrix} K_{11} & K_{12} \\ & K_{22} \end{bmatrix}}_{\text{3 components}} \quad (6.4.7)$$

3) if $K_{ij} = K$ (a constant) for $i = j$ and $K_{ij} = 0$,

$$\begin{bmatrix} K & 0 & 0 \\ 0 & K & 0 \\ 0 & 0 & K \end{bmatrix} \Rightarrow \underset{\substack{\downarrow \\ \textit{Kronecker delta}}}{K\delta_{ij}} \Rightarrow \underset{\substack{\uparrow \\ \textit{scalar}}}{K} \begin{bmatrix} 1 & 0 & 0 \\ 0 & 1 & 0 \\ 0 & 0 & 1 \end{bmatrix} \quad (6.4.8)$$

Equation (6.4.8) means that hydraulic conductivity of the medium is isotropic; that is, the hydraulic conductivity value of the medium does not vary with the direction.

4) when $K_{ij} \neq$ constant for $i = j$, and $K_{ij} \neq 0$ for $i \neq j$ (unless in the principal directions, which are discussed later), the hydraulic conductivity tensor of the medium is called **anisotropic.** That is,

$$\begin{bmatrix} K_{11} & K_{12} & K_{13} \\ K_{21} & K_{22} & K_{23} \\ K_{31} & K_{32} & K_{33} \end{bmatrix} \text{ or } \begin{bmatrix} K_{11} & 0 & 0 \\ 0 & K_{22} & 0 \\ 0 & 0 & K_{33} \end{bmatrix} \tag{6.4.9}$$

In the case where our coordinates align with the principal directions, the diagonal terms in the tensor are zero. Generally speaking, the principal directions for an anisotropic medium composed of alternating layers are the directions parallel to the bedding planes and those perpendicular to the bedding planes.

5) The transformations from any x, y coordinates to the principal x', y' coordinates can be obtained by using the following relationships (2-D case)

$$K_{x'x'} \quad or \quad K_{y'y'} = \frac{K_{xx} + K_{yy}}{2} \pm \left[\left(\frac{K_{xx} - K_{yy}}{2} \right)^2 + K_{xy}^{\ 2} \right]^{1/2} \tag{6.4.10}$$

Likewise, if the principal directions and the principal hydraulic conductivity values are known, the following rule can be used to derive the hydraulic conductivity values in any given x and y coordinates:

$$\begin{aligned} K_{xx} \quad or \quad K_{yy} &= \frac{K_{x'x'} + K_{y'y'}}{2} \pm \frac{K_{x'x'} - K_{y'y'}}{2} \cos 2\theta \\ \text{and } K_{xy} &= \frac{K_{x'x'} - K_{y'y'}}{2} \sin 2\theta \end{aligned} \tag{6.4.11}$$

where θ is the angle between the principal directions of the x-y coordinate system. These transformation rules are derived from the rule of coordinate transformation, which is discussed next.

Now, let us examine how the hydraulic conductivity tensor can resolve the issue of noncollinear relationship between the hydraulic gradient and the specific discharge vector in cases where the hydraulic gradient is at an angle with a series of layers (Figure 6.6). Recall the 2-D Darcy's law for the case where the coordinate system is not aligned with the principal directions:

$$\begin{aligned} q_x &= K_{xx} J_x + K_{xy} J_y \\ q_y &= K_{xy} J_x + K_{yy} J_y \end{aligned} \tag{6.4.12}$$

According to Figure 6.6b, the hydraulic gradient exists in the x direction only. That is, $J_x \neq 0$ and $J_y = 0$. Therefore, the system of equations (6.4.12) becomes

$$\begin{aligned} q_x &= K_{xx} J_x \\ q_y &= K_{xy} J_x \end{aligned} \tag{6.4.13}$$

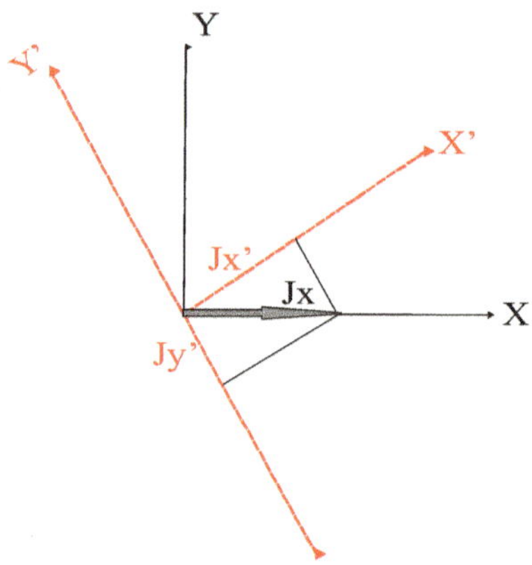

Figure 6.10. A schematic illustration of components of a hydraulic gradient in different coordinate systems.

On the basis of these equations, the specific discharge in the y direction will be nonzero since $K_{xy} \neq 0$ and $J_x \neq 0$. As a consequence, the $\mathbf{q}$ vector will not align with J_x.

Suppose we choose the principal directions (x', y') as our coordinate system (Figure 6.10). The hydraulic gradient, J_x, can be decomposed into $J_{x'}, J_{y'}$, and Darcy's law takes the following form:

$$\begin{aligned} q_{x'} &= K_{x'x'} J_{x'} \\ q_{y'} &= K_{y'y'} J_{y'} \end{aligned} \tag{6.4.14}$$

The $\mathbf{q}$ vector will be in the same direction as that of J_x if $K_{x'x'} = K_{y'y'}$; that is, the medium is isotropic. Since $K_{x'x'} \neq K_{y'y'}$, the $\mathbf{q}$ vector will not align with J_x. Therefore, the hydraulic conductivity tensor approach resolves the noncollinearity issue discussed previously. In addition, this analysis illustrates that the off-diagonal components of a hydraulic conductivity tensor, that is, K_{xy} and K_{yx}, are mathematical artifacts due to the coordinate transformation.

6.4.3 Coordinate Transformation

Suppose there is a point P whose coordinates are x and y, defined according to an x-y coordinate system. If we use a new coordinate system (i.e., x'–y' system) that is at angle θ with respect to the x-y coordinate system, the location of the point is given in terms of x' and y'. We can determine the relationship between x and x' as well as that between y and y'. According to the geometric relationship between $P(x, y)$ and $P(x', y')$, shown in Figure 6.11, the formula to convert the x' and y' coordinates of point P to the x and y coordinates of the same point P are

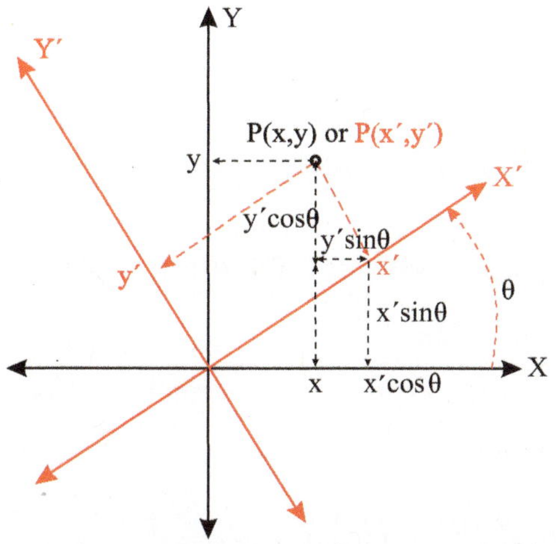

Figure 6.11. A schematic illustration of the geometric relationship of coordinates of the same point based on two different coordinate systems.

$$x = x' \cos\theta - y' \sin\theta \qquad \begin{bmatrix} x \\ y \end{bmatrix} = \begin{bmatrix} \cos\theta & -\sin\theta \\ \sin\theta & \cos\theta \end{bmatrix} \begin{bmatrix} x' \\ y' \end{bmatrix} \tag{6.4.15}$$
$$y = x' \sin\theta + y' \cos\theta, \quad \text{or}$$

Similarly, if we transform the x and y coordinates of the point P to the new coordinates x', y', the transformation rule is

$$x' = x\cos\theta + y\sin\theta \qquad \begin{bmatrix} x' \\ y' \end{bmatrix} = \begin{bmatrix} \cos\theta & \sin\theta \\ -\sin\theta & \cos\theta \end{bmatrix} \begin{bmatrix} x \\ y \end{bmatrix} \tag{6.4.16}$$
$$y' = -x\sin\theta + y\cos\theta, \quad \text{or}$$

The angle θ is measured from x to x' in a counterclockwise sense, and it is positive in that direction. Furthermore, cosθ and sinθ are called the “directional cosines.”

6.4.4 Hydraulic Conductivity Tensor Rotation

The tensor transformation rule is given by

$$[K] = [D][K'][D]^T \tag{6.4.17}$$

where $[K]$ is the hydraulic conductivity tensor (a matrix) in the x-y coordinates system, $[D]$ is the directional cosine matrix, $[K']$ is the hydraulic conductivity tensor (matrix) in the x'–y' coordinates system, and $[D]^T$ denotes the transpose of the matrix $[D]$. For example, a two-dimensional hydraulic conductivity tensor can be transformed from the x'–y' coordinate system to the x-y coordinate system by

$$\begin{bmatrix} K_{xx} & K_{xy} \\ K_{yx} & K_{yy} \end{bmatrix} = \overbrace{\begin{pmatrix} \cos\theta & -\sin\theta \\ \sin\theta & \cos\theta \end{pmatrix}}^{D} \overbrace{\begin{bmatrix} K_{x'x'} & 0 \\ 0 & K_{y'y'} \end{bmatrix}}^{\substack{\text{Principal} \\ Ks}} \overbrace{\begin{bmatrix} \cos\theta & \sin\theta \\ -\sin\theta & \cos\theta \end{bmatrix}}^{D^T}$$

$$= \begin{bmatrix} K_{x'x'}\cos\theta & -K_{y'y'}\sin\theta \\ K_{x'x'}\sin\theta & K_{y'y'}\cos\theta \end{bmatrix} \begin{bmatrix} \cos\theta & \sin\theta \\ -\sin\theta & \cos\theta \end{bmatrix}$$

$$= \begin{bmatrix} K_{x'x'}\cos^2\theta + K_{y'y'}\sin^2\theta & K_{x'x'}\cos\theta\sin\theta - K_{y'y'}\sin\theta\cos\theta \\ K_{x'x'}\sin\theta\cos\theta - K_{y'y'}\cos\theta\sin\theta & K_{x'x'}\sin^2\theta + K_{y'y'}\cos^2\theta \end{bmatrix} \quad (6.4.18)$$

Recall the following relationships:

$$\sin^2\theta = \frac{1}{2}(-\cos 2\theta + 1) \text{ and } \cos^2\theta = \frac{1}{2}(\cos 2\theta + 1) \quad (6.4.19)$$

The components in the transformed hydraulic conductivity tensor become

$$\begin{aligned} K_{xx} &= K_{x'x'}\cos^2\theta + K_{y'y'}\sin^2\theta \\ &= \frac{K_{x'x'}}{2}\cos 2\theta + \frac{K_{x'x'}}{2} + \frac{-K_{y'y'}}{2}\cos 2\theta + \frac{K_{y'y'}}{2} \\ &= \frac{K_{x'x'} + K_{y'y'}}{2} + \frac{(K_{x'x'} - K_{y'y'})}{2}\cos 2\theta \end{aligned} \quad (6.4.20)$$

And

$$\begin{aligned} K_{yy} &= K_{x'x'}\sin^2\theta + K_{y'y'}\cos^2\theta \\ &= \frac{-K_{x'x'}}{2}\cos 2\theta + \frac{K_{x'x'}}{2} + \frac{K_{y'y'}}{2}\cos 2\theta + \frac{K_{y'y'}}{2} \\ &= \frac{K_{x'x'} + K_{y'y'}}{2} - \frac{(K_{x'x'} - K_{y'y'})}{2}\cos 2\theta \end{aligned} \quad (6.4.21)$$

The off-diagonal terms are

$$K_{yx} = K_{x'x'}\sin\theta\cos\theta - K_{y'y'}\cos\theta\sin\theta = K_{xy} \quad (6.4.22)$$

since $\sin 2\theta = 2\sin\theta\cos\theta$,

$$K_{yx} = \frac{K_{x'x'} - K_{y'y'}}{2}\sin 2\theta = K_{xy} \quad (6.4.23)$$

Note that according to equation (6.4.23), K_{xy} and K_{yx} can have negative values, depending on the values of $K_{x'x'}$ and $K_{y'y'}$. This implies that the mixed or off-diagonal terms (K_{xy} and K_{yx}) are mathematical artifacts of the coordinate transformation, and they do not bear any physical meanings.

At this moment, several questions arise: 1) Why is the hydraulic conductivity tensor symmetrical? 2) How is it related to the concept of REV or reversibility of effective hydraulic properties? 3) Are all these related to the diffusive nature of the flow process?

6.5 Equations for Flow through Equivalent Homogeneous Media

For equivalent homogeneous anisotropic porous media under variably saturated conditions, the governing flow equation takes the following form:

$$\frac{\partial}{\partial x}\left[K_{xx}(h)\frac{\partial h}{\partial x}\right]+\frac{\partial}{\partial y}\left[K_{yy}(h)\frac{\partial h}{\partial y}\right]+\frac{\partial}{\partial z}\left[K_{zz}(h)\left(\frac{\partial h}{\partial z}+1\right)\right]=(C(h)+\beta S_s)\frac{\partial h}{\partial t} \tag{6.5.1}$$

where h is the pressure head. It is positive if the medium is fully saturated, and it is negative if the medium is unsaturated. A saturation index, β, is used to control the storage property of the porous medium. For fully saturated media, β is set to 1; otherwise its value is 0. $K_{xx}(h)$, $K_{yy}(h)$, and $K_{zz}(h)$ are pressure head–dependent effective hydraulic conductivity functions in the principal directions, x, y, and z (see Sections 6.2 and 6.3). Here, we assume that the coordinate system for the flow equation is aligned with that of the principal directions of the hydraulic conductivity tensor. Otherwise, terms related to the nine components of the hydraulic conductivity tensor must be specified. The homogeneity here implies that the mathematical models (e.g., the Gardner or MVG model; see Chapter 3) and associated parameters for hydraulic conductivity–pressure head and moisture content–pressure head relationships, as well as the specific storage term, are the same at every location of the medium. However, the actual value of the conductivity or the moisture capacity term [$C(h)$] may vary spatially, depending on pressure head distribution. If the medium is fully saturated, the effective hydraulic conductivity functions are effective saturated hydraulic conductivities, which are independent of locations and thus can be moved out of the partial derivatives.

Unlike the effective hydraulic conductivities, the effective moisture capacity [$C(h)$] and specific storage terms [S_s] in equation (6.5.1) are scalars and independent of flow directions. Nevertheless, the effective hydraulic conductivity, its anisotropy, specific storage, and moisture capacity term may evolve with time until the flow process encounters sufficient heterogeneity (see Chapter 7, Section 7.4). They are also flow dependent due to the effects of volume averaging as illustrated in Equation (5.9.21) in Chapter 5.

6.6 Equations for Flow through Equivalent Homogeneous Fractured Media

Generally speaking, equivalent homogeneous conceptual models of flow through fractured rocks fall into two broad categories: 1) an equivalent porous media (more

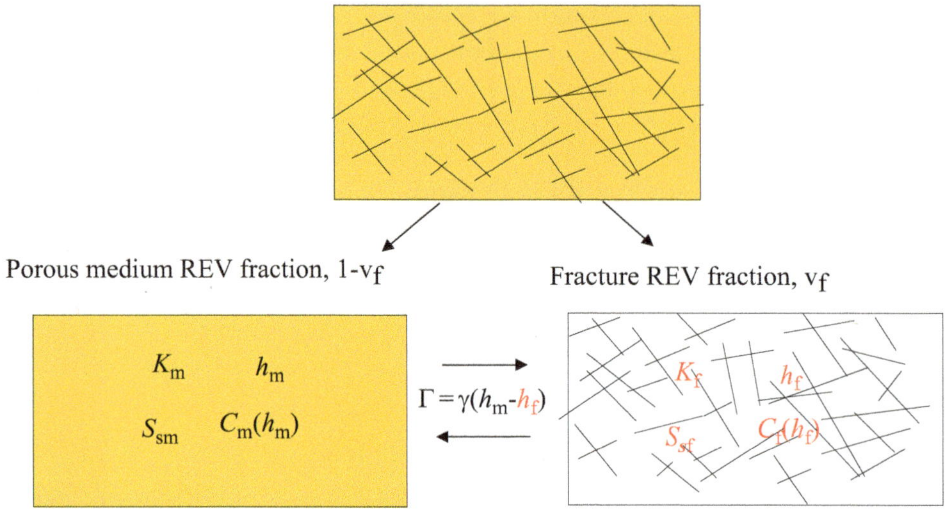

Figure 6.12. A schematic illustration of an equivalent homogeneous, dual continuum conceptual model for variably saturated flow in fractured rocks.

appropriately, a single continuum) approach (Bear, 1972) that treats a rock as a single homogeneous porous continuum, representing the fracture network and the rock matrix itself and 2) a dual continuum approach that treats the rock as two overlapping continua of matrix and fracture (Barenblatt et al., 1960; Duguid and Lee, 1977; Pruess and Narasimhan, 1985; Gerke and van Genuchten, 1993a, b). Both models either assume that fracture density is sufficiently high such that spatial REVs exist or they merely adopt the ensemble REV concept.

The governing partial differential equation for the single continuum approach is identical to equation (6.5.1), with the exception that the effective properties are heavily weighted toward hydraulic characteristics of the fractures themselves. Additionally, the hydraulic exchange or mass transfer between rock matrix and fractures is generally assumed to be at equilibrium at all times. This implies that any mass exchange between the fractures and the rock matrix is instantaneous, and this may not be the case, in general.

Consideration of the effects of the contrast in hydraulic properties of the two void populations (fracture and matrix) and the effects of the slow mixing process between processes in fractures and matrix has led to the use of a dual porosity or continua/mass transfer model. Such a dual continua model has been employed to explain delayed release of water from matrix to the fracture in fractured aquifers, which is similar to that of a leaky or an unconfined aquifer during a pumping test.

The dual porosity or continua conceptual model is illustrated in Figure 6.12. This conceptual model partitions the spatial or ensemble REV for the entire fractured medium into fracture REV and matrix (porous medium) REV, according to the volume fraction of the fracture. The hydraulic properties of each REV are spatially invariant.

Governing equations for this model consist of two equations. First, for the flow through fractures, the governing equation is

$$\frac{\partial}{\partial x_i}\left[K_f(h_f)\frac{\partial(h_f+z)}{\partial x_j}\right]-\frac{\Gamma}{v_f}=\left(C_f(h_f)+\beta S_{sf}\right)\frac{\partial h_f}{\partial t} \tag{6.6.1}$$

where K_f, h_f, v_f, C_f, S_{sf}, and β_f are the hydraulic conductivity, pressure head, volume fraction of fractures in the REV, moisture capacity term, specific storage, and saturation index of the fractures. Further, the term Γ is the exchange flux between the fractures and the porous matrix. The other equation for the flow through porous matrix is given by

$$\frac{\partial}{\partial x_i}\left[K_m(h_m)\frac{\partial(h_m+z)}{\partial x_j}\right]-\frac{\Gamma}{1-v_f}=(C_m(h_m)+\beta S_{sm})\frac{\partial h_m}{\partial t} \tag{6.6.2}$$

The definitions of the variables in equation (6.6.2) are the same as those in equation (6.6.1), but as indicated by the subscript m, these variables are now referring to the porous matrix. The interaction between fracture and porous matrix is modeled by a linear reservoir assumption, which can be expressed as

$$\Gamma=\gamma(h_f-h_m) \tag{6.6.3}$$

in which γ is the transfer coefficient, which dictates the rate of flow exchange between the fracture and matrix. Additionally, the hydraulic conductivity anisotropy can be included in these equations.

The major issue confronting this dual continua approach are the physical meanings of the predicted head in fracture (h_f) and in matrix (h_m) in equations (6.6.1), (6.6.2), and (6.6.3). Theoretically, they are averages over their corresponding CV or REV. But how do they relate to our observations, since our head observation often includes the heads from both the fracture and the matrix?

Finally, as discussed in the previous chapter (Chapter 5), a more realistic and general model for fractured media is the heterogeneous conceptual model, which treats flow in the fractured rocks as flow in a highly heterogeneous porous medium, where interaction between fractures and the matrix is inherent. However, in this approach, the major challenge is mapping of all heterogeneities in the medium.

6.7 Conceptual Models, REV, Observation Scale, and Resolution

In Chapter 5, we introduced different conceptual models, i.e., highly parameterized, geologic or zonation, and equivalent homogeneous models. We later formulated governing equations for 3-D and 2-D flow for the highly parameterized conceptual model, which includes the geologic or zonation conceptual model. In this chapter, we focused on the governing equations for the equivalent homogeneous conceptual models. After discussing all these conceptual models and their associated mathematical equations, we present below an overall discussion of these conceptual models and equations in terms of CV, REV, observation scale, and predictive resolution.

The equivalent homogeneous conceptual model is built upon the assumption that a spatial REV exists in a porous medium. Otherwise, the model adopts the ensemble REV concept. In a medium where a spatial REV exists, the heterogeneous property field of the medium is a realization of a stationary random field. As such, the spatial statistics about the spatial variability of the medium property can be inferred from the samples of a small portion of the medium, that is, a sample volume covering several correlation scales (see Chapter 4), and the spatial REV exists. Because the size of this spatial REV is much smaller than the size of the entire medium, and the head propagation is diffusive (see Section 2.5.2), the head observed by our sampling device (such as an observation well) will likely experience the same heterogeneity within the spatial REV (see Section 2.5.3). The observed head will thus agree with the simulated head on the basis of the property defined over the REV. If it does not, the difference ought to be small. The homogeneous model therefore often yields satisfactory flow predictions.

However, most field problems involve nonstationary or statistically heterogeneous geologic media. Such geologic media do not have any spatial REV but an ensemble REV (i.e., homogeneity exists only in an ensemble sense). Under these circumstances, the ensemble REV is the entire medium. This implies that the variance of the property for the ensemble REV is likely large, since it includes both trends and perturbations at different scales. As discussed previously in Section 6.2.3, the predicted head based on the effective property is the ensemble average of all possible heads corresponding to all possible properties over the ensemble. With such a large variation in the property, our observed head, which is an average over a small volume of the medium around the sampling device, will not experience all heterogeneity, as does the predicted head. Consequently, the observed head will be quite different from the predicted mean head on the basis of the effective properties for the equivalent homogeneous model.

On the other hand, the governing flow equations for the heterogeneous conceptual model, including the geologic (or zonation) model and highly parameterized model, are built upon many CVs. The CV is a volume where mass is balanced and a representative hydraulic property of this volume is defined. The head at a point in the space described by these equations consequently represents the average of all possible heads (ensemble average) at that point, resulting from heterogeneity within the CV.

Since CV (i.e., each block in Figure 5.1b) is small, many CVs allow the mathematical model to represent both large-scale trend and small-scale heterogeneity of a field site closely. For that reason, the CV encompasses only mild heterogeneity, and the simulated ensemble-averaged head at any location in the CV is expected to coincide with the corresponding observed, spatially averaged head.

In view of all these reasons, as long as the property of each block of the heterogeneous conceptual model is known, the simulated head field will capture the effects of the large-scale trend and small-scale heterogeneity. Accordingly, a highly parameterized, heterogeneous conceptual model can yield high-resolution predictions that mirror reality.

The resolution here refers to the degree of spatial and temporal details of the observed true flow process (e.g., head) that can be reproduced by the mathematical model of a conceptual model. To illustrate the resolution of a model, a scatterplot, which plots the predicted process on one of the axes of an *x-y* plot and the observed process on the other axis, is useful. A perfect resolution then means that pairs of observed and simulated heads in time and space perfectly align with the 45° line (i.e., perfect correlation). On the other hand, lower resolution refers to the situation where the pairs of data scatter around the 45° line (or the perfect correlation no longer exists).

Figure 6.13 illustrates the concept of the resolution using scatterplots for different conceptual models. If boundary and initial conditions, inflow, and outflow are known exactly, excluding possible systemic biases, the best resolution that an equivalent homogeneous model can yield is shown in Figure 6.13a, and that a geologic conceptual model can produce is plotted in Figure 6.13b. Finally, Figure 6.13c demonstrates the resolution that a highly parameterized heterogeneous model can offer. The degree of scattering reflects the degree of unresolved heterogeneity by the three conceptual models and our head measurement volume. Note that measurement errors are excluded in the preceding discussion. In addition, as the size of CV approaches the volume our head measurement represents, the resolution becomes higher.

6.8 Summary

- Homogenization creates effective hydraulic conductivities.
- Effective hydraulic conductivity for flow parallel to bedding is equivalent to the arithmetically weighted average of conductivity of each layer.
- Effective hydraulic conductivity for flow perpendicular to bedding is the harmonically weighted average of conductivity of each layer.
- Effective conductivities and anisotropy depend on the contrasts of heterogeneities and their structures within the REV.
- The hydraulic heads predicted by the equivalent homogeneous model are heads in some fictitious media, and they are the same as those observed at a single location.
- Effective hydraulic conductivity anisotropy is an artifact, resulting from the spatial or ensemble averaging of heterogeneities at scales smaller than CV or REV of porous media.

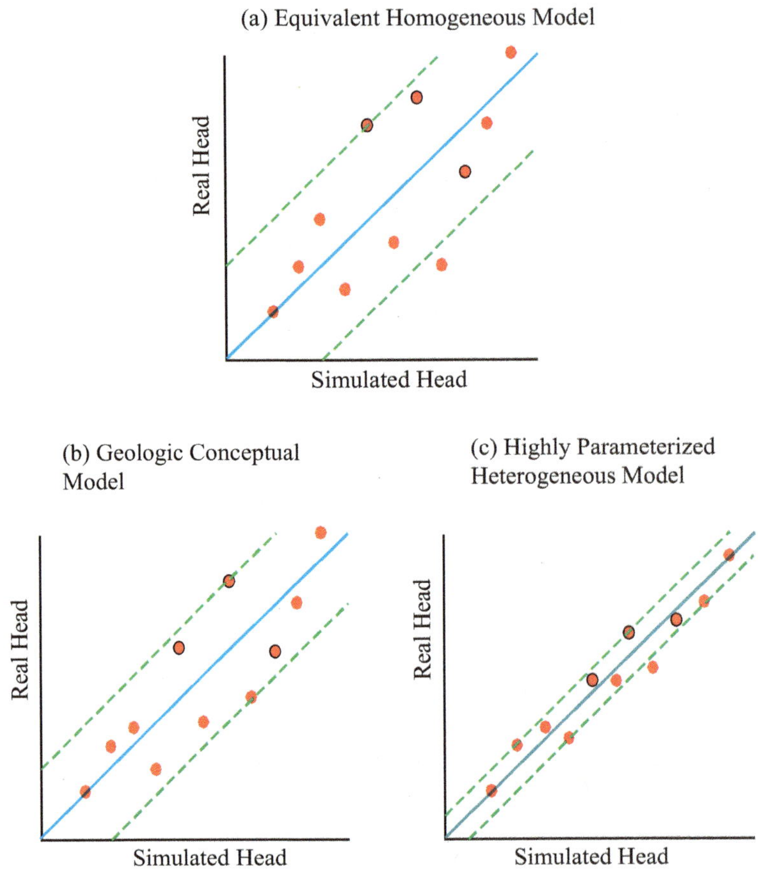

Figure 6.13. Illustrations of the predicted head resolutions and their deviations (upper and lower bounds – green lines) of the three conceptual models. The deviations will vary with the mean gradient (see equation 6.2.12).

6.9 Exercises

1. The hydraulic heads in three observation wells located in a homogeneous confined aquifer of constant transmissivity $T = 5000\ \text{m}^2\text{d}^{-1}$ are as follows:

$$\text{Well A} : x = 0\ \text{m}, y = 0\ \text{m}, h = 10.0\ \text{m}$$
$$\text{Well B} : x = 50\ \text{m}, y = 0\ \text{m}, h = 9.0\ \text{m}$$
$$\text{Well C} : x = 0\ \text{m}, y = 80\ \text{m}, h = 6.0\ \text{m}$$

(a) Draw the hydraulic head contour map surface and determine the hydraulic gradient vector.

(b) Determine the discharge through the aquifer (magnitude and direction) per unit width.

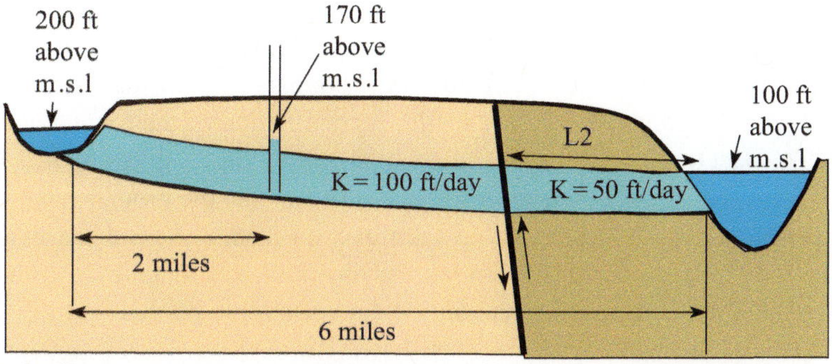

Figure 6.14.

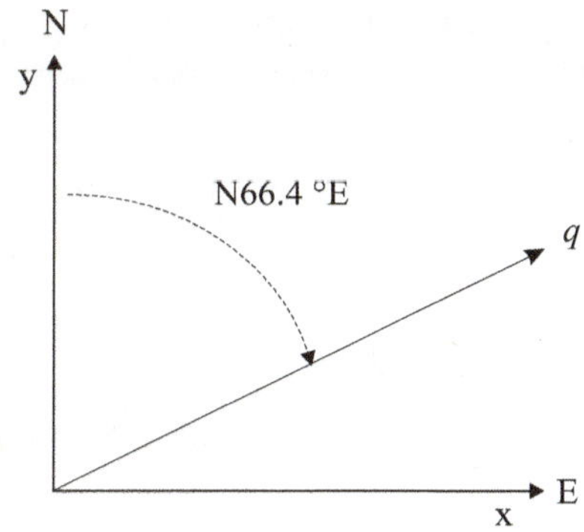

Figure 6.15.

2. Consider the aquifer in Problem 1 as anisotropic with $\overline{K}$ (md^{-1}) tensor given as

$$\begin{bmatrix} 30 & 8 \\ 8 & 10 \end{bmatrix}$$

The aquifer thickness is 50 m. Determine the hydraulic gradient and discharge through the aquifer (magnitude and direction) per unit width.

3. For the $\overline{K}$(m/d) tensor in Problem 2, determine the principal values and principal directions.
4. Let $K_x = 36$ m/d and $K_y = 16$ m/d be the principal values for the $\overline{K}$ tensor in an anisotropic aquifer, in the x and y directions, respectively, in a two-dimensional flow. The hydraulic gradient is 0.004 in a direction making an angle of 30° with the x axis.
   (a) Draw the Darcian flux (magnitude and direction).
   (b) Determine the angle between the hydraulic gradient and the Darcian flux vectors.
5. A confined aquifer receives water from a lake and discharges into an adjacent river valley, as shown in Figure 6.14. The aquifer was shifted as a result of a normal fault so that the upstream portion of the aquifer has a hydraulic conductivity K_1: 100 ft/d, and the downstream portion has a lower hydraulic conductivity $K_2 = 50$ ft/d. Given that the aquifer

thickness is 50 ft, calculate (A) the flow rate into the aquifer from the lake per unit length of shore line (B) the length of the low conductivity portion of the aquifer, L_2.

6. Consider uniform, steady, horizontal flow in a porous medium that is anisotropic with respect to hydraulic conductivity. The matrix representation of the hydraulic conductivity tensor with respect to the x–y system (shown in the following) is given by $K_{xx} = 10$ cm/sec, $K_{xy} = K_{yx}$: 3 cm/sec, $K_{yy} = 2$ cm/sec. A tracer test shows that the groundwater velocity (not specific discharge) is 0.0046 cm/sec in the direction $N66.4°$ E, and the porosity of the medium is 0.38 (Figure 6.15). Determine the direction and magnitude of the hydraulic gradient. What are the principal directions of the anisotropic porous media?
7. As we discussed in Chapter 5, the moisture-based Richards equation is not appropriate for heterogeneous soils. However, heterogeneity is a relative term. That is, any heterogeneous soil formation can always be envisioned as an equivalent homogeneous soil as shown in this chapter. As a result, the moisture-based Richards equation should be appropriate for any heterogeneous soils. Are these statements contradict to each other? Can you use the discussion in section 6.7 to explain this apparent paradox in science?

7

Flow toward a Well Due to Pumping (Part 1)

7.1 Introduction

Well hydraulics (radial flow to or from a well) is an important component of subsurface hydrology. Theories for well hydraulics developed over the past few decades have been widely adopted to evaluate and develop groundwater resources, to design aquifer tests, and, most importantly, to estimate aquifer properties. This well hydraulics topic will be discussed in this chapter and Chapter 8. In this chapter, we formulate first the governing partial differential equations for steady flow and transient radial flow in a 2-D, depth-averaged confined aquifer. We then develop analytical solutions to these equations for predicting drawdowns during pumping or recovery tests in unbounded aquifers. Subsequently, we also examine several widely utilized methods to estimate isotropic or anisotropic transmissivity and storage coefficient (i.e., analysis of aquifer tests). At the end of this chapter, we discuss fundamental limitations of these theories as they are applied to real-world problems, and then we propose alternative methods for aquifer characterization. Chapter 8 then extends the analysis to bounded confined aquifers, leaky aquifers, and unconfined aquifers.

7.2 Steady-State Radial Flow to a Well in a Confined Aquifer

7.2.1 Homogeneous and Isotropic Aquifers

Consider the groundwater flow to a well, induced by pumping in a confined aquifer. To simplify the mathematical analysis, we will first assume that the aquifer is perfectly horizontal, bounded by an aquitard above and below, has uniform thickness and infinite lateral extent, and is homogeneous and isotropic. We further assume that the pumping well fully penetrates the aquifer and is screened from the bottom to the top of the aquifer, and it is pumped at a constant rate, Q, over time (Figure 7.1).

From Chapter 5, the governing equation for two-dimensional, depth-averaged, steady groundwater flow takes the following form:

$$T\frac{\partial^2 H}{\partial x^2} + T\frac{\partial^2 H}{\partial y^2} = 0 \quad \text{or} \quad \frac{\partial^2 H}{\partial x^2} + \frac{\partial^2 H}{\partial y^2} = 0 \tag{7.2.1}$$

where H is the hydraulic head, and T is the transmissivity, which is constant in space and time. The flow toward a well under these assumptions, as stated previously, is an axisymmetric, converging radial flow. As a consequence, all streamlines toward the well are identical.

The axisymmetric flow thus leads to the idea of transformation of equation (7.2.1) into the polar coordinate system. In the polar coordinate system, $x = r\cos\theta$ and $y = r\sin\theta$, where $r = \sqrt{x^2 + y^2}$ is the radial distance of the location (x, y) from the pumping well (located at the origin in Figure 7.2), and θ is the angle between the r

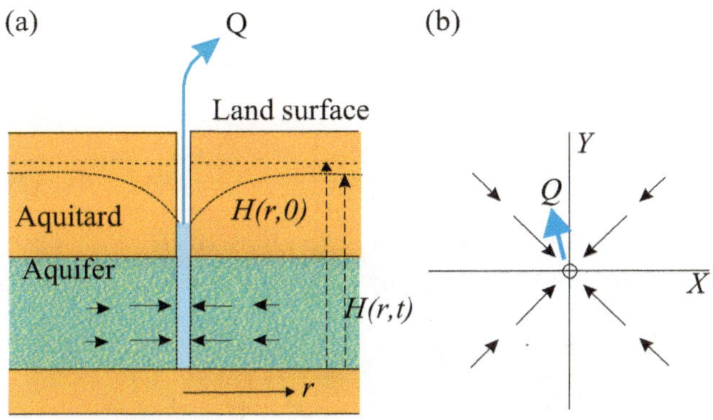

Figure 7.1. a) A cross-sectional view of flow to a well in a confined aquifer. Q is the well discharge or pumping rate, r is the distance from the pumping well, H (r, 0) is the total head at distance r and time zero, and $H(r, t)$ is the total head at distance r and time t. Velocity vectors show flow direction toward the well, and increasing velocity due to the steep hydraulic gradient near the well. The dotted line illustrates the total head profile. b) A plan view of groundwater flow toward a well: axisymmetric (radial) flow due to homogeneity and isotropy assumptions.

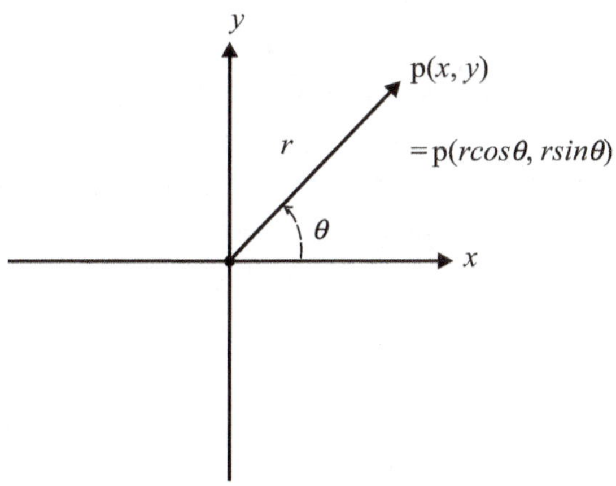

Figure 7.2. An illustration of the relationship between the Cartesian coordinate system and a polar coordinate system.

vector and the x axis shown in Figure 7.2. The derivative of r with respect to x then becomes

$$\frac{\partial r}{\partial x} = \frac{x}{r} \tag{7.2.2}$$

Using the chain rule from calculus, the second derivative of H with respect to x can also be transformed into the polar coordinates as

$$\frac{\partial^2 H}{\partial x^2} = \frac{d}{dr}\left(\frac{dH}{dr}\frac{\partial r}{\partial x}\right)\frac{\partial r}{\partial x} = \frac{d}{dr}\left(\frac{dH}{dr}\frac{x}{r}\right)\frac{x}{r}$$

$$= \frac{d^2 H}{dr^2}\left(\frac{x}{r}\right)^2 + \frac{x}{r}\frac{dH}{dr}\frac{\partial (x/r)}{\partial r} = \frac{d^2 H}{dr^2}\left(\frac{x}{r}\right)^2 + \frac{dH}{dr}\left(\frac{1}{r} - \frac{x^2}{r^3}\right) \quad (7.2.3)$$

Using the same procedure, the second derivative of H with respect to y can be transformed to the following expression:

$$\frac{\partial^2 H}{\partial y^2} = \frac{d^2 H}{dr^2}\left(\frac{y}{r}\right)^2 + \frac{dH}{dr}\left(\frac{1}{r} - \frac{y^2}{r^3}\right) \quad (7.2.4)$$

Substitution of equations (7.2.3) and (7.2.4) into equation (7.2.1) yields

$$\frac{\partial^2 H}{\partial x^2} + \frac{\partial^2 H}{\partial y^2} = \left(\frac{x^2}{r^2} + \frac{y^2}{r^2}\right)\frac{d^2 H}{dr^2} + \frac{dH}{dr}\left(\frac{2}{r} - \frac{x^2}{r^3} - \frac{y^2}{r^3}\right) = \frac{d^2 H}{dr^2} + \frac{1}{r}\frac{dH}{dr} = 0 \quad (7.2.5)$$

Note that $x^2 + y^2 = r^2$. Therefore, the governing partial differential equation for steady radial flow in the polar coordinate system is

$$\frac{d^2 H}{dr^2} + \frac{1}{r}\frac{dH}{dr} = 0 \quad (7.2.6)$$

Notice that the polar coordinate transformation of the governing flow equation reduces a two-dimensional partial differential equation to a one-dimensional ordinary differential equation, expressed in terms of r, instead of x and y. As a result, deriving a solution of the governing flow equation becomes less cumbersome.

Since equation (7.2.6) is a second-order, homogeneous, ordinary differential equation, a solution can be obtained by the method of integration. Before carrying out the integration, we will define a new variable: $\Phi = dH/dr$. Using this new variable, equation (7.2.6) becomes

$$\frac{d\Phi}{dr} + \frac{1}{r}\Phi = 0 \quad (7.2.7)$$

which has a general solution:

$$r\Phi = c_1 \quad \text{or} \quad r\frac{dH}{dr} = c_1 \quad (7.2.8)$$

Thus, the general solution for H in terms of r is

$$H(r) = c_1 \ln r + c_2 \quad (7.2.9)$$

where c_1 and c_2 are unknown integration constants. To evaluate the constant c_1, consider the total inflow, Q, to a cylindrical surface around the pumping well at any distance r. In terms of Darcy's law, it is

$$Q = KIA = K\frac{dH}{dr}2\pi rb = 2\pi rT\frac{dH}{dr} \tag{7.2.10}$$

where I is the hydraulic gradient, A is the surface area of the cylindrical surface (equals to $2\pi rb$), and T is the transmissivity, which is equal to the product of the hydraulic conductivity, K, and the thickness of the aquifer, b. Substitution of equation (7.2.10) into equation (7.2.8) yields

$$c_1 = r\frac{dH}{dr} = \frac{Q}{2\pi T} \tag{7.2.11}$$

Then, substitution of equation (7.2.11) into equation (7.2.9) yields

$$H(r) = \frac{Q}{2\pi T}\ln r + c_2 \tag{7.2.12}$$

Next, we will determine c_2. If we say H at some radial distance, $r = r_0$, is H_0, equation (7.2.12) becomes

$$H(r_0) = H_0 = \frac{Q}{2\pi T}\ln r_0 + c_2 \tag{7.2.13}$$

Therefore, we have

$$c_2 = H_0 - \frac{Q}{2\pi T}\ln r_0 \tag{7.2.14}$$

After substituting equations (7.2.11) and (7.2.14) into equation (7.2.9), a particular solution takes the following form:

$$H(r) = \frac{Q}{2\pi T}\ln r + H_0 - \frac{Q}{2\pi T}\ln r_0 \tag{7.2.15}$$

Thus, the drawdown, $s(r) = H_0 - H(r)$, the difference between the initial head and the head observed during steady flow at any radial distance r, is

$$s(r) = H_0 - H(r) = \frac{Q}{2\pi T}\ln\left(\frac{r_0}{r}\right) \tag{7.2.16}$$

This is the so-called **Thiem's equation.** The r_0 in equation (7.2.16) is called the radius of influence, which is defined as the radial distance at which the drawdown due to pumping is insignificant under a steady-state flow situation.

We next derive a solution to equation (7.2.16) using a different boundary condition, say, $H(r_w) = H_w$ at $r = r_w$, where r_w is the radius of the well, and H_w denotes the water level at the well radius. Imposing this boundary condition on equation (7.2.12), we have

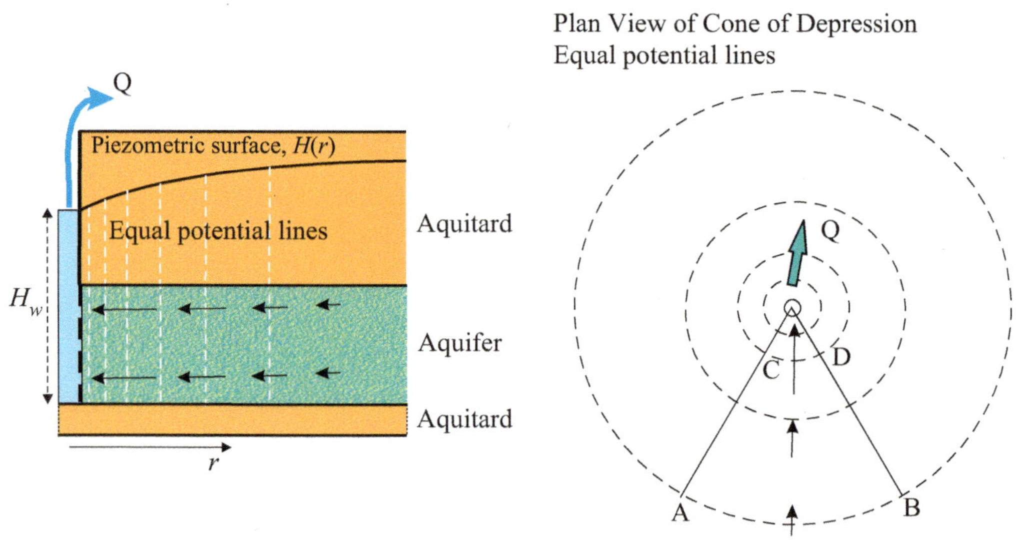

Figure 7.3. A cross-sectional view of steady flow toward a pumping well (left); plan view of the cone of depression (head contours) created by the pumping well (right).

$$H_w = \frac{Q}{2\pi T} \ln r_w + c_2 \tag{7.2.17}$$

which leads to an expression for c_2

$$c_2 = H_w - \frac{Q}{2\pi T} \ln r_w \tag{7.2.18}$$

The particular solution to equation (7.2.9) subject to this new boundary condition then becomes

$$H(r) = H_w + \frac{Q}{2\pi T} \ln\left(\frac{r}{r_w}\right) \tag{7.2.19}$$

These solutions to Thiem's equation lead to the following two conclusions:

(1) Spatial distribution of the hydraulic head during a steady flow condition, induced by pumping at a well, is logarithmic. That is, the head decreases logarithmically toward the pumping well as illustrated in Figure 7.3.

(2) The drawdown distribution in three dimensions forms a cone of depression with a logarithmic profile. Creation of this cone of depression can be easily understood if we consider the steady flow field as illustrated in Figure (7.3). Since the flow is converging toward the pumping well, the surface area of a cylinder, where the groundwater flows toward the well, becomes small, as flow approaches the well (length or area along AB >> CD). The velocity must be greater near the well than far away from the well, in order to maintain a constant flow rate (Q) along any stream tube. Therefore, large gradients must occur near the well.

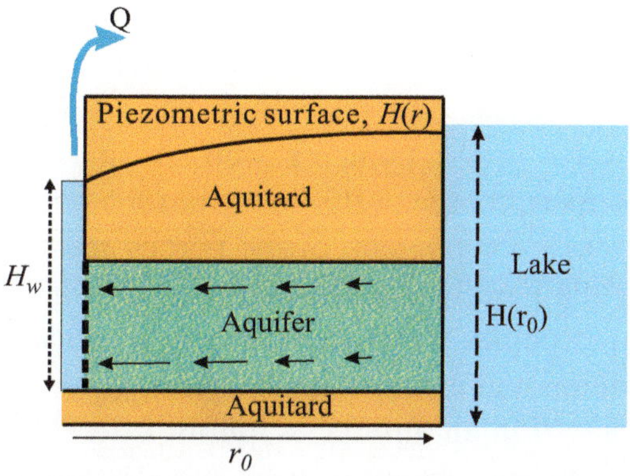

Figure 7.4. A schematic illustration of a cross-sectional view of steady flow due to pumping in a confined aquifer that is surrounded by a lake.

While the preceding two conclusions are useful, we also find that according to the Thiem equation, the head at $r = \infty$ must be infinitely large. That is,

$$H(r) = H_w + \frac{Q}{2\pi T}\ln\left(\frac{r}{r_w}\right) \quad \text{Let} \quad r = \infty \quad H(\infty) = H_w + \infty = \infty \tag{7.2.20}$$

This is physically impossible. Why does Thiem's equation produce such an erroneous solution? In order to answer this question, we ask ourselves, Is the assumption of steady flow to a well in a confined aquifer of infinite lateral extent reasonable? Let us examine three cases of flow to a well and use common sense to reason out the answer.

Case (1): Flow to a well in a circular aquifer surrounded by a lake. In this case (Figure 7.4), the head at the lake, H_0, (or $H(r_0)$ as indicated in the figure) must be constant if the lake is large enough to supply water throughout the time of pumping. As the pumping starts, the cone of depression grows outward to the lake. During this period, all the pumped water must flow from the aquifer storage (i.e., compaction of the aquifer material and expansion of water). Once the cone of depression reaches the lake, water discharged from the well is siphoned directly from the lake. The aquifer no longer has to release water from the storage, and thus the pressure head in the aquifer must reach a steady state. In addition, the head in the aquifer at the lake must be equal to the water level in the lake. Therefore, a steady-state flow condition exists in this case. Then, the head distribution along r, based on Thiem's equation, is

$$H(r) = H_0 + \frac{Q}{2\pi T}\ln\left(\frac{r}{r_0}\right) \tag{7.2.21}$$

At $r = r_0$, $H(r_0) = H_0$. The problem of an infinite head value as mentioned earlier no longer exists.

Case (2): Flow to a well in a circular confined aquifer surrounded by an impermeable boundary. Since this aquifer is surrounded by an impermeable boundary, when the drawdown reaches the impermeable boundary, no external source is available to supply water to the pumping well. As a result, the aquifer is not replenished and water must continuously flow from the aquifer storage as a result of expansion of water and compression of the porous medium. This will result in eventually dewatering the pore space. Therefore, drawdown will continue until the groundwater reservoir becomes unsaturated and is practically fully depleted. No steady-state flow condition will exist in this aquifer.

Case (3): Flow to a well in an aquifer of infinite lateral extent. In this case, the cone of depression grows (or change in head propagates) continuously in lateral directions to release water from aquifer storage and to supply water to the pumping well. As the cone of depression grows, the volume of the aquifer that releases water increases, and the amount of pressure drop required to sustain the pumpage at the far field thus becomes smaller. In practice, the drawdown at the far field may be small and negligible, but theoretically, the drawdown exists. Consequently, the cone of depression continuously grows outward, and a steady-state flow condition never exists in this case. This should explain why the head predicted by Thiem's equation is physically impossible at large distances.

Remarks

Thiem's equation is not valid for confined aquifers with infinite lateral extent. However, it can approximate the drawdown in the close proximity of a well, where steady flow conditions seem present.

7.2.2 Homogeneous and Anisotropic Aquifers

The governing partial differential equation for steady radial flow toward a well in an anisotropic aquifer is

$$T_x \frac{\partial^2 H}{\partial x^2} + T_y \frac{\partial^2 H}{\partial y^2} = 0 \tag{7.2.22}$$

Note that equation (7.2.22) assumes that our x-y coordinate system aligns with the principal directions (see Chapter 6) of the transmissivity. Before discussing the mathematical solution to this equation, let us guess what the solution will be. on the basis of your common sense, you would probably say that the cone of depression will not be concentric and circular; rather it will be elliptical in shape (see Figure 7.5).

Well, can we visualize an elliptical cone of depression as a circular one? Sure, it is simple: Rescale the coordinates. That is,

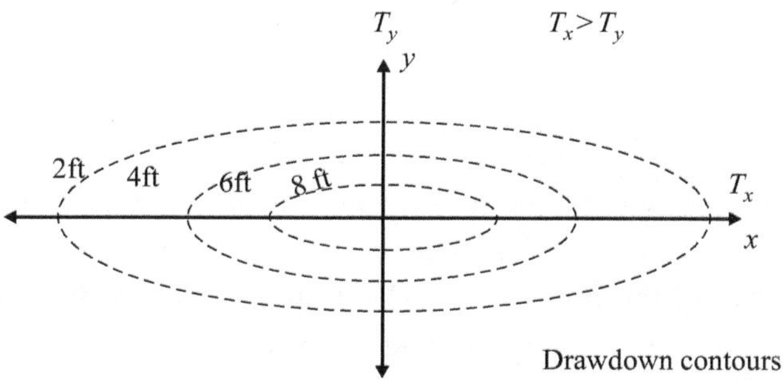

Figure 7.5. Cone of depression due to pumping in an anisotropic aquifer.

$$\text{If } T_x > T_y, \text{ we define a new coordinate } x' = x\sqrt{T_y/T_x} \tag{7.2.23}$$

This means that distances in the direction of the greater T are scaled down. Of course, you can choose to scale the y axis instead. Consequently, the spatial derivatives can be expressed in terms of the new coordinate system. That is,

$$\frac{\partial H}{\partial x} = \frac{\partial H}{\partial x'}\frac{\partial x'}{\partial x} = \sqrt{T_y/T_x}\frac{\partial H}{\partial x'} \tag{7.2.24}$$

$$\frac{\partial^2 H}{\partial x^2} = \sqrt{T_y/T_x}\frac{\partial}{\partial x}\left(\frac{\partial H}{\partial x'}\right) = \sqrt{T_y/T_x}\frac{\partial}{\partial x'}\left(\frac{\partial H}{\partial x'}\right)\frac{\partial x'}{\partial x} = \frac{T_y}{T_x}\frac{\partial^2 H}{\partial x'^2} \tag{7.2.25}$$

Therefore,

$$T_x\frac{\partial^2 H}{\partial x^2} = T_y\frac{\partial^2 H}{\partial x'^2} \tag{7.2.26}$$

The original equation in terms of T_x, T_y, x, and y becomes

$$T_y\frac{\partial^2 H}{\partial x'^2} + T_y\frac{\partial^2 H}{\partial y^2} = 0 \quad \text{or} \quad \frac{\partial^2 H}{\partial x'^2} + \frac{\partial^2 H}{\partial y^2} = 0 \tag{7.2.27}$$

This equation is the same as the governing equation for the steady radial flow in isotropic aquifers, which can be easily transformed into the polar coordinate system, and then it can be solved analytically. In fact, we have already derived the solution to this equation at the beginning of this chapter. However, notice that in equation (7.2.27) the x coordinate has been scaled. Questions: Can you derive an analytical expression for the solution for the steady-state flow to a well in an anisotropic aquifer now? Remember to scale the coordinates of the solution back to the original xy coordinate system.

7.3 Transient Radial Flow to a Well in a Confined Aquifer

7.3.1 Homogeneous and Isotropic Aquifers

7.3.1.1 Prediction of Drawdowns

Again, we assume that the aquifer is confined with an infinite lateral extent, it has uniform thickness, and aquifer properties are isotropic and homogeneous. As far as the discharging well is concerned, it is assumed that the well is fully penetrating and screened over the entire thickness of the aquifer, and it is discharging at a constant rate, Q. The 2-D depth-averaged governing equation in the polar coordinate system is

$$T\frac{\partial^2 H}{\partial r^2}+\frac{T}{r}\frac{\partial H}{\partial r}=S\frac{\partial H}{\partial t} \tag{7.3.1}$$

with boundary conditions

$$\lim_{r\to 0}\left(r\frac{\partial H}{\partial r}\right)=\frac{Q}{2\pi T}\text{ and }H(r=\infty,t)=H_0\text{ for all }ts \tag{7.3.2}$$

and initial condition

$$H(r,t=0)=H_0\text{ for all rs.} \tag{7.3.3}$$

These equations can be written in terms of drawdown $s(r,t)=H_0-H(r,t)$

$$\frac{\partial^2 s}{\partial r^2}+\frac{1}{r}\frac{\partial s}{\partial r}=\frac{S}{T}\frac{\partial s}{\partial t} \tag{7.3.4}$$

with boundary conditions

$$\lim_{r\to 0}\left(-r\frac{\partial s}{\partial r}\right)=\frac{Q}{2\pi T}\text{ and }s(r=\infty,t)=0\text{ for all }ts \tag{7.3.5}$$

and initial condition

$$s(r,t=0)=0\text{ for all }rs. \tag{7.3.6}$$

Note that lowercase s represents drawdown and capital S is the storage coefficient, which is dimensionless. Theis (1935) uses the Boltzman transformation to derive the analytical solution for the partial differential equation (7.3.5). The Boltzman transformation is

$$u=\frac{r^2 S}{4Tt} \tag{7.3.7}$$

which leads to

$$\frac{\partial u}{\partial r} = \frac{2\,rS}{4\,Tt} = \frac{2u}{r} \tag{7.3.8}$$

Using the new variable u, the first spatial derivatives in equation (7.3.4) can be expressed as

$$\frac{\partial s}{\partial r} = \frac{ds}{du}\frac{\partial u}{\partial r} = \frac{rS}{2Tt}\frac{ds}{du} \tag{7.3.9}$$

and the second spatial derivative is

$$\frac{\partial^2 s}{\partial r^2} = \frac{\partial}{\partial r}\left(\frac{\partial s}{\partial r}\right) = \frac{\partial}{\partial r}\left(\frac{rS}{2Tt}\frac{ds}{du}\right) = \frac{S}{2Tt}\frac{ds}{du} + \frac{rS}{2Tt}\frac{\partial}{\partial r}\left(\frac{ds}{du}\right)$$

$$= \frac{S}{2Tt}\frac{ds}{du} + \frac{rS}{2Tt}\frac{d\left(\frac{ds}{du}\right)}{du}\left(\frac{\partial u}{\partial r}\right) = \frac{S}{2Tt}\frac{ds}{du} + \frac{rS}{2Tt}\left(\frac{rS}{2Tt}\right)\frac{d^2 s}{du^2} \tag{7.3.10}$$

It can be simplified to

$$\frac{\partial^2 s}{\partial r^2} = \frac{S}{2Tt}\frac{ds}{du} + \frac{r^2 S^2}{4T^2 t^2}\frac{d^2 s}{du^2} \tag{7.3.11}$$

Now, the time derivative in equation (7.3.1) becomes

$$\frac{\partial s}{\partial t} = \frac{ds}{du}\frac{\partial u}{\partial t} = -\frac{r^2 S}{4Tt^2}\frac{ds}{du} \tag{7.3.12}$$

Substitution of these expressions into equation (7.3.4) yields the following equation:

$$u\frac{d^2 s}{du^2} + (u+1)\frac{ds}{du} = 0 \tag{7.3.13}$$

Note that in equation (7.3.13) it is assumed that $u \neq 0$. Further notice that by using the transformation, we have removed the time derivative and reduced the partial differential equation to a second-order ordinary differential equation, in which s is a function of u only, with the following conditions

$$\lim_{u \to 0}\left(-u\frac{ds}{du}\right) = \frac{Q}{4\pi T} \tag{7.3.14}$$

and $s(\infty) = 0$ when $u \to \infty$. Finding a solution to this ordinary differential equation becomes relatively simple. To solve this equation, we rewrite the previous equation in a new form:

$$\frac{d}{du}\left(u\frac{ds}{du}\right) + u\frac{ds}{du} = 0 \tag{7.3.15}$$

and letting $\Phi = u\frac{ds}{du}$, the equation becomes

$$\frac{d\Phi}{du} = -\Phi \tag{7.3.16}$$

Integration of the equation leads to

$$ln\, \Phi = -u + c \qquad \Phi = e^{-u+c} = e^{-u}e^{c} \tag{7.3.17}$$

We then have

$$u\frac{ds}{du} = c_1 e^{-u} \tag{7.3.18}$$

Now apply the boundary condition to obtain c_1

$$\lim_{u \to 0}\left(-u\frac{ds}{du}\right) = c_1 = \frac{Q}{4\pi T} \tag{7.3.19}$$

and we now have

$$\frac{ds}{du} = \frac{Q}{4\pi T}\frac{e^{-u}}{u} \tag{7.3.20}$$

The final solution is obtained by integrating *ds/du*:

$$s = \frac{Q}{4\pi T}\left[\int_u^{\infty} \frac{e^{-x}}{x} dx\right] + c_2 \tag{7.3.21}$$

Using the boundary condition for s at $u = \infty$, the integration constant, c_2, can be determined; it is zero in this case. The integral in the solution is called an exponential integral and the final solution is expressed as

$$s(r,t) = \frac{Q}{4\pi T} W(u) \qquad \text{where } W(u) = \int_u^{\infty} \frac{e^{-x}}{x} dx \tag{7.3.22}$$

This final solution is called the **Theis solution**, and the function, $W(u)$, is called the well function for a nonleaky, uniform, confined aquifer, which can be expressed in an infinite series as

$$W(u) = \left[-0.5772 - \ln(u) + u - \frac{u^2}{2.2!} + \frac{u^3}{3.3!} + \cdots\right] \tag{7.3.23}$$

Numerical Recipes in Fortran 77 provides a Fortran program to evaluate the exponential integral of equation (7.3.22). Furthermore, polynomial and rational approximations of the $W(u)$ function can also be found in Abramowitz and Stegun, *Handbook of Mathematical Functions,* National Bureau of Standards (1964).

If we consider a constant head or zero drawdown boundary, rather than an infinite lateral boundary, say, $s(u_{r_e}) = 0$ at a radial distance, r_e, the solution then becomes

$$s(r,t) = \frac{Q}{4\pi T}\left[\int_u^\infty \frac{e^{-x}}{x}dx\right] - \frac{Q}{4\pi T}\left[\int_{u_{r_e}}^\infty \frac{e^{-x}}{x}dx\right] \tag{7.3.24}$$

Question:

The preceding analysis assumes that no regional gradient exists during the aquifer pumping test. If a regional gradient does exist, how will it affect the cone of depression?

Dimensionless drawdown analysis. From the previous analysis, we learn that

$$s(r,t) = \frac{Q}{4\pi T}W(u), \quad t = \frac{r^2 S}{4T}\left(\frac{1}{u}\right) \tag{7.3.25}$$

Next, we express the drawdown and time as dimensionless drawdown and time as

$$s^* = \frac{s(r,t)4\pi T}{Q} = W(u), \quad t^* = \frac{1}{u} = \frac{4Tt}{r^2 S} \tag{7.3.26}$$

That is to say, the well function, $W(u)$, and $1/u$ are dimensionless drawdown and time, respectively. According to equation (7.3.26), the real drawdown, s, is linearly proportional to $W(u)$, and time, t, is linearly proportional to $1/u$.

$$s = aW(u), t = b\left(\frac{1}{u}\right), \text{ and } a = \frac{Q}{4\pi T}, b = \frac{r^2 S}{4T} \tag{7.3.27}$$

If we take the log of the equations, then we have

$$\log_{10} s = \log_{10} a + \log_{10} W(u) \qquad \log_{10} t = \log_{10} b + \log_{10}(1/u) \tag{7.3.28}$$

These two equations imply that a plot of s versus t curve on a log-log paper is merely a displaced form of the plot of $W(u)$ versus $1/u$ by an amount of log a on the drawdown axis and log b on the time axis. This leads to the idea of type-curve matching using a log-log plot (see Figure 7.6).

Similarly, a plot of s versus $1/r^2$ on a log-log paper is a form of a displacement of a plot of $W(u)$ versus $1/u$ by an amount of log a on the drawdown axis and log b on the $1/r^2$ axis. However, the definition of b is $\frac{S}{4Tt}$, in this case.

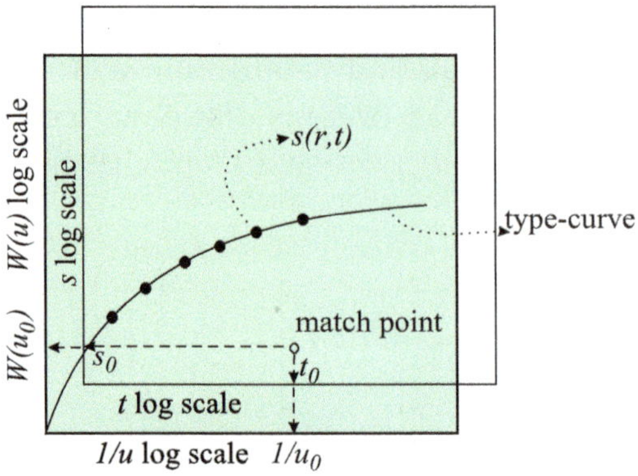

Figure 7.6. A schematic illustration of the concept of the type curve matching approach for estimating aquifer properties from drawdown-time data.

7.3.1.2 Aquifer Tests and Parameter Estimation (Inverse Modeling)

An **aquifer test** has been a standard practice used by hydrogeologists to evaluate the characteristics of an aquifer. The test "stimulates" the aquifer through a constant pumping rate at one well and observes the aquifer's "response" (drawdown) in a monitoring well and/or additional monitoring wells, or piezometers ("point" observation wells) – a cross-hole test. A monitoring well is simply a well, which is not being pumped but is used to monitor the hydraulic head in the aquifer. One of the reasons to use a monitoring well to observe aquifer responses arises from the fact that the water level in the pumping well is in general unstable and fluctuating. Typically, both monitoring and pumping wells are screened at the same depth interval across the same aquifers.

The drawdown data collected at different time intervals during an aquifer test are often referred to as a drawdown-time curve or well hydrography. A well hydrography is typically interpreted by using an analytical model of groundwater flow in an aquifer (i.e., the Theis solution) to estimate aquifer properties: that is, solving the Theis solution for T and S with a given pumping rate, locations of pumping and monitoring wells, and the observed drawdown-time curve (i.e., **inverse modeling**, see Chapter 9). Traditional methods discussed in the following have been widely used for analyzing hydrographs from aquifer tests.

1) Type-Curve Method

This type-curve method evolves from the previously discussed dimensionless analysis. This method can be applied to a) drawdown-time data observed from one well or b) drawdown data at a given time observed at many wells at different locations.

A) The drawdown-time data fitting involves the following steps:

1. Plot the type curve on a log-log paper.
2. Plot drawdown versus time data on a transparent paper with the same length of each logarithmic cycle as that for the type curve in step 1.

3. Overlap the data sheet onto the type curve, holding the coordinate axes of the sheet and the type curve in such a way that the data best fit the type curve.
4. At any common point, the match point, arbitrarily chosen on the overlapping part of the curves, determines mutual values of s and $W(u)$ and t and $1/u$ at the match point.
5. Determine T and S using the following equations:

$$T = \frac{Q}{4\pi s} W(u) \text{ and } S = \frac{4Ttu}{r^2} \tag{7.3.29}$$

Notice that this approach and these formulae are valid for any consistent units for every variable. In general, the visual match between the type curve and data is somewhat subjective. As a result, the estimate values may vary from one analysis to another, although the estimates are unique in theory.

B) Drawdown-distance fitting. The same procedure explained earlier can be applied to s-r (drawdown-distance) data. That is, we can plot observed drawdowns versus the reciprocal of the square of the distance (or $1/r^2$), as we did in Figure 7.6. According to the previous dimensional analysis, the type curve ($W(u)$ vs. $1/u$) can also be employed to find the match point, and, subsequently, T and S can be estimated.

Applications of this approach require drawdowns at a selected time from at least two wells (the more the better!) at different distances. The drawdown-distance method should be first applied to the late-time drawdown data (see cross-correlation in Chapter 9) to estimate the transmissivity. Then, estimation of the storage coefficient is carried out using the early-time data by fixing the value of the transmissivity estimated in the previous step.

2) Automated Curve Fitting Method (Method of Least Squares)

The type curve approach is merely a curve fitting exercise, which can be accomplished manually or through a nonlinear optimization routine (using some search algorithm; see Press et al. (1992)). This routine automatically searches for the best parameter values that produce simulated drawdowns in agreement with what is observed. That is, it successively adjusts estimated T and S to minimize the following objective function

$$\sum_{i=1}^{n} \left[s^*(r,t_i) - s(r,t_i)\right]^2 = \text{minimal} \tag{7.3.30}$$

where $s^*(r, t_i)$ is the observed drawdown at a location, with a radial distance r from the pumping well at time t_i $s(r, t_i)$ is the corresponding drawdown calculated using the Theis solution, with a set of estimated T and S values. A more detailed discussion of this topic is given in Chapter 9, where the inverse modeling techniques are discussed.

At this point, it may appear that the transient drawdown data are primarily controlled by the ratio of S to T (diffusivity) according to equation (7.3.4). This implies that it may be difficult to separate S or T from the diffusivity term, using only a short record of the drawdown-time data set. Actually, this conjecture is mathematically false. As explained in Section 9.2 in Chapter 9, since the flux along any streamline to the well is known (i.e., $Q/2\pi rb$, where b is the thickness of the aquifer) and the analytical form of the gradient (the Theis solution) is given, a unique solution can always be obtained, as long as heads at two times (e.g., one late and one early time; see Cooper-Jacob's straight-line method, to be discussed next, for definitions of the early and late times) are given, although noisy data could lead to incorrect solutions for T and S.

3) Cooper-Jacob's straight-line method

Cooper-Jacob's straight-line method (1946) can be applied to either the drawdown-time or drawdown-distance data sets:

A) Straight-line plot of *s* versus log *t*. Recall the well function:

$$W(u) = -0.5772 - \ln(u) + u - \frac{u^2}{2.2!} + \frac{u^3}{3.3!} - \frac{u^4}{4.4!} + \cdots \tag{7.3.31}$$

For small values of u, say, $u \leq 0.01$, the infinite series can be approximated by its first two terms only, that is,

$$s(r,t) = \frac{Q}{4\pi T}(-0.5772 - \ln(u)) \tag{7.3.32}$$

The error involved in this approximation is less than 6 percent. The solution then becomes

$$\begin{aligned} s(r,t) &= \frac{Q}{4\pi T}\left(\ln\frac{4Tt}{r^2 S} - \ln 1.78\right) = \frac{Q}{4\pi T}\ln\left[\frac{2.25Tt}{r^2 S}\right] \\ &= \frac{0.183Q}{T}\log_{10}\left[\frac{2.25Tt}{r^2 S}\right] \end{aligned} \tag{7.3.33}$$

This equation implies that s is linearly proportional to log t. That is, if s is plotted on a linear scale and t is on a log-scale, the drawdown time data then will follow a straight line when $u \leq 0.01$ or when r is small or t is large (representing a late time, see Figure 7.7).

Further, for $t = t_1$, the drawdown at a given distance r is

$$s(r,t_1) = \frac{0.183Q}{T}\log_{10}\left[\frac{2.25Tt_1}{r^2 S}\right] \tag{7.3.34}$$

For $t = t_2$ at r, the drawdown is

$$s(r,t_2) = \frac{0.183Q}{T}\log_{10}\left[\frac{2.25Tt_2}{r^2 S}\right] \tag{7.3.35}$$

The difference in the drawdown at the two times becomes

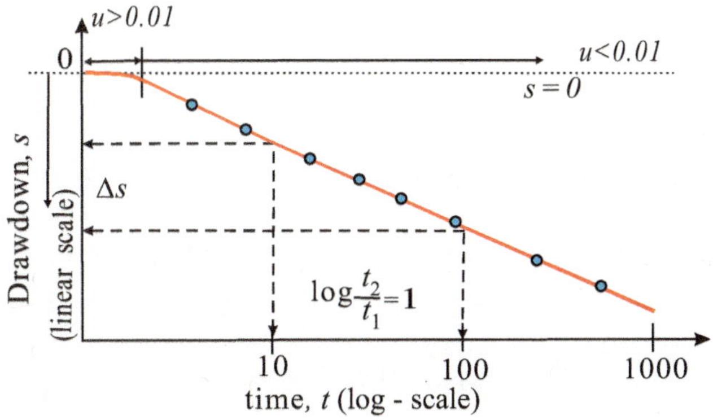

Figure 7.7. Drawdown-time plot on a linear-log scale.

$$\Delta s = s(r,t_2) - s(r,t_1) = \frac{0.183Q}{T}\log_{10}\left[\frac{t_2}{t_1}\right] \tag{7.3.36}$$

If Δs is the drawdown difference over one log-cycle time interval, we then have

$$\Delta s = \frac{0.183Q}{T}\log_{10}\left[\frac{100}{10}\right] = \frac{0.183Q}{T} \tag{7.3.37}$$

Thus, T can be estimated from the preceding equation as

$$T = \frac{0.183Q}{\Delta s} \tag{7.3.38}$$

Cooper-Jacob's straight line method provides not only a convenient tool for estimating transmissivity, but also a means to illustrate that the rate of change in drawdown at the late time is mainly controlled by the aquifer transmissivity. A practical implication of the results for aquifer test analysis is that the late-time drawdown data should be used to estimate the transmissivity first and the early time data should then be used to estimate the storage coefficient of a confined aquifer.

In order to estimate S, we extrapolate the straight line portion of the hydrograph to intercept the horizontal line that represents zero drawdown.

From the intercept, we determine t_0 from the time axis at which the linear approximation of drawdown (shown by a dotted line in Figure 7.8) is zero. On the basis of this t_0, we can estimate S. That is, consider the equation

$$s(r,t) = \frac{0.183Q}{T}\log_{10}\left(\frac{2.25Tt}{r^2S}\right) \tag{7.3.39}$$

At t_0 when $s = 0$, equation (7.3.39) becomes

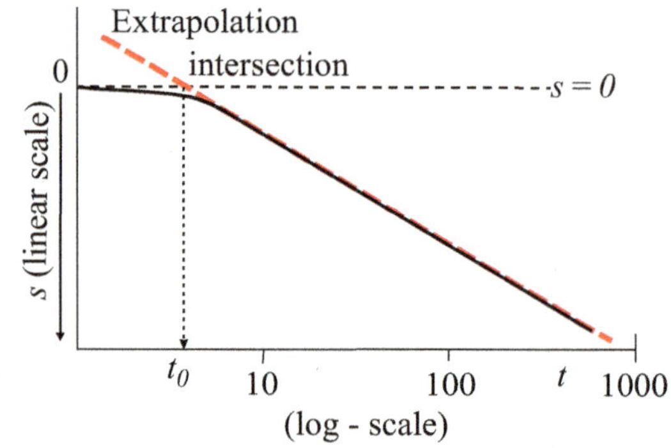

Figure 7.8. A drawdown-time plot on a linear-log scale to illustrate the time at which the drawdown is considered to be virtually zero.

$$\log_{10}\left(\frac{2.25Tt_0}{r^2S}\right) = 0 \tag{7.3.40}$$

which leads to

$$\frac{2.25Tt_0}{r^2S} = 1 \tag{7.3.41}$$

and

$$S = \frac{2.25Tt_0}{r^2} \tag{7.3.42}$$

which yields an estimate of the storage coefficient of the aquifer. In addition to estimating the storage coefficient, the approach suggests that the arrival of a noticeable drawdown at the observation well at this early time is mainly controlled by the ratio of the storage coefficient to the transmissivity (the inverse of the water diffusivity, $D = T/S$). That is,

$$t_0 = \frac{Sr^2}{2.25T} \tag{7.3.43}$$

The differences in drawdowns recorded at different times at late time (i.e., after t_0) are mainly controlled by the transmissivity, as illustrated in equation (7.3.38).

B) Straight-line plot of *s* versus log *r*. Cooper-Jacob's approximation can also be used with drawdown-distance plots. Consider two observation wells at distances r_1 and r_2 from the pumping well and drawdowns at the two wells of s_1 and s_2, respectively. According to the first-order approximation of the Theis solution, the drawdowns can be written as

$$s(r_1,t) = \frac{Q}{4\pi T}\left(-0.5772 - \ln(u_1)\right)$$
$$\text{and } s(r_2,t) = \frac{Q}{4\pi T}\left(-0.5772 - \ln(u_2)\right) \tag{7.3.44}$$

The difference of the two drawdowns then becomes

$$s(r_1, t) - s(r_2, t) = \frac{Q}{4\pi T} \ln\left(\frac{u_2}{u_1}\right) \tag{7.3.45}$$

which leads to the following expression

$$\Delta s = \frac{Q}{4\pi T} \ln \frac{\left(\frac{r_2^2 S}{4Tt}\right)}{\left(\frac{r_1^2 S}{4Tt}\right)} = \frac{Q}{4\pi T} \ln\left(\frac{r_2^2}{r_1^2}\right) \tag{7.3.46}$$

Thus,

$$\Delta s = \frac{Q}{2\pi T} \ln\left(\frac{r_2}{r_1}\right) \tag{7.3.47}$$

Please note that Δs here is defined as $s(r_1, t) - s(r_2, t)$. Furthermore, after expressing this equation in terms of log base 10, we have

$$\Delta s = \frac{2.303Q}{2\pi T} \log_{10}\left(\frac{r_2}{r_1}\right) \tag{7.3.48}$$

If we find the difference in drawdown between two points, separated by 1 log cycle, we have

$$\Delta s = \frac{2.303Q}{2\pi T} \tag{7.3.49}$$

Therefore, the estimate of the transmissivity is

$$T = \frac{2.303Q}{2\pi \Delta s} \tag{7.3.50}$$

To determine the storage coefficient, we extrapolate the straight line to intercept the log r axis (see Figure 7.9). Now, we know that at r_0 the drawdown (s) is zero. According to the linear approximate solution, we have

$$0.5772 = -\ln(u) = \ln(u)^{-1} = \ln\left(\frac{1}{u}\right) \tag{7.3.51}$$

Substitution of the expression for u into equation (7.3.51) leads to

$$\frac{4Tt}{r_0^2 S} = e^{0.5772} = 1.78 \tag{7.3.52}$$

Therefore, the estimate of S is given as

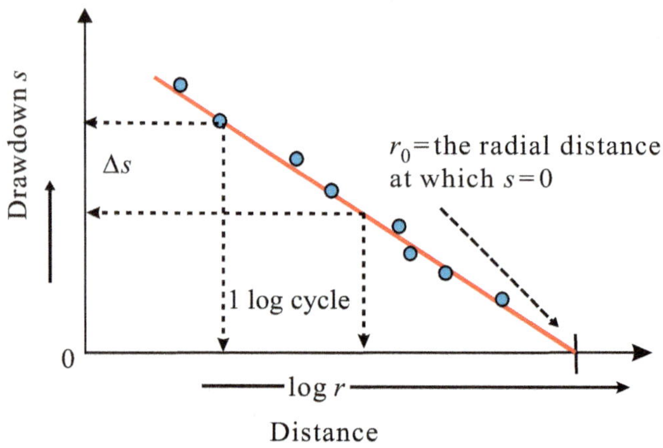

Figure 7.9. A drawdown-distance plot on a linear-log scale to illustrate the distance at which the drawdown is zero.

$$S = \frac{2.25Tt}{r_0^2} \tag{7.3.53}$$

7.3.2 Homogeneous and Anisotropic Aquifers

7.3.2.1 Prediction of Drawdowns

Next, we examine the analytical solution for flow to a well in a homogeneous and anisotropic aquifer. The governing partial differential equation for transient radial flow to a well in a homogeneous and anisotropic, two-dimensional, horizontal aquifer can be written as

$$T_x \frac{\partial^2 s}{\partial x^2} + 2T_{xy} \frac{\partial^2 s}{\partial x \partial y} + T_y \frac{\partial^2 s}{\partial y^2} = S \frac{\partial s}{\partial t} \tag{7.3.54}$$

The preceding equation is written for the scenario in which the principal directions of the transmissivity are not aligned with the x-y coordinates of the governing equation. To derive an analytical solution to this equation, we first carry out a coordinate transformation such that the x and y axes align with the principal directions of the transmissivity, indicated by the new coordinates x' and y'.

$$T_{x'} \frac{\partial^2 s}{\partial x'^2} + T_{y'} \frac{\partial^2 s}{\partial y'^2} = S \frac{\partial s}{\partial t} \tag{7.3.55}$$

Now, we can convert the equation to the one for an isotropic aquifer by scaling the axes.

$$x'' = \sqrt{T_{y'}}x' \text{ and } y'' = \sqrt{T_{x'}}y' \tag{7.3.56}$$

Then the governing equation becomes

$$(T_{x'}T_{y'})\left(\frac{\partial^2 s}{\partial x''^2}+\frac{\partial^2 s}{\partial y''^2}\right)=S\frac{\partial s}{\partial t} \quad \text{or}$$
$$(\sqrt{T_{x'}T_{y'}})\left(\frac{\partial^2 s}{\partial x''^2}+\frac{\partial^2 s}{\partial y''^2}\right)=\left(\frac{S}{\sqrt{T_{x'}T_{y'}}}\right)\frac{\partial s}{\partial t} \tag{7.3.57}$$

which can be rewritten as

$$T_e\left(\frac{\partial^2 s}{\partial x''^2}+\frac{\partial^2 s}{\partial y''^2}\right)=S_e\frac{\partial s}{\partial t} \tag{7.3.58}$$

where

$$T_e=\sqrt{T_{x'}T_{y'}} \quad \text{and} \quad S_e=\frac{S}{\sqrt{T_{x'}T_{y'}}}=\frac{S}{T_e} \tag{7.3.59}$$

Now, the aquifer is isotropic in the x'' and y'' coordinate system. If we change the equation to a radial coordinate system (also known as a polar coordinate system), we then have

$$\frac{\partial^2 s}{\partial r_e^2}+\frac{1}{r_e}\frac{\partial s}{\partial r_e}=\frac{S_e}{T_e}\frac{\partial s}{\partial t} \tag{7.3.60}$$

where

$$r_e^2=x''^2+y''^2=T_{y'}x'^2+T_{x'}y'^2 \tag{7.3.61}$$

Using the coordinate transformation rule,

$$\begin{aligned} x' &= x\cos\theta+y\sin\theta \\ y' &= -x\sin\theta+y\cos\theta \end{aligned} \tag{7.3.62}$$

we have

$$r_e^2=T_{y'}(x\cos\theta+y\sin\theta)^2+T_{x'}(-x\sin\theta+y\cos\theta)^2 \tag{7.3.63}$$

And

$$r_e^2=T_xy^2+T_yx^2-2T_{xy}xy \tag{7.3.64}$$

$$T_e=\sqrt{T_{x'}T_{y'}} \quad \text{or} \quad T_e^2=T_{x'}T_{y'} \tag{7.3.65}$$

recall

$$T_{x'} = \frac{T_x + T_y}{2} + \left[\left(\frac{T_x - T_y}{2}\right)^2 + T_{xy}^2\right]^{\frac{1}{2}}$$
$$T_{y'} = \frac{T_x + T_y}{2} - \left[\left(\frac{T_x - T_y}{2}\right)^2 + T_{xy}^2\right]^{\frac{1}{2}} \tag{7.3.66}$$

$$T_e^2 = T_x T_y - T_{xy}^2 \tag{7.3.67}$$

Solution to equation (7.3.58) takes the form of the Theis solution

$$s(r_e, t) = \frac{Q}{4\pi T_e} W(u_e) \tag{7.3.68}$$

where

$$u_e = \frac{r_e^2 S_e}{4T_e t} = \frac{S(T_x y^2 + T_y x^2 - 2T_{xy} xy)}{4\left(T_x T_y - T_{xy}^2\right)t} \tag{7.3.69}$$

While the form of the solution is the same as in the homogeneous and isotropic aquifer case, the flow is no longer axisymmetric. There exist steep gradients in the direction of low transmissivity and gentle gradients in the direction of high transmissivity.

7.3.2.2 Aquifer Tests and Parameter Estimation

For estimating the hydraulic properties of an anisotropic aquifer, one must recognize that there are two unknown transmissivity parameters (T_x and T_y), if the principal directions of transmissivity are known, and there are three unknown parameters (T_x, T_y, and T_{xy}), otherwise. This implies that three hydrographs observed at three different directions are needed in order to determine transmissivity, since principal directions are generally unknown. Similarly, one can use either type curve matching or Cooper-Jacob's straight-line approach to estimate the properties as one can for isotropic aquifers.

1) Type Curve Matching This approach requires plotting drawdown-time data of three observation wells (Wells 1, 2, and 3) on a log-log paper. The coordinates of the three wells are (x_1, y_1), (x_2, y_2), and (x_3, y_3), respectively. As we have shown previously, the solution to the governing equation for an anisotropic aquifer, after some transformation, takes the same form as that of the equation for an isotropic aquifer.

$$s(r_e, t) = \frac{Q}{4\pi T_e} W(u_e) \tag{7.3.70}$$

This implies that one can match the drawdown-time curve with the Thies solution to obtain the dimensionless drawdown and time in terms of transformed variables. That is, we can find the match points (s_1, t_1) and $(u_{e_1}, W(u_{e_1}))$, (s_2, t_2) and $(u_{e_2}, W(u_{e_2}))$, and (s_3, t_3) and $(u_{e_3}, W(u_{e_3}))$ for the observed hydrographs at wells 1, 2, and 3, respectively.

Subsequently, using each matching point, we can obtain a T_e value for each hydrograph using the following equation:

$$T_e = \frac{Q}{4\pi s(r_e, t)} W(u_e) \tag{7.3.71}$$

These three T_e values, in theory, should be identical if the medium is homogeneous. If not, an average of the three can be used as an estimate.

According to the previously employed transformation procedure and equation (7.3.59), we have

$$u_e = \frac{r_e^2 S_e}{4T_e t} = \frac{r_e^2}{4t} \frac{S}{T_e^2} \tag{7.3.72}$$

which leads to

$$r_e^2 S = 4tT_e^2 u_e \tag{7.3.73}$$

Substitution of the expression for r_e^2 from equation (7.3.64) into the preceding equation yields

$$S\left(T_x y^2 + T_y x^2 - 2T_{xy} xy\right) = 4u_e t T_e^2 \tag{7.3.74}$$

Since we have three hydrographs from three wells, and we have already used them to obtain values of $W(u_{e1})$, $W(u_{e2})$, and $W(u_{e3})$ for each hydrograph from the type curve matching procedure, we can now form a system of equations that consists of three equations with three unknowns. That is,

$$\begin{bmatrix} y_1^2 & x_1^2 & -2x_1y_1 \\ y_2^2 & x_2^2 & -2x_2y_2 \\ y_3^2 & x_3^2 & -2x_3y_3 \end{bmatrix} \begin{bmatrix} ST_x \\ ST_y \\ ST_{xy} \end{bmatrix} = \begin{bmatrix} 4u_{e1}t_1T_e^2 \\ 4u_{e2}t_2T_e^2 \\ 4u_{e3}t_3T_e^2 \end{bmatrix} \tag{7.3.75}$$

The left-hand side of equation (7.3.75) involves the known coordinates of the wells, and the right-hand side of the equation is known from the type curve matching. Thus, this system of equations can be solved for ST_x, ST_y, and ST_{xy}. Afterward, using the relationship in equation (7.3.67), we solve for S by using the following formula:

$$S^2 = \left(ST_x ST_y - S^2 T_{xy}^2\right) / T_e^2 \tag{7.3.76}$$

Once the estimate of S is obtained, T_x, T_y, and T_{xy} can be determined.

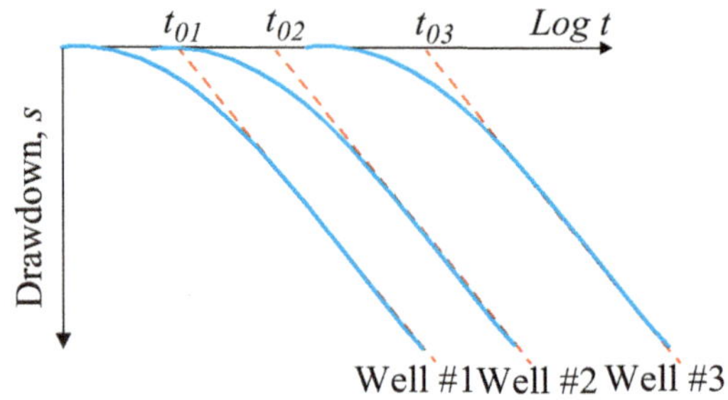

Figure 7.10. Illustration of drawdown-time curves and corresponding Cooper-Jacob's straight line approximations at three observation wells in an anisotropic aquifer.

2) Cooper-Jacob's straight line method Of course, Cooper-Jacob's approximation approach can be applied here to estimate properties for an anisotropic aquifer. However, to do so, again three hydrographs from three different wells are needed. We then plot them on a semilog paper as shown in Figure 7.10.

The Cooper-Jacob approximation corresponding to equation (7.3.70) is

$$s(x, y, t) \cong \frac{2.303Q}{4\pi T_e} \log_{10}\left[\frac{2.25T_e^2 t}{r_e^2 S}\right] \tag{7.3.77}$$

Because $r_e^2 = \left(T_x y^2 + T_y x^2 - 2T_{xy} xy\right)$, equation (7.3.77) can be rewritten as

$$s(x, y, t) \cong \frac{2.303Q}{4\pi T_e} \log_{10}\left[\frac{2.25T_e^2 t}{\left(T_x y^2 + T_y x^2 - 2T_{xy} xy\right)S}\right] \tag{7.3.78}$$

To estimate the properties of an anisotropic aquifer using equation (7.3.78), we need to estimate T_e from the *s–log t* plots. This procedure is the same as that for an isotropic aquifer. That is,

$$T_e = \frac{0.183Q}{\Delta s} \tag{7.3.79}$$

where Δs is the drawdown difference over one log cycle on the time axis. Again, in theory, the estimate of T_e should be identical for the observation wells.

During the next step, we find t_0 (where $s = 0$) for each hydrograph by extrapolation, as we did for an isotropic aquifer. Applying these values to Cooper-Jacob's approximate solution,

$$s(x, y, t_0) = 0 = \log_{10}\left[\frac{2.25T_e^2 t_0}{\left(T_x y^2 + T_y x^2 - 2T_{xy} xy\right)S}\right] \tag{7.3.80}$$

we then have three equations for the three hydrographs:

$$\begin{aligned}(T_x y_1^2 + T_y x_1^2 - 2T_{xy} x_1 y_1)S &= 2.25T_e^2 t_{01} \\ (T_x y_2^2 + T_y x_2^2 - 2T_{xy} x_2 y_2)S &= 2.25T_e^2 t_{02} \\ (T_x y_3^2 + T_y x_3^2 - 2T_{xy} x_3 y_3)S &= 2.25T_e^2 t_{03}\end{aligned} \tag{7.3.81}$$

where subscripts 1, 2, and 3 denote the well number associated with each hydrograph. We then solve equation (7.3.81) for ST_x, ST_y, and ST_{xy}. We next use the following equation to determine the value of S.

$$S^2 = \left(ST_x ST_y - S^2 T_{xy}^2\right)/T_e^2 \tag{7.3.82}$$

Then, T_x, T_y, and T_{xy} can be determined.

While these analytic approaches for aquifer analysis in anisotropic aquifers are appealing, they often yield unreasonable values of T_x, T_y, T_{xy}, and S since identifying the values of T_x, and T_y is inherently difficult in field problems. Although this can be attributed to noise or measurement errors in the drawdown data, the major reason lies in the fact that a hydrograph is a point sample of the response of a heterogeneous aquifer. This hydrograph is not the same as the hydrograph calculated from models that assume homogeneity. In other words, the problem arises from the inconsistency between the scale of the conceptual model and that of head measurement as discussed in Chapters 2, 5, and 6. To alleviate this problem, an optimization approach based on the least squares of the differences between the measured and calculated drawdowns at many observation wells would be most appropriate.

7.3.3 Residual Drawdown of Recovery Tests

Aquifer tests and analyses, discussed previously, focus on the responses of an aquifer subject to pumping water from an aquifer or injection of water into an aquifer. However, as the excitation ceases, the head field of the aquifer recovers. The change of the well hydrograph during the recovery phase is called the **residual drawdown.** The rate and time of recovery of the drawdown in the aquifer certainly depend on the hydraulic properties of the aquifer, and the size of the cone of depression created before the pumping operation ceases.

In this case, numerical models also can be employed to analyze the residual drawdown hydrograph to estimate the hydraulic properties of the aquifer. Now, instead of using that approach, we will employ the Theis solution and a superposition technique to derive an analytical solution for these purposes because of the ingenuity and elegance of this approach.

Since the Theis solution assumes continuous pumping or injection in time, it certainly cannot be applied directly to the recovery test, in which the excitation lasts for only a period of time. To apply the Theis solution to this situation, a simple but

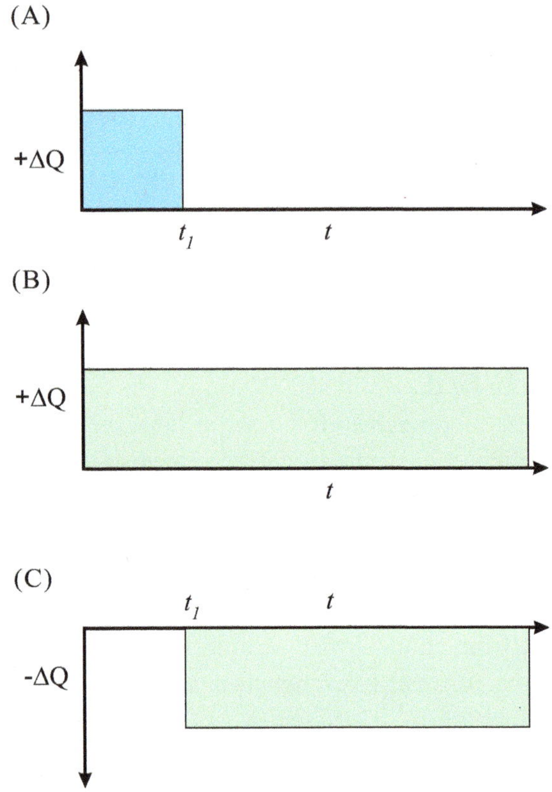

Figure 7.11. Schematic illustration of the simple fact that a period of discharge from $t = 0$ to t_1 (A) can be derived by superposition of a continuous discharge (B) and a continuous injection (C) starting at t_1.

elegant trick has been employed. The trick is to replace a step (pumping over a finite amount of time) pumping operation

$$Q = \begin{cases} \Delta Q & if \;\; 0 \leq t \leq t_1 \\ 0 & if \;\; t \geq t_1 \end{cases} \tag{7.3.83}$$

by a sum of a continuous and constant pumping operation ($Q = \Delta Q$ for $t \geq 0$) and a continuous and constant injection operation, $Q_1 = -\Delta Q$ at $t \geq t_1$. These two operations are offset by a period of t_1 as illustrated in Figure 7.11.

Since the drawdown, due to the continuous pumping, and the buildup, due to the continuous injection, can be calculated by the Theis solution individually, the residual drawdown then is modeled as the sum of the Theis solution for the pumping, $s(r, t)$, and that for the injection, $s_i(r, t - t_1)$. That is,

$$s_r(r,t) = s(r,t) + s_i(r,t-t_1) \tag{7.3.84}$$

or

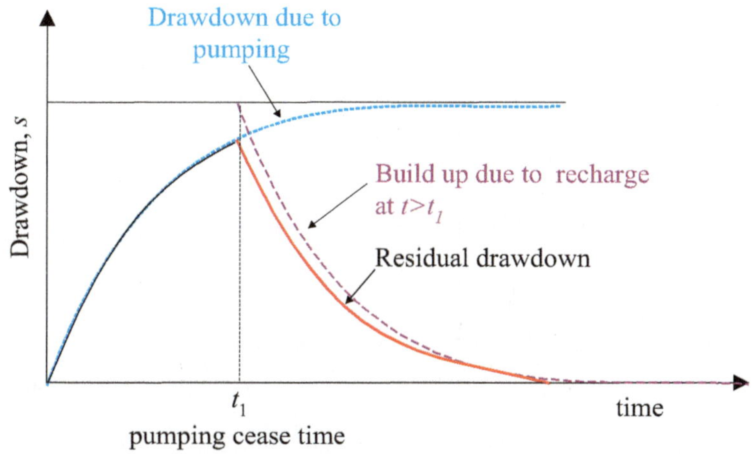

Figure 7.12. Illustration of the superposition approach for deriving the residual drawdown.

$$s_r(r,t) = \frac{\Delta Q}{4\pi T}\left[W\left(\frac{r^2 S}{4Tt}\right) - W\left(\frac{r^2 S}{4T(t-t_1)}\right)\right] \tag{7.3.85}$$

The second term on the right-hand side of the equation will become effective only when $t > t_1$.

The short dashed blue line represents drawdown behavior corresponding to the pumping event (B) and the long-dashed purple line denotes the buildup due to the injection event (C) in Figure 7.12. The residual drawdown is depicted by the solid red line.

Suppose that $u < 0.01$; then equation (7.3.85) can be approximated using Cooper-Jacob's approximation:

$$\begin{aligned} s_r &= \frac{\Delta Q}{4\pi T}\left(\ln\frac{2.25Tt}{r^2 S} - \ln\frac{2.25T(t-t_1)}{r^2 S}\right) \\ &= \frac{\Delta Q}{4\pi T}\ln\left(\frac{t}{t-t_1}\right) = \frac{2.303\Delta Q}{4\pi T}\log_{10}\left(\frac{t}{t-t_1}\right) \qquad \text{where } t > t_1 \end{aligned} \tag{7.3.86}$$

Equation (7.3.86) implies that if the residual drawdown during the recovery is plotted as a function of the log of time, the residual drawdown versus the log of time relationship after the recovery begins is linear, as illustrated in Figure 7.13.

As a consequence, we can choose the drawdown difference of one log-time cycle and estimate the transmissivity:

$$T = \frac{2.303Q}{4\pi \Delta s_r} \tag{7.3.87}$$

Then, the storage coefficient, S, can be determined from the drawdown at the time of the shutdown of the pumping using the Theis curve approach. That is,

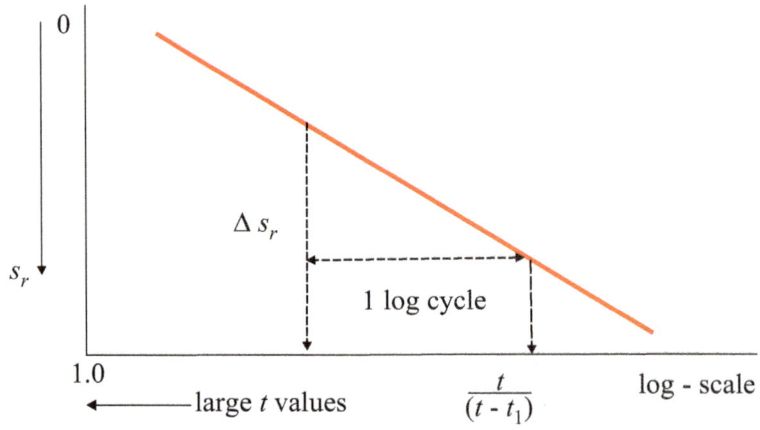

Figure 7.13. Plot of residual drawdowns as a function of log time after the pumping ceases.

$$W(u_{t1}) = \frac{4\pi T}{Q} s(r, t_1) \tag{7.3.88}$$

Then, using a $W(u)$ versus u table to find u, u for the corresponding value of $W(u)$ can be estimated. Once u is known, S can be determined as

$$u = \frac{r^2 S}{4Tt_1} \quad \Rightarrow \quad S = \frac{4Tt_1 u}{r^2} \tag{7.3.89}$$

where r can be r_w, the well radius.

7.3.4 Effects of Partial Penetration of the Pumping Well

Effects of partial penetration of a pumping well can be easily understood if we examine Figure 7.14. In an idealized confined aquifer, a partially penetrating pumping well will yield a three-dimensional converging flow field as opposed to a two-dimensional flow field. Since the partially penetrating well will produce vertically upward flow components in the vicinity of the well, the flow paths must be longer than those in the fully penetrating well, which produces essentially a two-dimensional horizontal flow regime. Therefore, we can argue that the average drawdown in a fully penetrating observation well, due to pumping from a partially penetrating well, will be more than that in the case of a fully penetrating pumping well, if both wells discharge the same amount of water. The effect of the partially penetrating well will be small if the aquifer is highly anisotropic. To minimize the effect of the partial penetration, the observation well should be installed at a distance far away from the pumping well.

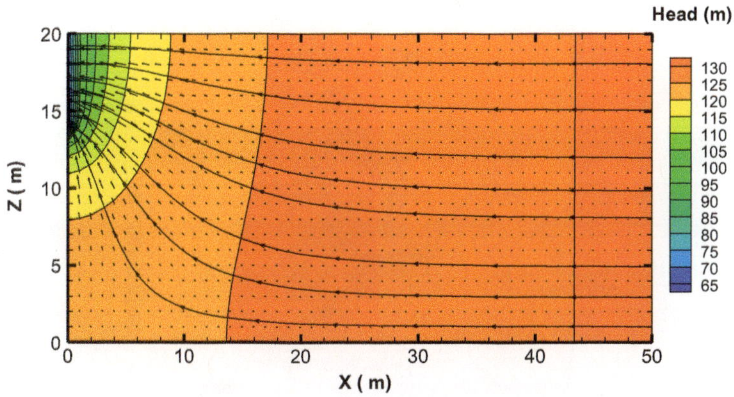

Figure 7.14. Equipotential lines and streamlines created by a partial penetrating pumping well.

7.4 Limitations of the Conventional Aquifer Tests in a Heterogeneous Aquifer

Aquifers are inherently heterogeneous at a multiplicity of scales. The Theis solution, on the other hand, assumes aquifer homogeneity, horizontal flow, a fully penetrating and perforated well, and many other characteristics for mathematical convenience. The Theis solution, in essence, lumps a three-dimensional flow process into a two-dimensional, depth-averaged flow and then adopts the ensemble REV concept to justify the homogeneity assumption. The Theis solution thus predicts some ensemble-averaged drawdown distribution (see Chapter 6), which could be quite different from the observed head field or drawdown. For example, Figure 7.15 shows an irregular drawdown distribution (colored contour) induced by pumping at a well in a heterogeneous aquifer. It also shows a concentric cone of depression derived from the Theis solution, which assumes aquifer homogeneity. The two drawdown distributions are different, and the drawdown values at the same observation location should not be the same. Therefore, one should only expect that the predicted drawdown distribution based on the Theis solution fits the observed drawdown distribution in the mean square error (or spatial average) sense. Mathematically, this implies the following objective function should be met:

$$\sum_{i=1}^{n}[s^*(r_i,t)-s(r_i,t)]^2 = \min \tag{7.3.90}$$

Specifically, at a given time, t, the difference between the drawdown predicted by the Theis solution $s(r_i, t)$ and the observed drawdown $s^*(r_i, t)$ everywhere (for all r_i s) in the aquifer should be minimal. This suggests that a drawdown-distance analysis (Section 7.3.1.2) for estimating aquifer properties is consistent with the Theis theory. However, a large number of observation wells is required to conduct this analysis

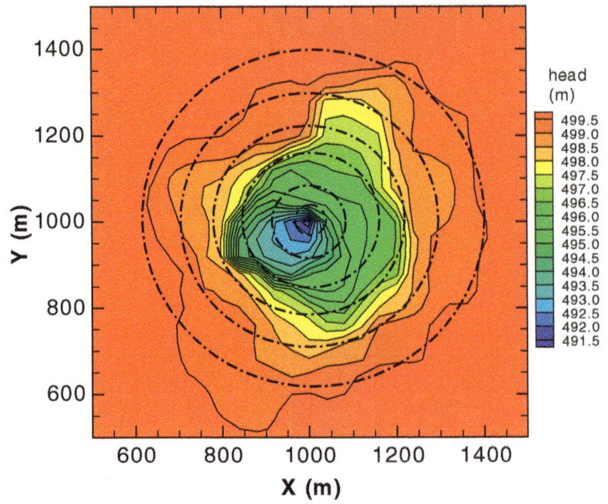

Figure 7.15. Contours of the simulated cone of depression (head in meters) due to pumping at the center of a heterogeneous aquifer at some given time, and the concentric cone of depression calculated by using the Theis solution (dashed-dot black line), assuming aquifer homogeneity.

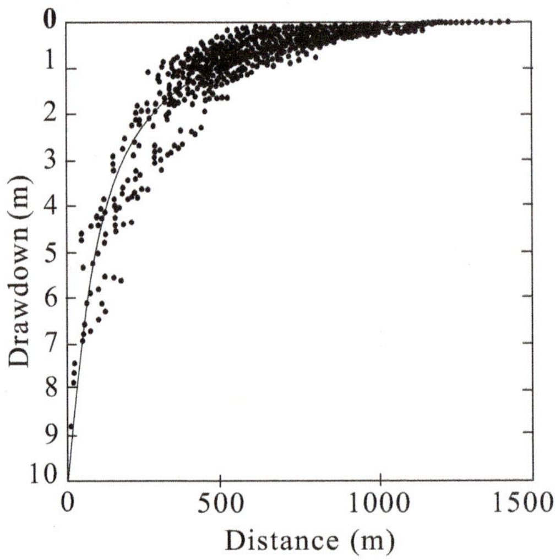

Figure 7.16. A drawdown-distance plot using many observation wells during a pumping test in a synthetic heterogeneous aquifer. Units are meters and minutes.

properly (see Figure 7.16) such that the spatial average of the drawdowns approaches the ensemble average. Then, the question is how many wells are needed and how long the pumping test has to last such that the cone of depression encounters enough heterogeneity. On the other hand, one may argue that a heterogeneous conceptual model approach, which attempts to delineate aquifer heterogeneity, may be more appropriate if there is a large number of observation wells (Wu et al., 2005; Wen et al., 2010).

Most field situations often have only one pumping well and one observation well, and aquifers are heterogeneous. Therefore, it is likely that a pumping well may be located in a highly permeable zone of an aquifer, while an observation well is located in the low permeable zone. It is then logical to question the common practice, which fits the observed drawdown in a heterogeneous aquifer to the Theis solution to estimate T and S of the aquifer. Does the common practice compare apples and oranges? More importantly, we must ask what these estimated T and S represent. Do they represent averaged hydraulic properties over the cone of depression or something else?

This question and other related questions will be discussed using the cross-correlation between the observed head perturbation (h) at the observation well and the T (or ln$T$) variation (y) at every part of the aquifer, ρ_{hy}, and the cross-correlation between the observed head perturbation (h) at the observation well and the S (or ln$S$) variation (z) at every part of the aquifer, ρ_{hz}. As stated in Sun et al. (2013), a cross-correlation contour map shows that the degree of the head perturbation at the observation location at a given time is influenced by the ln$T$ or ln$S$ perturbation at every location of the aquifer as does a sensitivity map but in a statistical sense. The cross-correlation value ranges between +1 (fully positively correlated) to −1 (fully negatively correlated). Mathematical formation of the cross-correlation is given in Section 9.3 of Chapter 9.

Contour maps of the cross-correlation values between the head at an observation well and T perturbation (i.e., ρ_{hy}) everywhere in a 2-D horizontal and heterogeneous aquifer and the head field during a pumping test are plotted in Figures 7.17a and 7.17c for two early times (i.e., dimensionless time $t^* = 0.4$ and 2.2), and Figure 7.18a and 7.18c for late time and steady state (i.e., $t^* = 6.2$ and 41.4), respectively. The dimensionless time t^*, which equals $1/u$, was evaluated by using the mean values of T and S of the aquifer. The corresponding cross-correlation maps for the head and the S field (ρ_{hz}) are shown in Figures 7.17b and 7.17d for the early times and 7.18b and 7.18d for late times. Note that the head fields are the mean heads, which were calculated using the mean T and S values of the heterogeneous aquifer.

At a very early time ($t^* = 0.4$, $t = 200$ s), the ρ_{hy} values are significantly negative over a large area and have the highest correlation (i.e., around −0.6) between the pumping well and observation well (Figure 7.17a). Meanwhile, the cross-correlation ρ_{hz} ranges from 0.1 to 0.4 (Figure 7.17b). The high positive values (i.e., around 0.4) are confined to the small circular area between the pumping well and the observation well. Within the aquifer domain, the nonzero velocities at this very early time are limited to the vicinity of the pumping well and the drawdown has not propagated to the observation well yet. As a result, the high cross-correlation at this time stage is not useful for relating the hydraulic properties to the head at the observation well and beyond.

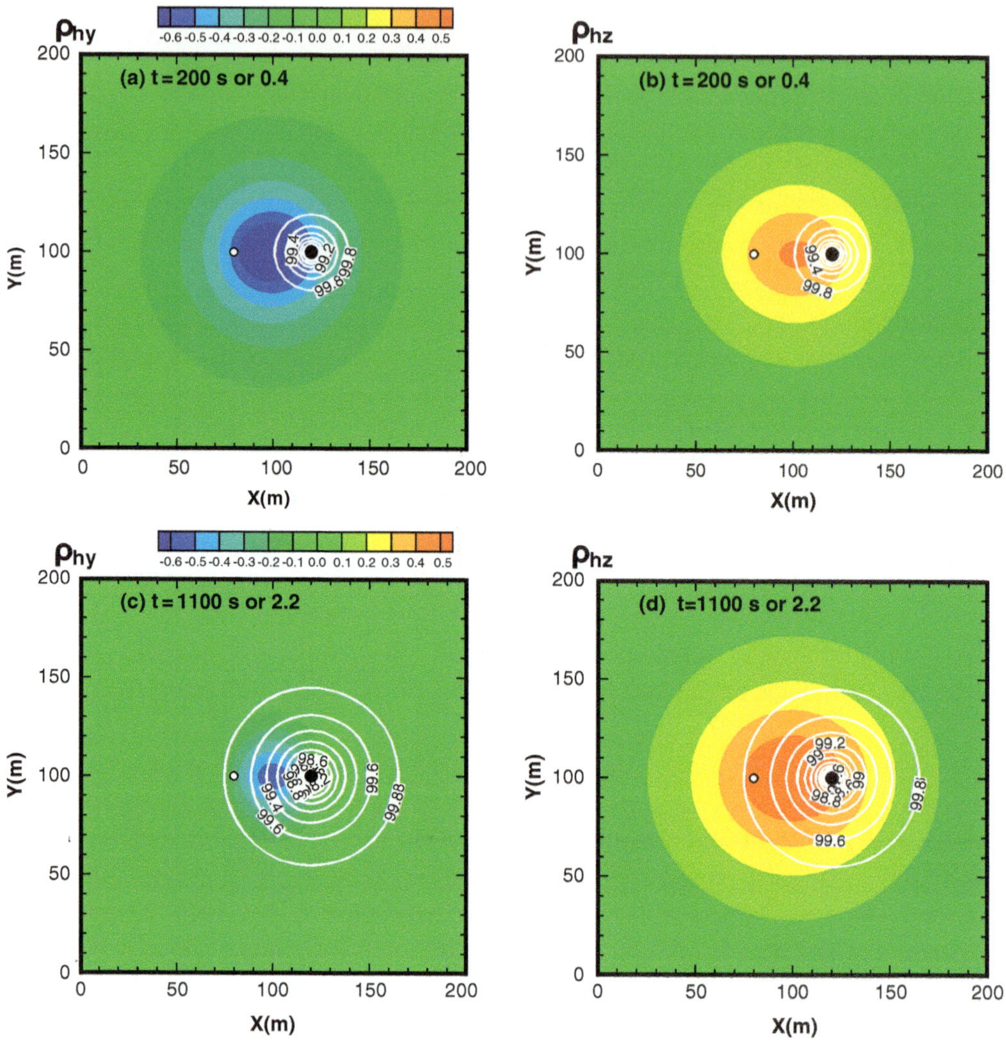

Figure 7.17. a) and c) Contour maps of cross-correlation between the head at the observation well (the white circle) and ln$T$ (ρ_{hy}) everywhere in the aquifer at early dimensionless time 0.4, and 2.2 after pumping started at the pumping well (black circle). b) and d) Cross-correlation maps between head and ln$S$ (ρ_{hz}) for the two periods. The aquifer is surrounded by constant heads of 100 m. White contours are equipotential lines in meters. (Sun et al., 2013)

At an early time $t^* = 2.2$ or $t = 1100$s (approximately equal to t_0 time), the noticeable drawdown has expanded to the observation well, but only to some limited areal extents (Figures 7.17c and 7.17d). The ρ_{hy} values are close to zero everywhere except the small circular area centered between the pumping well and the observation well, where the values range from −0.5 to 0 (Figure 7.17c). The negative correlation means that if the head at the observation well in a heterogeneous aquifer is higher than the head calculated from the mean T and S values of the aquifer, the T values downstream of the observation are likely to be lower than the mean T value. Meanwhile, ρ_{hz} the values are ranging from 0.1 to 0.4 over the area, with the highest correlation in a small circular area between the pumping well and the observation well (Figure 7.17d). The positive cross-correlation implies that if the observed head in a heterogeneous aquifer is high in comparison to the head derived from the mean T and S of the aquifer, the S values in the area between the pumping well and the observation well are likely to be higher than the mean S value. Notice that the area of significant ρ_{hz} values is much larger than that of ρ_{hy}. These high cross-correlation values suggest that the head at the observation well at this time is largely influenced by the S heterogeneity within the small red circle with ρ_{hz} = 0.4 enclosing the pumping and observation wells, and the area outside the circle also has some influence.

At t^* = 6.2 (intermediate time, t = 3000 s), as shown in Figures 7.18a and 7.18b, the edge of the cone of depression has reached the right-hand side of the constant head boundary. The shapes of ρ_{hz} remain similar to that at t^* = 2.2 but the values drop below 0.4 (Figure 7.18b), indicating that the effect of storage coefficient on head is diminishing and flow is moving toward a steady-state condition. At this time, ρ_{hy} has evolved to positive values everywhere (see Figure 7.18a). The cross-correlation pattern forms two kidney-shaped humps: one at the upstream side of the observation well (i.e., from the left-hand side boundary to the observation well) and the other at the upstream side of the pumping well (i.e., from the right-hand side boundary to the pumping well). The values in the area between the two wells are smaller.

When the flow reaches the steady state (t^* = 41.4 or t = 20,000s), the shape of the spatial distribution of ρ_{hy} (Figure 7.18c) is similar to that at t^* = 6.2 (Figure 7.18a), but their magnitudes everywhere in the aquifer are elevated. Meanwhile, the ρ_{hz} values in every part of the aquifer diminish to zero, reflecting the fact that the steady hydraulic head is not related to S (Figure 7.18d).

The cross-correlation patterns in these figures at different stages of the pumping test suggest that by using the drawdown early, the effective T and S estimated from the Theis solution likely reflect the geology between the pumping and the observation well.

As the time progresses, the effective T estimate becomes an average of the heterogeneities within the cone of depression, which is heavily influenced by the heterogeneity of the two kidney-shaped regions upstream of the observation and the pumping wells.

Overall, the estimates of the effective T and S not only evolve with time, but also change with the locations of the pumping and observation wells, implying that they are scenario-dependent parameters or their values change with flow fields.

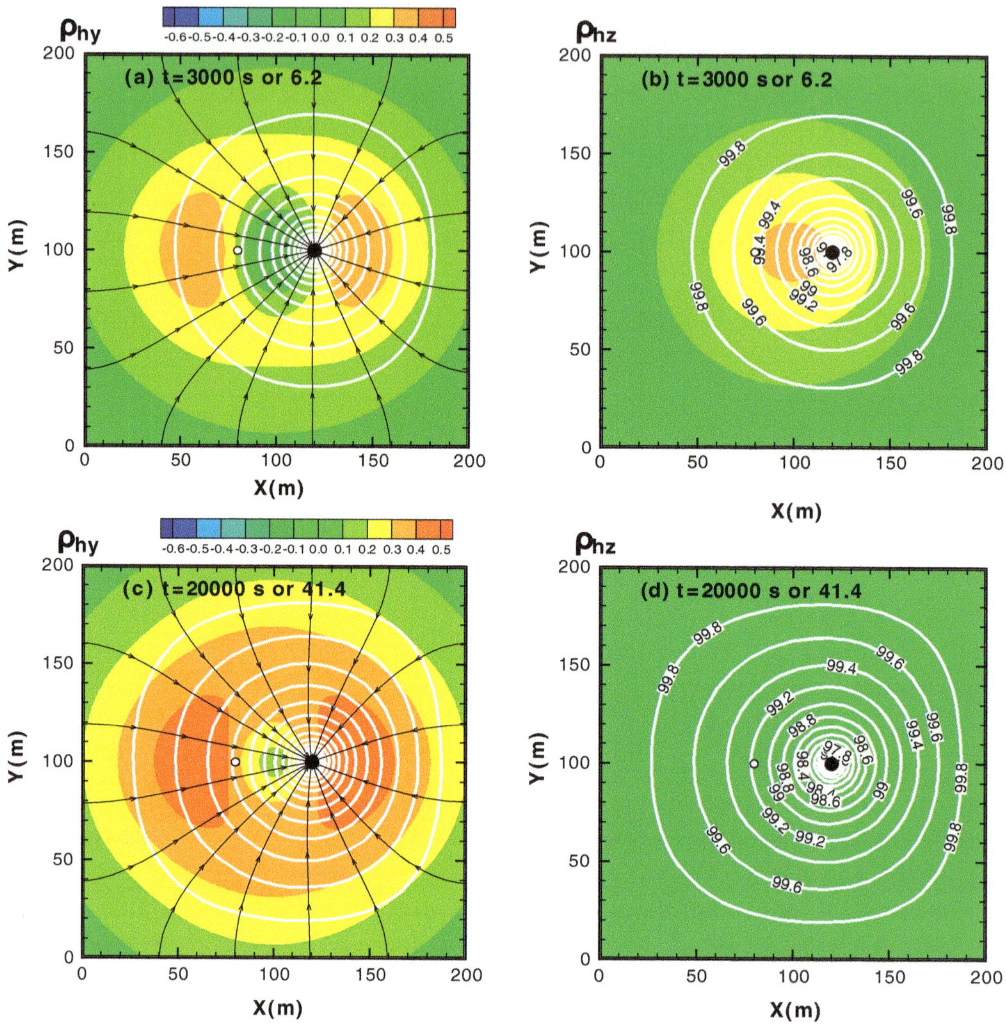

Figure 7.18. a) and c) Contour maps of cross-correlation between the head at the observation well (the white circle) and ln*T* (ρ_{hy}) everywhere in the aquifer at late dimensionless time 6.2 and 41.4 after pumping started at the pumping well (the black circle). b) and d) Cross-correlation maps between head and ln*S* (ρ_{hz}) for the two periods. The white contours represent the head distribution and black arrow lines are streamlines. (Sun et al., 2013)

Wu et al. (2005) showed that at a large time, the effective *T* estimated for an equivalent homogeneous aquifer approaches the geometric mean of all *T*s of statistically homogeneous aquifers (see Chapter 4), and the effective *S* is close to the arithmetic mean of all *S*s, although these estimates are still scenario-dependent (i.e., varying with the pumping location in this case). Analysis of field pumping test data by Straface et al. (2007), Huang et al. (2011), and Wen et al. (2010) corroborates these findings. Wen et al. (2010) further showed that the principal directions and magnitudes of anisotropy of the effective T also evolve with time and stabilize at very large times.

7.5 Summary

- The homogeneity assumption leads to an axisymmetric flow field toward or away from a well. Thus, application of the polar coordinate transformation to a two-dimensional plan-view governing equation transforms it to a one-dimensional equation in the radial coordinate system, for which an analytical solution can be easily derived.
- Anisotropic transmissivity elongates the cone of depression: sparse contours in the direction of high transmissivity, and vice versa.
- Theis solution:

$$s(r,t) = \frac{Q}{4\pi T} W(u), \quad u = \frac{r^2 S}{4Tt}$$

- Cooper-Jacob's approximate solutions:

$$s(r,t) = \frac{Q}{4\pi T} \ln\left[\frac{2.25\,Tt}{r^2 S}\right] = \frac{0.183Q}{T} \log_{10}\left[\frac{2.25\,Tt}{r^2 S}\right]$$

- These two formulae are often used in aquifer test analyses for estimating T and S. Cooper-Jacob's solution shows that the drawdown, at a large time or small u, increases linearly as a function of log time.
- Theis's or Cooper-Jacob's solution predicts the ensemble- (or spatial-) averaged heads or drawdowns, which are likely different from the observed drawdowns at an observation well.
- In order to obtain representative T and S estimates using Theis's or Cooper-Jacob's solution, drawdown-time data from a dense well field are necessary.
- Using drawdown-time data from an observation well induced by pumping at another well in a heterogeneous aquifer, T and S estimates based on the Theis or the Cooper-Jacob approach likely vary with the pumping time. They are scenario-dependent parameters.

7.6 Exercises

1. A nonleaky, isotropic artesian aquifer, with an infinite areal extent, is 100 feet thick. A production well fully penetrating the aquifer was pumped at a constant rate of 1,500 gpm for a period of 1 day. The drawdowns were observed at a distance of 300 feet from the pumping well in a fully penetrating observation well. They are listed in Table 7.1. Compute the transmissivity, hydraulic conductivity, and storage coefficient of the aquifer.
2. The transmissivity and storage coefficient of a nonleaky artesian aquifer with an infinite areal extent are 1×10^5 gpd/ft and 3×10^{-4}, respectively. A fully penetrating production well has been discharging water at a constant rate of 1,000 gpm for a period of 1 year. Compute the drawdowns at a distance of 5,000, 10,000, 15,000 feet from the production well for pumping periods of 10 days, 50 days, 1 year; determine the first arrival times of

Table 7.1. *Temporal drawdown at an observation well located 300 feet from the pumping well*

Time after Pumping Started (min)	**Drawdown (ft)**
1	0.45
2	0.74
3	0.91
4	1.05
5	1.13
6	1.22
7	1.28
8	1.36
9	1.38
10	1.45
21	1.90
30	2.02
40	2.17
50	2.30
60	2.34
70	2.41
80	2.50
90	2.60
100	2.67
200	2.96
300	3.11
400	3.25
500	3.32
600	3.41
700	3.46
800	3.50
900	3.55
1,000	3.60
1,440	3.81

drawdowns (t_0 of Cooper-Jacob's solution) at these locations; and estimate the effective radius of the cone of depression for a pumping period of 1 year.

3. Explain the reason the one-dimensional equation for flow to a well is different from a 1-D equation for uniform flow.
4. What is the difference between the contours of the head distributions resulting from the two flow situations in Problem 3?
5. Derive the dimensionless sensitivity of the drawdown (equation 7.3.22) in an unbounded aquifer with respect to $\ln T$ and $\ln S$. Plot them as a function of dimensionless time ($1/u$). Discuss their implications for parameter estimation from aquifer test data.
6. How would the aquifer heterogeneity affect the approach to estimating anisotropy (Section 7.3.2.2)? Would the estimated anisotropy evolve with time?
7. What are the necessary conditions (e.g., well locations) such that the estimates of transmissivity, anisotropy, and storage coefficient are unique?

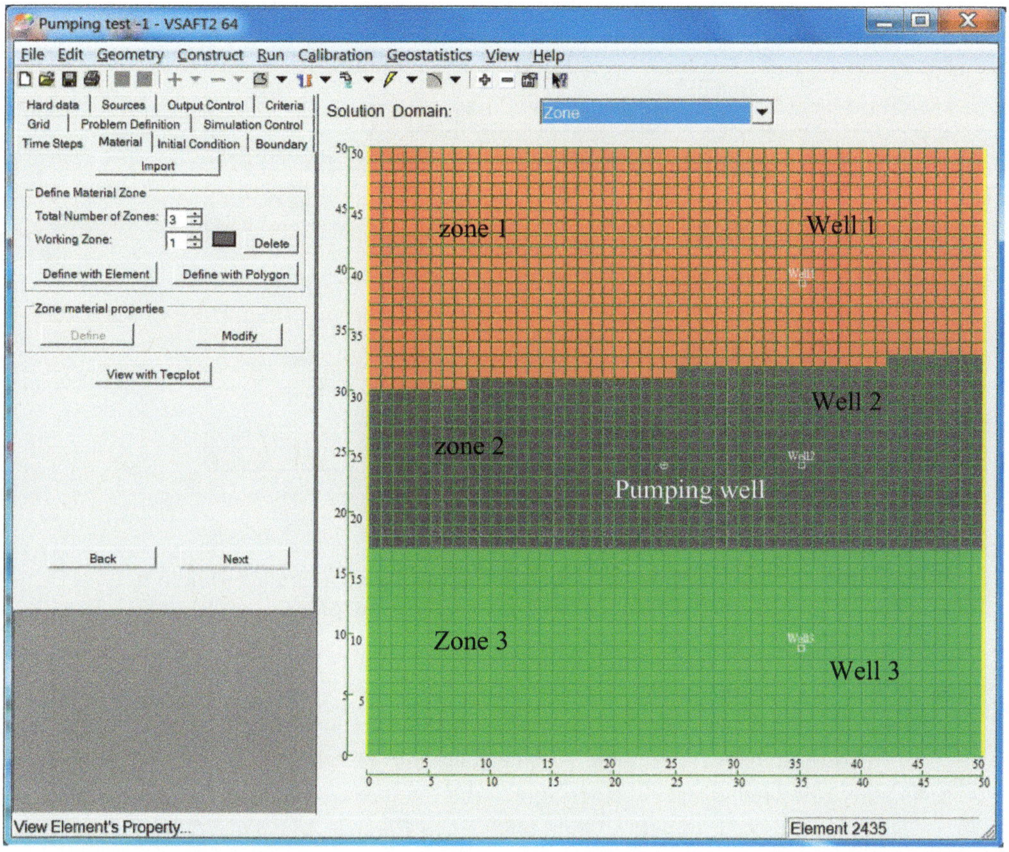

Figure 7.19. The locations of the pumping well and the observation wells in a synthetic aquifer for problem 9.

8. On the basis of Figures 7.17 and 7.18, speculate what the cross-correlation map of head and transmissivity and of head and storage would look like if the distance between the observation well and the pumping well decreased. Also discuss what this new correlation map tells you about the relationship between the aquifer heterogeneity and the T and S estimates based on the head data observed at the pumping well.
9. A program (VSAFT2, available at www.hwr.arizona.edu/yeh/download or contact stochasticfusion@gmail.com) was used to simulate drawdowns due to pumping at a pumping well (x = 24 m, y = 24 m) in a two-dimensional, depth-averaged confined aquifer. The dimension of the aquifer is 50 m × 50 m, and the left-hand side and the right-hand side of the aquifer are constant head of 100 m, and the other two sides are no flow boundaries (Figure 7.19). There are three geologic zones: The T value of zone 1 is 1 m^2/d, zone 2 is 1000 m^2/d, and zone 3 is 10 m^2/d. All three zones are assumed to have the same S values (0.001). The pumping rate is 100 m^3/d. Observed heads as a function of time at observation well 1, 2, and 3 are illustrated in Figure 7.20.

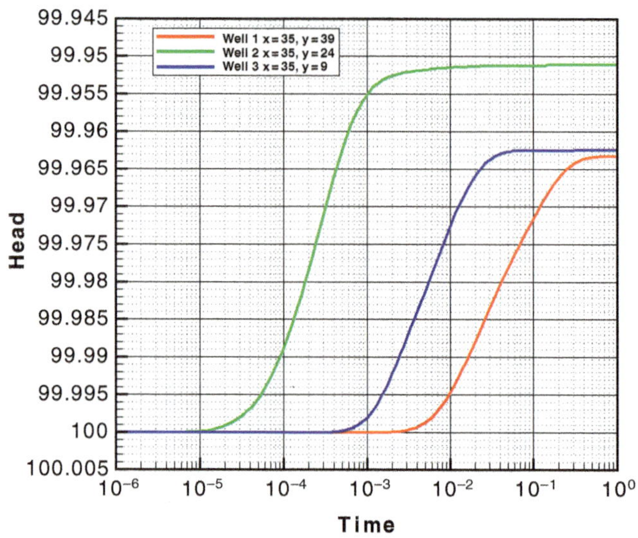

Figure 7.20. Simulated drawdown-time curves in the aquifer for problem 9.

Use Cooper-Jacob's straight-line approach to estimate *T* and *S* values based on the three observed drawdown-log time curves.

a. Are the slopes of the straight-line portion of the drawdown-log time curves the same? In terms of the slopes, estimate *T* value. Can you tell which zone is more permeable?
b. Do the estimated *T* and S values agree with the true ones?
c. Use the results of cross-correlation analysis in Figures 7.17 and 7.18 to discuss your estimated *T* and *S* values. What does this pumping test data set tell us about the aquifer heterogeneity?

8

Flow toward a Well Due to Pumping (Part 2)

8.1 Introduction

This chapter is an extension of the analytical models of radial flow toward a well, introduced in Chapter 7, which assumes that aquifers are confined and unbounded (i.e., infinitely extended aquifers). Here we derive analytical solutions for flow toward a well considering either impermeable or recharge boundaries at some distance from the well. Afterward, we discuss analytical models for leaky confined aquifers and unconfined aquifers. At last, we discuss the necessity of using the variably saturated flow model developed in Chapter 5 for analysis of pumping tests in unconfined aquifers.

8.2 Boundary Effects in Confined Aquifers

Aquifers are generally bounded by geologic features, such as faults, bedrocks of surrounding mountains, and rivers. As a result, effects of such boundaries on the observed well hydrographs are expected during an aquifer test or prediction of groundwater flow. Therefore, a good understanding of these effects is essential for interpretation of any aquifer tests and/or forecasts of water levels and groundwater flow in aquifers.

Rapid advances in numerical modeling techniques, such as three-dimensional finite difference or finite element numerical models, are most suitable for investigating the boundary effects, because they can realistically incorporate the three-dimensional configuration of these geologic boundaries. Nevertheless, the method of images has been employed for decades to implement these boundaries in analytical models for radial flow toward a well in a plan view. It remains an important topic in subsurface hydrology. This is attributed to the fact that the logic and idea behind the method are innovative, elegant, and practical. This section, therefore, aims not only to investigate the effects of geologic boundaries, but also to introduce the method of images.

Geologic or natural boundaries generally are irregular and complex in shape and may exhibit gradual transitions. In order to incorporate such boundaries in any analytical model(s), we will need to simplify their complexity first. Specifically, we will assume that all the boundaries exhibit abrupt or sharp discontinuity, and the boundaries are straight vertical planes, cutting through the aquifer. While these assumptions may be unrealistic, they facilitate analytical solutions, which reveal useful information about general effects of boundaries on the drawdown due to pumping in an aquifer. Two hydrologic boundaries will be examined in this section, namely, 1) impermeable and 2) recharge boundaries.

8.2.1 Impermeable Boundary

This type of boundary represents geologic or hydrologic features that prohibit groundwater flow from entering or leaving an aquifer (such as bedrock or

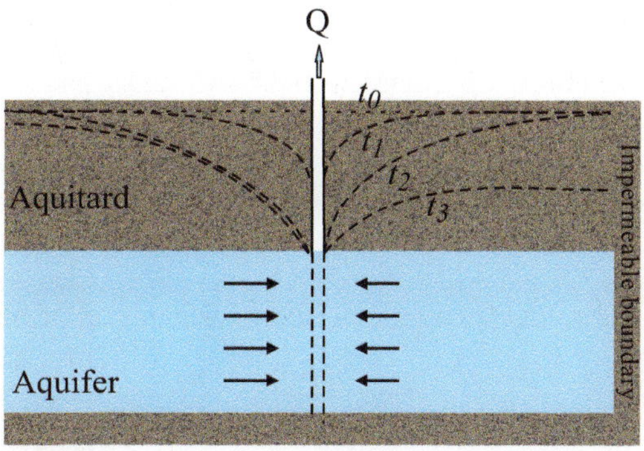

Figure 8.1. A schematic plot of a cross section depicting possible head distribution due to pumping near an impermeable boundary.

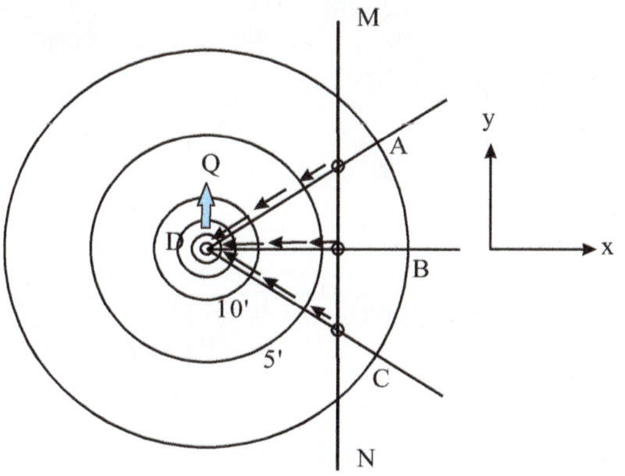

Figure 8.2. Illustration of the plan view of the cone of depression in an aquifer induced by pumping at location D, depicted by the Theis solution. The line M-N represents the location of an impermeable boundary.

groundwater divides) Intuitively, when the cone of depression induced by pumping reaches an impermeable boundary, the drawdown has to increase to reflect the fact that less water is available to replenish the aquifer, as compared with the situation when the impermeable boundary does not exist (see Figure 8.1). While this intuition is correct, quantification of this intuition is a challenge. Many methods can be used to take on this challenge; the method of images, which takes advantage of the Theis solution for unbounded aquifers, will be presented in this chapter.

Before developing the analytical solution for radial flow in a 2-D depth-averaged aquifer with an impermeable boundary, we first explain the rationale of application of the method of images to this problem. Let us start with Figure 8.2, which illustrates the drawdown distribution computed by using the Theis solution (see Chapter 7) at a given time due to pumping at the well located at point D.

Suppose that the line MN in Figure 8.2 represents an impermeable boundary. We first ask, What will be the specific discharge vectors at points A, B, and C for two

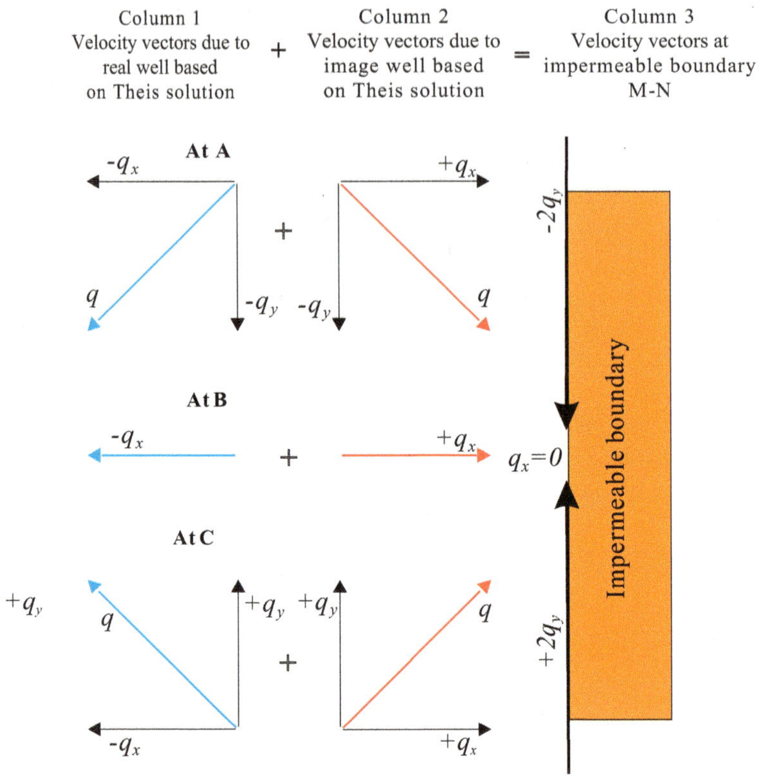

Figure 8.3. Specific discharge vectors at various positions in Figure 8.2 to elucidate the method of images.

possible scenarios: 1) if the impermeable boundary does not exist and 2) if the boundary is present. The answers to these questions are elucidated graphically in Figure 8.3.

In the first column of Figure 8.3, we plot the specific discharge vectors and it's the components in the x and y directions at locations A, B, and C, induced by pumping at point D in an aquifer of infinite lateral extent. In the third column, we plot the specific discharge vectors at the same locations, which now represent the interface between the aquifer and an impermeable boundary, according to principles of groundwater flow. That is, groundwater flow direction along the impermeable boundary must be parallel to this boundary, and no flow exists perpendicular to this boundary.

Suppose that the specific discharge vectors at points A, B, and C of the impermeable boundary are incorrectly given as those in the first column; what should be done in order to make them conform to the aforementioned principles of flow along an impermeable boundary? Adding mirror images of the specific discharge vectors of the first column (as shown in the second column) to themselves is one possibility. This implies that an imaginary well with the same discharge as the real well should be placed on the other side of the boundary, and the distance between the imaginary well

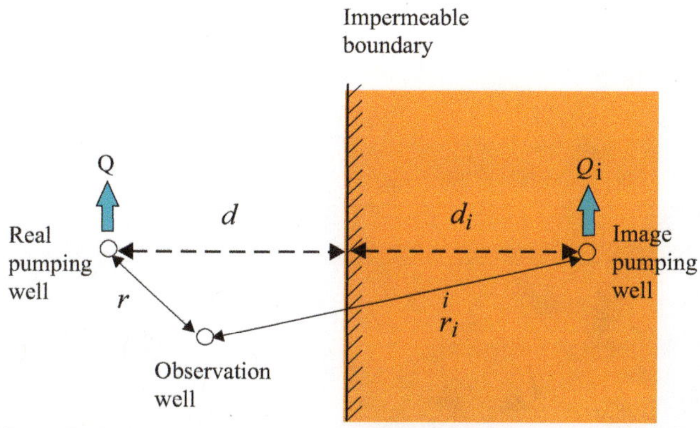

Figure 8.4. Illustration of the concept of the method of images. Q and d are discharge, the distance between the pumping well and the impermeable boundary, and r is the distance between the pumping well and the observation well. The discharge of the image is Q_i, the distance between the image well and the impermeable boundary is d_i, and the distance between the image well and the observation well is denoted by r_i. Note that $d = d_i$.

and the boundary should be the same as that between the impermeable boundary and the real well (point D in Figure 8.2). Mathematically, this suggests that one can use the Theis solution for an aquifer of an infinite areal extent to calculate the drawdown caused by the image well. Then, one combines the drawdowns induced by the real and image wells for the unbounded aquifer to yield the drawdown in the aquifer with the impermeable boundary (see Figure 8.4). This mathematical procedure is called the method of images, and it is derived next.

8.2.1.1 Method of Images

As illustrated in Figure 8.4, the distance between the real pumping well and an observation well is r, whereas the distance between the same observation well and image well is r_i. According to the method of images, the actual drawdown observed at the observation well at time t, $s_a(r, t)$, should be the sum of the drawdowns, resulting from the Theis solution for pumping at the real well, $s(r, t)$, and the Theis drawdown due to pumping at the image well $s_i(r, t)$. That is,

$$s_a(r,t) = s(r,t) + s_i(r_i,t) = \frac{Q}{4\pi T} W(u) + \frac{Q}{4\pi T} W(u_i) = \frac{Q}{4\pi T} [W(u) + W(u_i)] \tag{8.2.1}$$

where $u = \frac{r^2 S}{4Tt}$, $u_i = \frac{r_i^2 S}{4Tt}$. Figures 8.5 and 8.6 illustrate the resultant drawdown due to superposition of the drawdown from the image well on the drawdown from the real well.

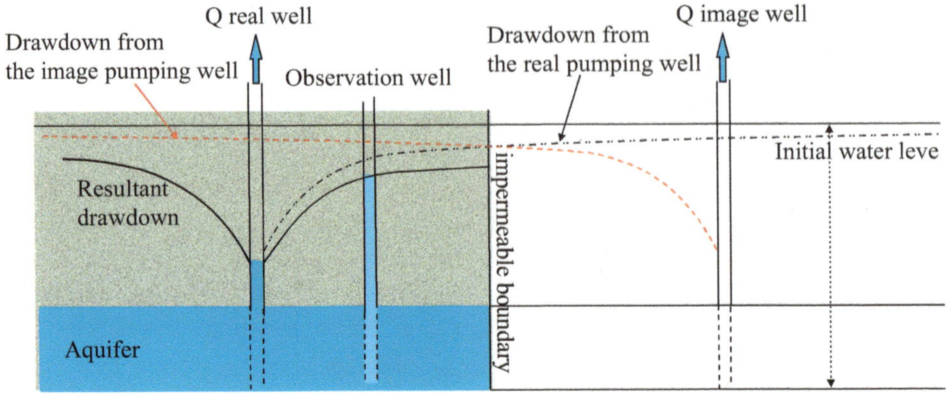

Figure 8.5. A cross-sectional view of the drawdown distributions derived from the Theis solution for the real and the image well in an unbounded aquifer (the black dashed line and the red dashed line, respectively). The resultant drawdown distribution after superposition of the drawdown from the two wells is shown as a solid black line, which reflects the effect of the impermeable boundary.

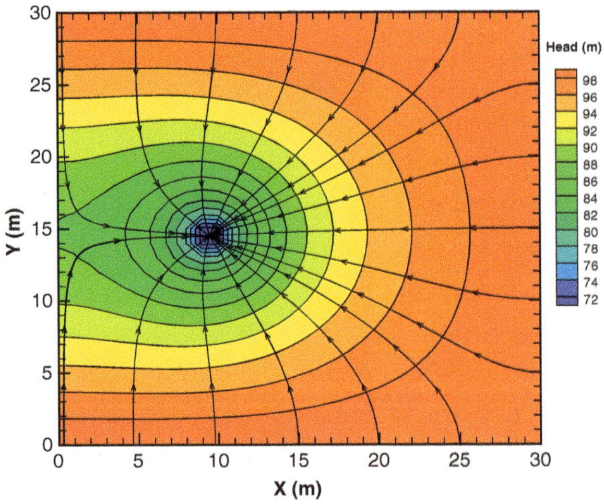

Figure 8.6. A plan view of the cone of depression resulting from the superposition of the two drawdowns, which reflects the effect of presence of an impermeable boundary on the left side of the aquifer. The other three sides are constant heads. The lines with arrows denote the streamlines.

For situations where $u \leq 0.01$, Copper-Jacob's approximation for drawdown in an aquifer of infinite lateral extent can also be employed. Therefore,

$$s(r,t) = \frac{Q}{4\pi T}\ln\left(\frac{2.25Tt}{r^2S}\right) \tag{8.2.2}$$

Now, if we define the variable β such that

$$\beta = \frac{2.25T}{r^2 S} \tag{8.2.3}$$

Equation (8.2.2) then can be rewritten as

$$s(r,t) = \frac{Q}{4\pi T} \ln \beta t = \frac{Q}{4\pi T} \ln \beta + \frac{Q}{4\pi T} \ln t \tag{8.2.4}$$

The drawdown due to the presence of a linear impermeable boundary, $s_a(r, t)$, then becomes

$$\begin{aligned} s_a(r,t) &= s(r,t) + s_i(r_i,t) \\ &= \frac{Q}{4\pi T}(\ln \beta + \ln t) + \frac{Q}{4\pi T}(\ln \beta_i + \ln t) \\ &= \frac{Q}{4\pi T}(\ln \beta\beta_i + 2 \ln t) \end{aligned} \tag{8.2.5}$$

or

$$s_a(r,t) = \frac{Q}{4\pi T} \ln \beta\beta_i + \frac{2Q}{4\pi T} \ln t \tag{8.2.6}$$

While equation (8.2.6) is an approximation of equation (8.2.1), it provides some very useful insights into the effects of an impermeable boundary. A comparison of equation (8.2.6) and equation (8.2.4) leads to the following conclusion. The straight-line portion of a drawdown-log time plot due to pumping in an aquifer with the presence of an impermeable boundary will have a slope twice that of the drawdown in an infinite aquifer (see Figure 8.7), confirming our intuition at the beginning of this section that the drawdown will be greater than that in an unbounded aquifer.

8.2.1.2 Location of the Boundary

In the previous section, we presented a methodology for determining the drawdown-time behavior in an aquifer with an impermeable boundary, whose location is known. We next investigate the possibility of using observed hydrographs to determine the location of the boundary. To do so, we examine Figure 8.4. It tells us that the impermeable boundary is located at a half of the distance between the real and the image pumping well, if the location of the image well is known. Therefore, in order to determine the location of the image well, we next introduce **the law of times** to calculate r_i (the distance between the observation well and the image well).

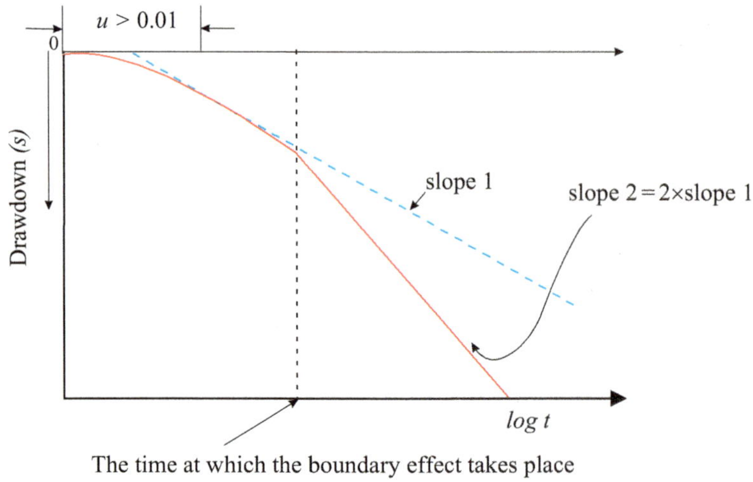

Figure 8.7. Drawdown-log time plot showing the change in slope of the drawdown versus log time plot due to the presence of an impermeable boundary (red solid line), compared with that without the boundary (blue dashed line).

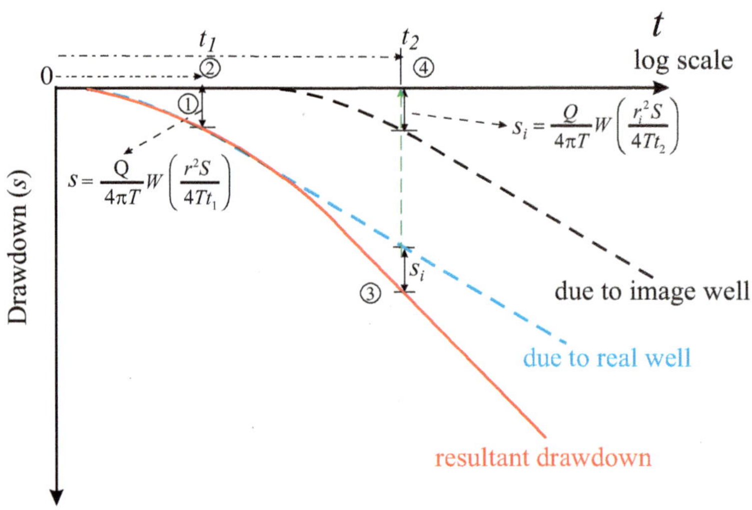

Figure 8.8. Illustration of the explanation of the law of times. The red solid line denotes the drawdown time curve in an aquifer with an impermeable boundary. The blue dashed line represents the drawdown-time curve for an unbounded aquifer. The drawdown-time curve created by the image well is represented by the black dashed line.

Figure 8.8 shows a drawdown-log time curve (solid line) that reflects the effect of a linear impermeable boundary and the long dashed curve for the case where such an impermeable boundary is absent. The deviation of the two curves starts at the time when the drawdown reaches the impermeable boundary as illustrated in the figure. Using Figure 8.8, we will derive the law of times in the following steps.

Steps:

1. Plot the drawdown and time data on a semilog paper (i.e., the red solid line) and select a drawdown value, s, from the straight-line portion of the plot, before the boundary effects are observed.
2. Then, find the time, t_1, associated with this selected drawdown in step 1.
3. Extend the straight-line portion beyond the time when the boundary effect occurs (the blue dashed line in Figure 8.8). At the second segment of the drawdown-time data (the red solid line) where the impermeable boundary effects prevail, find a drawdown, s_i (difference between the blue dashed line and the red solid line in Figure 8.8), that is equal to the magnitude of the drawdown determined from step 1 (s).
4. Then find the time, t_2, associated with s_i (the green dashed arrow line).

On the basis of the principle of superposition for calculating the drawdown due to the presence of an impermeable boundary, we know that the drawdown, s_i, is the drawdown calculated from the Theis solution for the image well at this particular time, t_2 (the black dashed line in Figure 8.8). Since s at t_1 equals s_i at t_2, the following equality must hold:

$$\frac{Q}{4\pi T}W\left(\frac{r^2S}{4Tt_1}\right) = \frac{Q}{4\pi T}W\left(\frac{r_i^2S}{4Tt_2}\right) \quad \text{or} \quad \frac{r^2S}{4Tt_1} = \frac{r_i^2S}{4Tt_2} \tag{8.2.7}$$

Therefore, we conclude that the following relationship exists:

$$\frac{r^2}{t_1} = \frac{r_i^2}{t_2} \tag{8.2.8}$$

Equation (8.2.8) is called the law of times. It states that the ratio of the square of the distance between the pumping well and the observation well to the time at which a given drawdown is observed will equal the ratio of the square of the distance between the image well and the observation well to the time at which the same drawdown due to pumping at the image well is observed. This relationship allows us to find r_i, the distance between the image well and the observation well, if r, t_1, and t_2 are known.

Here, one must recognize that the knowledge of r_i only indicates that the image well is located at a radial distance of r_i from the observation well. That is, there are an infinite number of possible locations for the image well. In order to pinpoint its location, at least hydrographs from three observation wells (say, OBW1, OBW2, and OBW3 in Figure 8.9) are needed. From these three hydrographs and using the law of times, three radii (r_{i1}, r_{i2}, and r_{i3}, representing the distance between the image well and the observation wells, OBW1, OBW2, and OBW3, respectively) thus can be determined. We next draw three circles with the radius of r_{i1}, r_{i2}, and r_{i3} using the locations of OBW1, OBW2, and OBW3, respectively, as their centers (see the red, blue, and green dashed lines in Figure 8.9). The intersection of the three circles then determines the location of the image well. Finally, the impermeable boundary

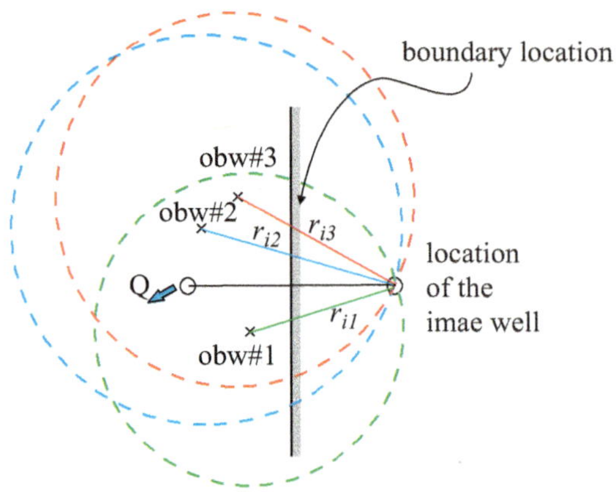

Figure 8.9. Illustration of the method to locate the image well and the impermeable boundary. The green, blue, and red dashed circles represent all possible image well locations based on the drawdown observed at the three observation wells (OBW1, OBW2, and OBW3, respectively). The intercept of the three circles is the only possible location where the added image well can reproduce the drawdowns at the three wells.

location is at half of the distance between the image well and the real pumping well, as illustrated in Figure 8.9.

An alternative to the graphical solution to find the location of the image well is to solve a system of following three equations:

$$\begin{aligned}(x_1-x_i)^2+(y_1-y_i)^2&=r_{i1}^2\\(x_2-x_i)^2+(y_2-y_i)^2&=r_{i2}^2\\(x_3-x_i)^2+(y_3-y_i)^2&=r_{i3}^2\end{aligned} \tag{8.2.9}$$

where (x_1,y_1), (x_2, y_2), and (x_3, y_3) are the coordinates of the observation wells OBW1, OBW2, and OBW3, respectively; (x_i, y_i) is the coordinate of the image well, which is to be determined.

Discussion points:

1. What factors can make the three circles fail to intersect at a point when you apply the approach in a field situation?
2. Can you use a similar approach to locate the boundary between two geologic media with different hydraulic properties?

8.2.1.3 Effects of More than One Boundary

Aquifers are often bounded by more than one boundary. For instance, an aquifer formed by sediment fills in a graben basin will likely have at least two impermeable boundaries. The method of images is applicable to such situations as well. However, an infinite series of image wells must be used for this scenario during superposition to reflect the effect of one impermeable boundary on the other. That is, the effect of the one impermeable boundary on the image wells for a selected impermeable boundary must be considered. The effects of additional image wells nevertheless rapidly decrease as their locations become more remote from the real well. Generally, this superposition approach is complicated and tedious. Therefore, computer models are more convenient tools to simulate the impact of these more complicated boundary types.

8.2.1.4 Vertical Irregularity of the Boundary

Geologic boundaries are also irregular in shape, deviating from the straight-line boundary assumed in the previous analysis. Application of the analytical approach to such situations is possible, but it will be quite complicated. Rather than deriving the solutions, we will use what you have already learned so far to postulate the effects of irregular boundaries on drawdown qualitatively. As an example, we will consider a fully penetrating well tapping into an aquifer that pinches out away from the well as illustrated in Figure 8.10.

To simplify the smoothly varying impermeable boundary, we will replace the boundary with another boundary, consisting of many steps, and then assume that each step represents a single layer. In addition, we will assume that the flow within each layer is essentially horizontal so that the drawdown can be obtained by using the Theis solution and the image well technique, as discussed previously. To illustrate the approach, we will replace the pinched-out part of the aquifer with a two-layer aquifer as illustrated in Figure 8.10.

Again, we will assume a horizontal flow in each layer. This assumption, however, is not valid in this case, but we will start from here and elaborate upon the effects of vertical flow between the two layers later. As pumping starts, drawdown in both layers should act as drawdown in a single layer, assuming the pumping well is fully screened over the two layers. The drawdown reaches first the impermeable boundary of the bottom layer, and it increases at a rate smaller than that of the case where the impermeable boundary is perfectly vertical (see Figure 8.11), because water in the second (upper) layer continues to supplement water to the pumping well. Once the drawdown reaches the second impermeable boundary at the upper layer, the drawdown will increase again. It will produce either the same drawdown, as the

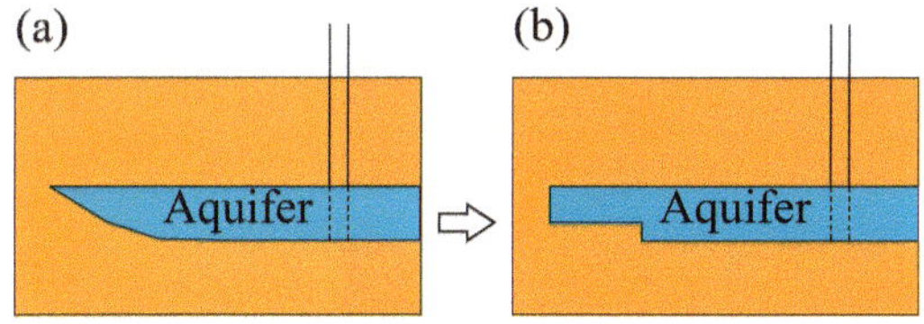

Figure 8.10. Illustration of a conceptualization for a vertically irregularly shaped impermeable boundary (a) by a two-step boundary (b).

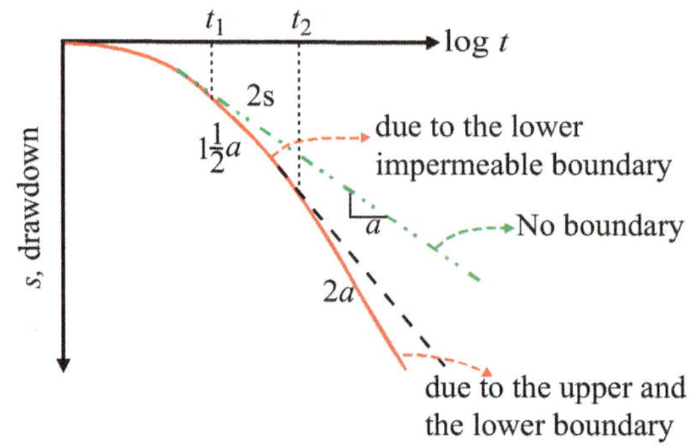

Figure 8.11. Conceptualized drawdown-log time curve (the red solid line) in an aquifer with a two-step impermeable boundary. The green dash-dot-dot line represents the drawdown-time curve for an unbounded aquifer with a slope of a. The red line segment between t_1 and t_2 (with a slope of approximately $1.5a$ and the black dashed line beyond is the drawdown behavior after the lower boundary is felt. The red line (with a slope of $2a$) beyond t_2 denotes the drawdown that reflects the effects of both upper and lower boundaries.

drawdown in the case where the boundary is fully vertical or a greater drawdown, since the vertical flow between layers now creates longer flow paths. You should verify this speculation using a numerical model. Following this logic, the drawdown behavior in the case of a curved impermeable boundary can be easily and intuitively drawn.

8.2.2 Recharge Boundary

In this section, surface water features (such as rivers, lakes, and reservoirs) will be considered as a hydrologic recharge boundary to an aquifer, although they may also represent a sink (discharge area) for the aquifer. Here, we will consider a pumping well and an observation well, which are located close to a river (see Figure 8.12). In the following analysis, we will assume that the river cuts through the entire thickness of the aquifer and it is a straight line in a plan view.

As we have already shown, once the drawdown due to pumping reaches the boundary, the effect of the boundary begins to impact the cone of depression. Instead of increasing as much as in the case of an impermeable boundary, the drawdown in this case will be less than the one predicted by Theis's solution for an unbounded aquifer. Furthermore, the drawdown will stabilize – reach a steady state due to continuous supplement of water from the river. To quantify this effect using the method of images, an injection well is placed on the other side of the river boundary, rather than a discharge well, as in the case of an impermeable boundary. Therefore, the drawdown at the observation well can be determined by

$$s(r,t) = \frac{Q}{4\pi T}[W(u) - W(u_i)] \tag{8.2.10}$$

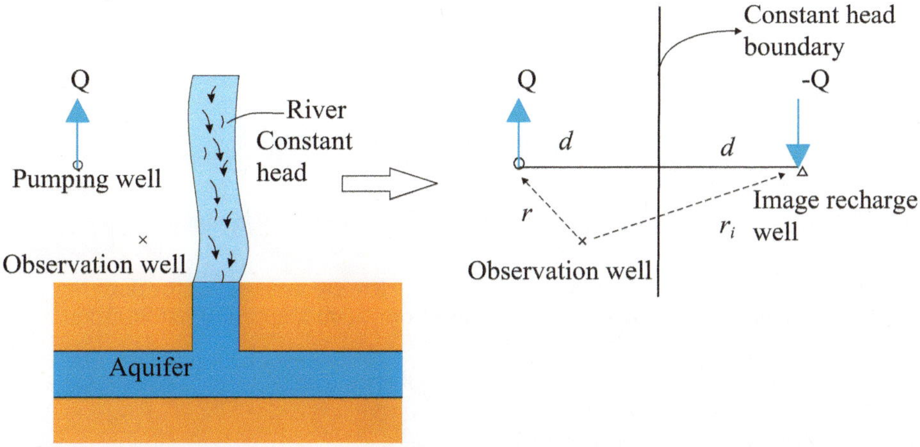

Figure 8.12. Illustration of a pumping well and an observation well near a constant-head river boundary, and the method of image approach for representation of the effect of a constant head boundary on the drawdown at an observation well due to extracting Q amount of water from the pumping well.

If we assume $u \leq 0.01$, we can then use Jacob's approximation for the well function:

$$W(u) \simeq -0.577 - \ln u + u \tag{8.2.11}$$

The drawdown in equation (8.2.10) then becomes

$$\begin{aligned} s(r,t) &\cong \frac{Q}{4\pi T}[(-0.577 - \ln u + u) - (-0.577 - \ln u_i + u_i)] \\ &= \frac{Q}{4\pi T}\left[\ln\left(\frac{u_i}{u}\right) + u - u_i\right] \end{aligned} \tag{8.2.12}$$

where $u = \frac{r^2 S}{4Tt}$ and $u_i = \frac{r_i^2 S}{4Tt}$. We thus have

$$s(r,t) \cong \frac{Q}{4\pi T}\left[2\ln\left(\frac{r_i}{r}\right) + \frac{S(r^2 - r_i^2)}{4Tt}\right] \tag{8.2.13}$$

Notice that the first term inside the bracket on the right-hand side is independent of time (see the constant drawdown line in Figure 8.13.) As time approaches infinity, the drawdown becomes

$$s(r,\infty) \cong \frac{Q}{2\pi T}\ln\left(\frac{r_i}{r}\right) \tag{8.2.14}$$

This equation proves the existence of a steady flow field in this aquifer, which is sustained by a continuous supply of water from the river or the constant head boundary.

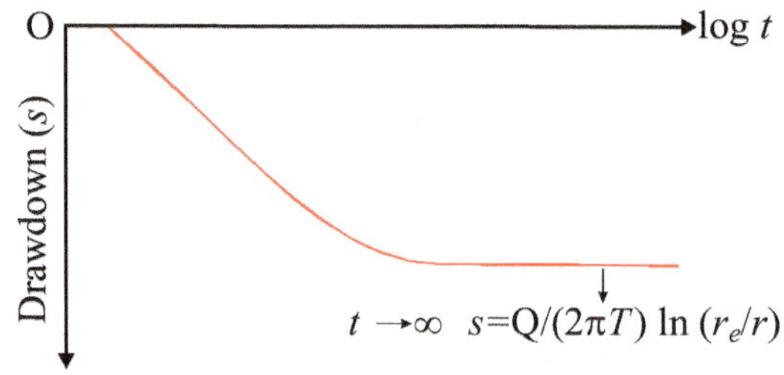

Figure 8.13. Schematic illustration of a drawdown-log time curve for pumping at a well near a recharge boundary. This is based on Jacob's approximate solution.

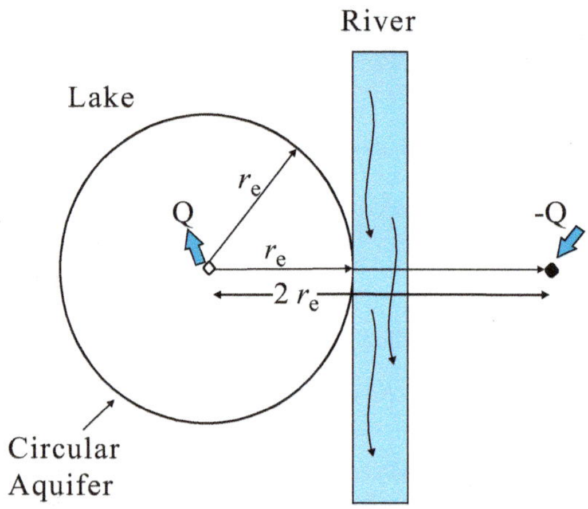

Figure 8.14. Schematic illustrations of an aquifer bounded by a circular lake at a distance r_e from the pumping well, and an aquifer with a straight-line recharge boundary, where the distance between the pumping well (which also serves as the observation well, the white circle) and the image well (the black circle) in equation 8.14 is 2 r_e.

8.2.2.1 Effects of the Shape of Recharge Boundary

The next question we would like to ask is how a meandering (instead of a straight-line) stream would affect the drawdown observed at a well. In order to answer this question, we will compare equation (8.2.14) with Thiem's equation for a circular aquifer surrounded by a constant depth lake (see equation 7.2.21), that is,

$$s = \frac{Q}{2\pi T} \ln\left(\frac{r_e}{r}\right) \tag{8.2.15}$$

The forms of the two equations, equations (8.2.14) and (8.2.15), are identical, with the exception that the distance between the pumping well and the image well, r_i, in equation (8.2.14) equals twice the radius of the circular island, r_e (see Figure 8.14) if the observation well is the same as the pumping well (i.e., $r = r_w$, the well radius).

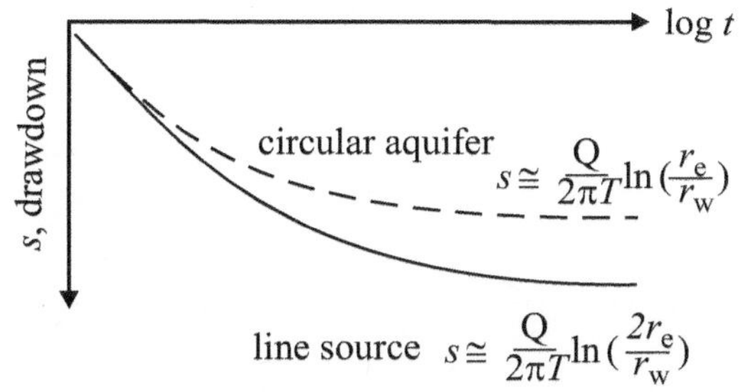

Figure 8.15. Conceptualized drawdown-log time curves for pumping at a well with drawdown influenced by two different configurations of recharge boundaries (a line source, the solid black line, and a circular aquifer, a dashed line). The line source is representative of a well near a river, and the circular aquifer represents an island aquifer surrounded by a lake or an ocean.

The drawdown-log time curves produced by equations (8.2.14) and (8.2.15) are illustrated in Figure 8.15. It shows that a well adjacent to a line source (e.g., a river) will take a longer time to reach a steady state than one surrounded by a circular lake. Furthermore, the drawdown at the well adjacent to a line source will be greater than that surrounded by a lake. In other words, the cone of depression has to travel longer distances to capture the same amount of water to reach a steady state than in the case where the well is surrounded by a circular lake. On the basis of this reasoning, try answering the following question: Supposing that a well is located near a convex portion of a meandering stream, will the drawdown be greater or smaller than that in an aquifer with a straight-line river?

8.3 Leaky Confined Aquifers

A leaky confined aquifer is generally referred to as an aquifer that is overlain by a semiimpermeable aquitard. Because it is semiimpermeable, water from aquifers above the aquitard flows downward through the aquitard and recharges the aquifers. A few analytical solutions were developed in the past for describing radial flow in a leaky confined aquifer. Each solution, however, relies on some simplifications of the true flow systems. In this section, we will first introduce a solution that omits the effect of storage in the aquitard (steady flow in the aquitard), and then a solution that considers the storage effect in that aquitard. The purpose of this section is not to review the solution technique or applications of the analytical solution. Rather, the goal is to elucidate the logic behind and the limitations of the mathematical formulation of the problem, as well as to illustrate the general behavior of a leaky aquifer.

8.3.1 Leaky Aquifer without Storativity in the Aquitard ($S'_s = 0$)

Consider a cross-sectional view of a geologic medium, which consists of a major aquifer (AQ_1) at the bottom, an aquitard (AT) in the middle, and another aquifer (AQ_2) overlying the top of the aquitard (Figure 8.16). The hydraulic conductivity, specific storage, and thickness of aquifer AQ_1 are K, S_s, and b, respectively. The vertical hydraulic conductivity of the aquitard is K'_z, the specific storage is S'_s, and the thickness is b'.

These geologic units are assumed to be homogeneous and isotropic, and they have uniform thickness and infinite lateral extents. In addition, the upper aquifer AQ_2 is considered to be unaffected by the pumping in the main aquifer (AQ_1), implying that the hydraulic head (h_0) in the upper aquifer remains constant at all times and in space. Furthermore, flow in the aquifers is assumed to be horizontal, and flow through the aquitard to the main aquifer is one-dimensional vertical steady flow, implying that the aquitard specific storage is zero, $S'_s = 0$. The pumping rate, Q, is constant in time and the initial head is the same everywhere in the main aquifer.

On the basis of these assumptions, the governing partial differential equation for the 2-D depth-averaged flow in the leaky confined aquifer, AQ_1, is

$$T\left(\frac{\partial^2 h}{\partial x^2} + \frac{\partial^2 h}{\partial y^2}\right) + q_Z = S\frac{\partial h}{\partial t} \tag{8.3.1}$$

where $T = Kb$ and $S = S_s b$; q_z represents the inflow from the aquitard above the main aquifer and it is defined as

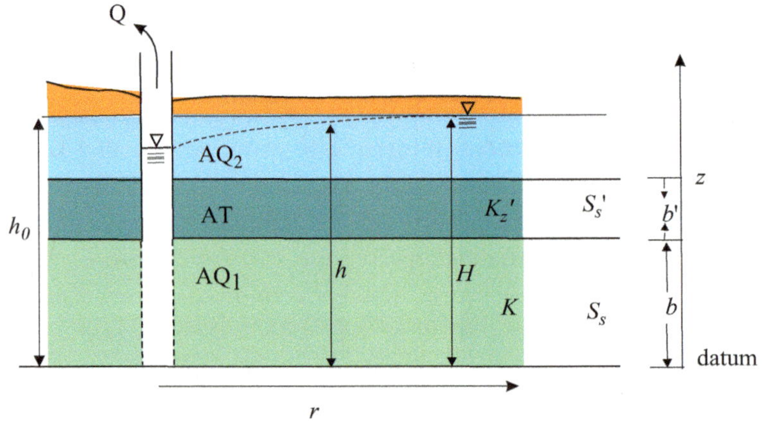

Figure 8.16. A schematic vertical geologic cross section consisting of three horizontal layers of material with different hydraulic conductivities including the main aquifer (AQ_1), an aquitard (AT), and the aquifer (AQ_2) above. K, S_s, and b are the hydraulic conductivity, specific storage, and thickness of AQ1. K'_z, S'_z, and b' are the vertical hydraulic conductivity, specific storage, and thickness of the aquitard.

$$q_Z = -K'_z \frac{(H-h)}{b'} \tag{8.3.2}$$

where K'_z and b' are the vertical hydraulic conductivity and the thickness of the aquitard (AT).

Combining equations (8.3.1) and (8.3.2), we obtain

$$\frac{\partial^2 h}{\partial x^2} + \frac{\partial^2 h}{\partial y^2} - \frac{K'_z(H-h)}{Tb'} = \frac{S}{T}\frac{\partial h}{\partial t} \tag{8.3.3}$$

Using the polar coordinate system and letting the drawdown $s = H - h$, equation (8.3.3) can be expressed as

$$\frac{\partial^2 s}{\partial r^2} + \frac{1}{r}\frac{\partial s}{\partial r} - \frac{s}{B^2} = \frac{S}{T}\frac{\partial s}{\partial t} \tag{8.3.4}$$

in which

$$B = \sqrt{Tb'/K'_z} \tag{8.3.5}$$

B is referred to as the leakage factor, which has a dimension. K'_z/b' is called the leakance or leakage coefficient, which has the dimension $[1/T]$. The boundary and initial conditions associated with this problem are

$$s(\infty, t) = 0, \ \lim_{r\to 0} r\frac{\partial s}{\partial r} = \frac{Q}{2\pi T}, \text{ and } s(r, 0) = 0 \tag{8.3.6}$$

8.3.1.1 Hantush–Jacob Solution (r/B solution)

Hantush-Jacob (1955) derived an analytical solution to equation (8.3.4) subjected to the boundary and initial conditions in equation (8.3.6). The solution is

$$s(r,t) = \frac{Q}{4\pi T} W\left(u, \frac{r}{B}\right) \tag{8.3.7}$$

where

$$W\left(u, \frac{r}{B}\right) = \int_u^\infty \frac{1}{y}\exp\left(-y - \frac{r^2}{4B^2 y}\right) dy \text{ and } u = \frac{r^2 S}{4Tt} \tag{8.3.8}$$

According to equation (8.3.5), as the ratio of K'_z to K approaches zero, $B = \sqrt{Kbb'/K'_z}$ approaches infinity; then r/B approaches zero, and thus $W(u, r/B)$ becomes $W(u)$, which is the Theis well function for a nonleaky confined aquifer. Figure 8.17 shows the behavior of the function $W(u, r/B)$.

From Figure 8.17, we can draw some conclusions:

1. Leakage from the aquitard stabilizes (i.e., it reaches a steady state) the drawdown in the main confined aquifer, causing its deviation from the Theis solution for the confined

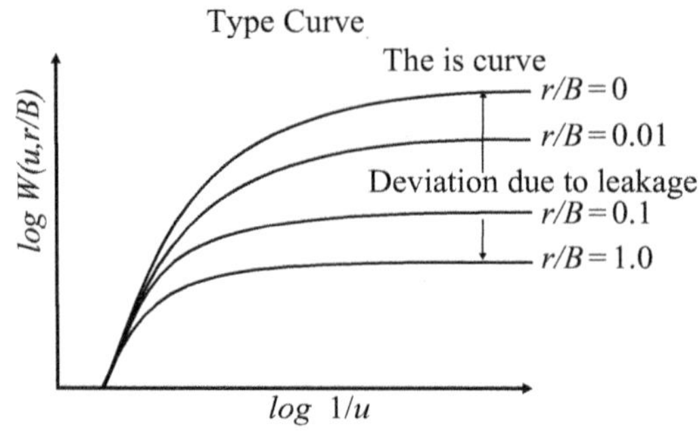

Figure 8.17. Hantush-Jacob type curve or a log-log plot between $W(u, r/B)$ and $1/u$ illustrating the behavior of $W(u, r/B)$ for different amounts of leakance (B).

aquifer. The drawdown behavior is similar to that of the case where the aquifer is connected with a recharge boundary.

2. At a fixed r, if B is small (i.e., K/K_z' is small or the vertical conductivity of the aquitard is relatively large) the drawdown s in the main aquifer stabilizes (or reaches a steady state) quickly.
3. As time goes on, more and more well discharge is derived from leakage, and ultimately the entire yield of the well is from leakage, and the flow becomes steady.

Estimation of T, S, and K_z': The type-curve approach again is one way to estimate the parameters for this type of leaky confined aquifer. This approach uses the type curves generated by using equation (8.3.8) to match the observed drawdown-time data in a way similar to the procedure previously mentioned for analyzing pumping test data from a confined aquifer (see Chapter 7 for more details). That is, from the curve matching, one finds a match point and the corresponding values of $s, t, W\left(u, \frac{r}{B}\right), u$ and $\frac{r}{B}$. Subsequently, one uses the following formulae to determine values of the parameters:

$$T = \frac{Q}{4\pi s} W\left(u, \frac{r}{B}\right), S = \frac{4Ttu}{r^2}, \text{ and } K_z' = \frac{Tb'\left(\frac{r}{B}\right)^2}{r^2} \tag{8.3.9}$$

Thus, to determine the vertical hydraulic conductivity (K'_z), one needs to know the thickness of the aquitard, b', which is generally estimated from the well-log data.

8.3.1.2 Steady-State Flow

For steady flow toward a well in a leaky aquifer, the governing equation becomes

$$\frac{\partial^2 s}{\partial r^2} + \frac{1}{r}\frac{\partial s}{\partial r} - \frac{s}{B^2} = 0 \tag{8.3.10}$$

with the same boundary conditions used in the previous unsteady-state case. The solution to equation (8.3.10) is

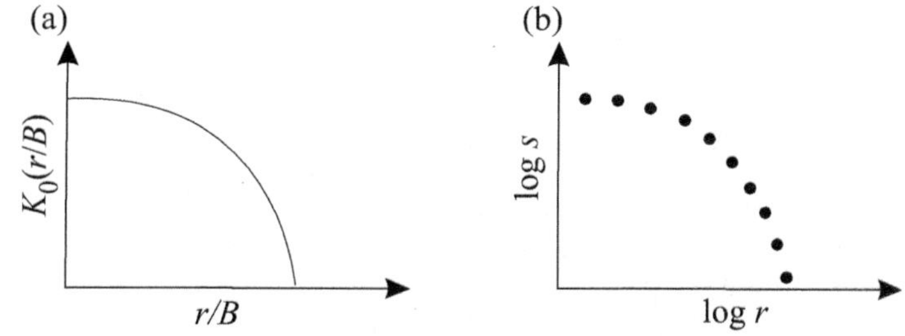

Figure 8.18. Plot of K_0 as a function of r/B (a) and the steady state drawdown as a function of distance (b).

$$s(r) = \frac{Q}{2\pi T} K_0\left(\frac{r}{B}\right) \tag{8.3.11}$$

where K_0 is the modified Bessel function of the second kind and of zero order. The behavior of the drawdown(s) as a function of distance (r) is shown in Figure 8.18.

Notice that the type curve can be constructed, and the parameters to be estimated (T and K'_z) are obtained by using the following formulae:

$$T = \frac{Q}{2\pi s} K_0\left(\frac{r}{B}\right) \text{ and } K'_z = \frac{Tb'(r/B)^2}{r^2} \tag{8.3.12}$$

Notice that we are dealing with steady-state flow, so that drawdown at the observation wells should not vary with time. In addition, the aquifer's S cannot be determined by this test, because the entire yield for the pumping well is derived from leakage.

8.3.2 Leaky Confined Aquifers with Storativity in the Aquitard ($S'_s \neq 0$)

In this analysis, transient (unsteady) leakage through the aquitard (Figure 8.16) is considered, and it is described by a one-dimensional vertical transient flow equation:

$$\frac{\partial^2 h'}{\partial z^2} = \frac{S'_s}{K'_z} \frac{\partial h'}{\partial t} \tag{8.3.13}$$

where h' is the head in the aquiftard at any time t and K'_z and S'_s are the vertical conductivity and storativity of the aquitard, respectively. The initial condition (I.C.) and boundary condition (B.C.) associated with the flow regime are

$$\begin{aligned} &\text{I.C.} \quad h'(z,0) = h_0 \quad \text{for} \quad b < z < b + b' \\ &\text{B.C.} \quad h'(b,t) = h(b,t) \\ &\qquad\quad h'(b+b',t) = h_0 \end{aligned} \tag{8.3.14}$$

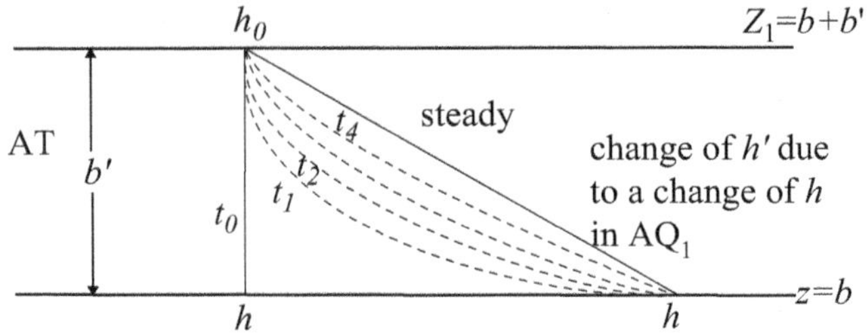

Figure 8.19. Temporal evolution of heads (dashed lines) to the steady-state (solid) head profile within the aquitard (AT), in response to a change in head (h_0 to h) in the underlying main aquifer (AQ_1).

The first boundary condition ensures that the hydraulic head at the bottom of the aquitard is equal to that of the main aquifer, AQ_1, at any time (i.e., an interface boundary condition). The second boundary condition states that the hydraulic head at the top of the aquitard is the same as the head in the top aquifer, AQ_2, which remains constant, and it is not affected by pumping in the main aquifer. As illustrated in Figure 8.19, while the head at the bottom of the aquitard changes from h_0 to h, the head distribution in the aquitard slowly adjusts and transitions to the steady-state profile as a result of the reduction in the storage effect in the aquitard with time.

Now, consider the equation for the flow in the main aquifer, which is described by a two-dimensional depth-averaged flow equation

$$T\left(\frac{\partial^2 h}{\partial r^2}+\frac{1}{r}\frac{\partial h}{\partial r}\right)+q_z=S\frac{\partial h}{\partial t} \tag{8.3.15}$$

with the following initial and boundary conditions

$$\begin{aligned} \text{I.C.}\quad & h(r,0)=h_0 \\ \text{B.C.}\quad & h(\infty,t)=h_0 \\ & \lim_{r\to 0} r\frac{\partial h}{\partial r}=\frac{Q}{2\pi T} \end{aligned} \tag{8.3.16}$$

where the leakage from the aquitard is given by

$$q_z=-K'_z\frac{\partial h'}{\partial z} \tag{8.3.17}$$

As you may imagine, the solution to these equations is rather complicated and only asymptotic analytical solutions are available (Hantush, 1964). Nevertheless, we will examine the general behavior of the flow regime in this type of aquifer-aquitard system using Figure 8.20.

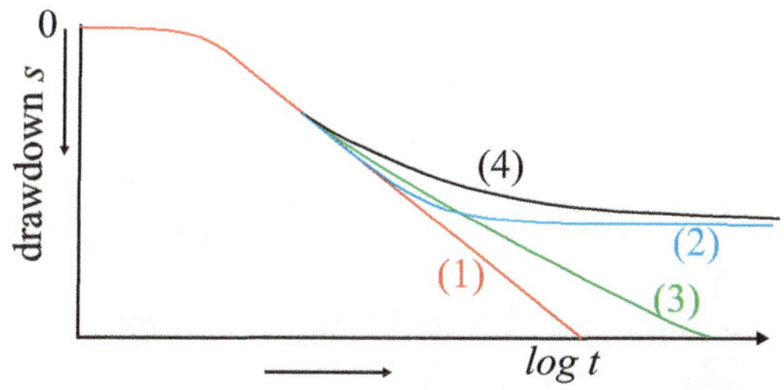

Figure 8.20. Schematic illustration of the drawdown-log time curves due to pumping at a well in four different types of nonleaky and leaky confined aquifers with infinite lateral extent.

Drawdown-time curves for four different types of leaky aquifers indicated by different color lines and numbers are illustrated in Figure 8.20, and they are explained in the following:

(1) In nonleaky aquifers with infinite lateral extent, drawdown increases continuously and steady state does not exist.
(2) In a confined aquifer with a leaky aquitard without storage in the aquitard of finite thickness, drawdown remains constant as soon as the discharge (well) equals the leakage from the aquitard (steady state).
(3) In a confined aquifer with a leaky aquitard with storage in the aquitard of infinite thickness, because of the infinite thickness, no steady leakage exists in the aquitard. Therefore, flow in the main aquifer will take a long time to reach a steady state.
(4) This case is for the confined aquifer with a leaky aquitard with storage in the aquitard of a finite thickness. This figure shows that steady-state flow in the aquifer will be reached when the leakage through the aquitard is equal to the well discharge. However, the time to reach a steady-state is delayed as a result of the storage effect of the aquitard. The time to the steady state (response time) depends mainly on K_Z, b', and S_s'.

Neuman and Witherspoon (1969) derived an analytical solution for a more realistic flow regime in a leaky aquifer system, which consists of an upper aquifer (AQ_2), an aquitard (AT), and the main aquifer (AQ_1) in which the pumping well is screened across (Figure 8.21). Their analysis treats 3-D flow in aquifers and 1-D vertical flow in the aquitard and allows heads in the upper aquifer and the aquitard to vary with the head in the bottom main aquifer. To simplify the analysis, they made the following assumptions:

- Flow is vertical and one-dimensional in the aquitard (AT).
- Only horizontal flows exist in both the lower and upper aquifers, namely, AQ_1, AQ_2, respectively.

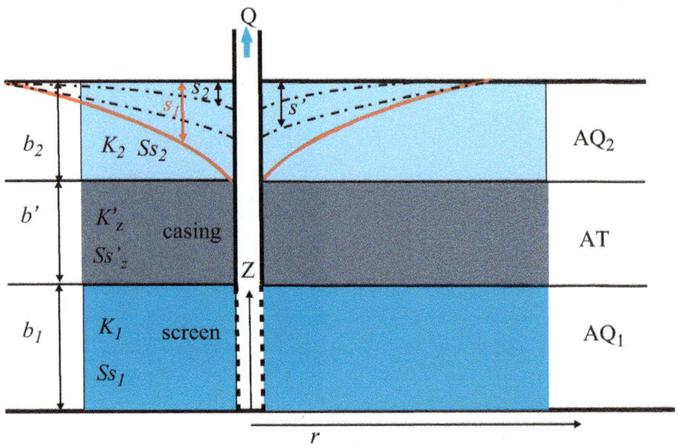

Figure 8.21. Schematic illustration of a vertical cross section of a leaky aquifer system used in the analysis by Neuman and Witherspoon (1969), which includes two aquifers (AQ_1 and AQ_2) and an aquitard (AT). Drawdowns in AQ_1, AQ_2, and AT are denoted as s_1, s_2, and s'. b_1, b_2, and b' denote the thickness of AQ1, AQ2, and AT. The same notation applies to the hydraulic properties K and S_s.

The governing partial differential equation for pumping from the lower aquifer, AQ_1, expressed in terms of drawdown is

$$\frac{\partial^2 s_1}{\partial r^2} + \frac{1}{r}\frac{\partial s_1}{\partial r} + \left(\frac{K_z{}'}{T_1}\frac{\partial s'}{\partial z}\right) = \frac{S_1}{T_1}\frac{\partial s_1}{\partial t} \tag{8.3.18}$$

I.C. and B.C. are

$$\begin{aligned} s_1(r,0) &= 0 \\ s_1(\infty,t) &= 0 \\ \lim_{r\to 0} r\frac{\partial s_1}{\partial r} &= \frac{Q}{2\pi T_1} \end{aligned} \tag{8.3.19}$$

where $T_1 = K_1 b_1$. The equation for the aquitard, AT, which is based on 1-D vertical flow assumption, is given by

$$\frac{\partial^2 s'}{\partial z^2} = \frac{S'_s}{K'_z}\frac{\partial s'}{\partial t} \tag{8.3.20}$$

with the following initial and boundary conditions:

$$\begin{aligned} \text{I.C.}\quad & s'(r,z,0) = 0 \\ \text{B.C.}\quad & s'(r,b_1,t) = s_1(r,t) \\ & s'(r,b_1+b',t) = s_2(r,t) \end{aligned} \tag{8.3.21}$$

The governing equation for the upper aquifer is

$$\frac{\partial^2 s_2}{\partial r^2}+\frac{1}{r}\frac{\partial s_2}{\partial r}-\frac{K_z'}{T_2}\frac{\partial s_z'}{\partial z}=\frac{S_2}{T_2}\frac{\partial s_2}{\partial t} \tag{8.3.22}$$

with the following initial and boundary conditions:

$$\begin{aligned} s_2(r,0)&=0\\ s_2(\infty,t)&=0\\ \lim_{r\to 0} r\frac{\partial s_2}{\partial r}&=0 \end{aligned} \tag{8.3.23}$$

Chen et al. (1986) have developed an analytical solution for more realistic flow situations than those examined by Neuman and Witherspoon (1969). In their analysis, the flow in the aquitard is allowed to be fully three-dimensional, but they assumed a two-dimensional horizontal flow regime for both the upper and lower aquifers. For details see "Exact Solution for the Problem of Cross-flow in a Bounded Two-Aquifer System with an Aquitard" by Chen et al. (1986).

8.3.3 Remarks

As we discussed in Chapters 5 and 6, leaky and nonleaky aquifers are merely simplified heterogeneous conceptual models such that analytical solutions of the models can be derived. However, rapid advances in numerical models and parameter estimation procedures nowadays allow us to analyze the pumping test using fully 3-D models with realistic descriptions of flow and geologic heterogeneity (see Chapter 9). In other words, the preceding analytical solutions are essentially very restricted and often deficient, as the implied assumptions do not always hold as the resolution of our interest becomes finer. They, nevertheless, help us to understand the flow mechanisms in leaky aquifers and their behavior during a pumping test. More importantly, they help us understand why they should be used with caution.

At last, we must emphasize that the behavior of a drawdown-time curve at one observation well is insufficient to diagnose the type of aquifer where the aquifer test is conducted. For example, the behavior of a drawdown-time curve in a leaky aquifer may be similar to that due to pumping in a confined aquifer with recharge boundaries. That is to say, a perfect fit of an analytical solution to an observed well hydrograph neither assures physical correctness of the solution nor suggests that this aquifer belongs to a nonleaky or leaky confined aquifer or an unconfined aquifer system.

8.4 Unconfined Aquifers (Water Table or Phreatic Aquifers)

Again, in order to derive analytical solutions for radial flow to a well in an unconfined aquifer (see Figure 8.22), simplified assumptions must be made. These include the following: The aquifer is isotropic and homogeneous; the effects of the unsaturated zone above the water table can be ignored; the pumping well discharge Q is constant;

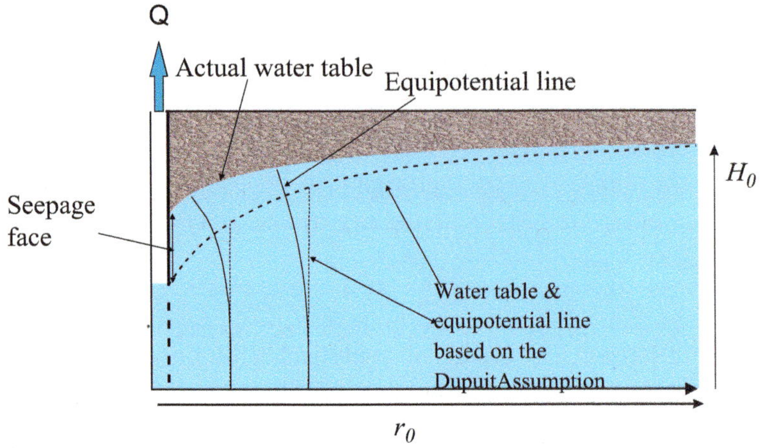

Figure 8.22. Schematic vertical cross section of an unconfined aquifer illustrating the Dupuit assumption and actual head behavior near a well. Q is the discharge of the well, and H_0 is the hydraulic head at a radial distance r_0 from the pumping well.

the bottom of the aquifer is horizontal and impermeable; Dupuit assumptions (horizontal flow and no seepage face above the water level in the well; see Chapter 5) are valid. With these assumptions, the Boussinesq equation (see Chapter 5) is deemed to be valid for flow to a pumping well in an unconfined aquifer. We will call these fictitious unconfined aquifers Dupuit aquifers.

8.4.1 Steady Radial Flow toward a Well in a Dupuit Aquifer

On the basis of the preceding assumptions, the governing partial differential equation for steady-state radial flow to a well in an unconfined aquifer in a plan view is

$$K\frac{\partial^2 h^2}{\partial x^2} + K\frac{\partial^2 h^2}{\partial y^2} = 0 \tag{8.4.1}$$

Equation (8.4.1) in the polar coordinate system takes the following form:

$$\frac{d^2 h^2}{dr^2} + \frac{1}{r}\frac{dh^2}{dr} = 0 \tag{8.4.2}$$

with the following boundary conditions:

$$h(r_0) = H_0 \quad \text{at} \quad r = r_0, \quad \lim_{r\to 0} r\frac{dh}{dr} = \frac{Q}{2\pi K h} \tag{8.4.3}$$

The particular solution to the preceding equation subjected to the boundary conditions in equation (8.4.3) is

$$H_0^2 - H^2 = \frac{Q}{\pi K}\ln\left(\frac{r_0}{r}\right) \tag{8.4.4}$$

Equation (8.4.2) is nonlinear in terms of h, but it is linear in terms of h^2. Because of the Dupuit assumption, equation (8.4.2) is not appropriate for describing the hydraulic head distribution near the pumping well. Using heads far away from the well and using equation (8.4.4), the hydraulic conductivity of the unconfined aquifer can be estimated as

$$K = \frac{Q}{\pi\left(H^2(r_2) - H^2(r_1)\right)}\ln\left(\frac{r_2}{r_1}\right) \tag{8.4.5}$$

if the hydraulic heads, $H_1(r_1)$ and $H_2(r_2)$, at distances r_1 and r_2 from a pumping well are available.

8.4.1.1 Jacob's Correction for Water Table Drawdown

If the drawdown at a well at a distance, r, from a pumping well, $s(r) = H_0(r) - H(r)$, is considered, the following relationships can be developed:

$$\begin{aligned} H_1^2 &= H_0^2 - 2H_0 s_1 + s_1^2 \\ H_2^2 &= H_0^2 - 2H_0 s_2 + s_2^2 \\ H_2^2 - H_1^2 &= 2H_0\left[\left(s_1 - \frac{s_1^2}{2H_0}\right) - \left(s_2 - \frac{s_2^2}{2H_0}\right)\right] \end{aligned} \tag{8.4.6}$$

If we let $s_c = s - \frac{s^2}{2H_0}$, which is called the corrected drawdown, then equation (8.4.5) becomes

$$T = KH_0 = \frac{Q}{2\pi(s_{c_1} - s_{c_2})}\ln\left(\frac{r_2}{r_1}\right) \tag{8.4.7}$$

If H_0 is much larger than s^2, observed drawdowns can be used directly in equation (8.4.7).

8.4.2 Transient Radial Flow toward a Well in a Dupuit Aquifer (Jacob, 1963a)

If the unconfined aquifer is conceptualized as a homogeneous and isotropic aquifer, then the Boussinesq equation for transient radial flow toward a well in a polar coordinate system is (see Chapter 5)

$$\frac{1}{r}\frac{\partial}{\partial r}\left(rKH\frac{\partial H}{\partial r}\right) = S_y\frac{\partial H}{\partial t} \text{ or } \frac{1}{r}\frac{\partial}{\partial r}\left(r\frac{\partial H^2}{\partial r}\right) = \frac{S_y}{KH}\frac{\partial H^2}{\partial t} \tag{8.4.8}$$

If we use the drawdown, $s = H_0 - H$ or $H = H_0 - s$, where H_0 is the initial hydraulic head, equation (8.4.8) can be rewritten in terms of the corrected drawdown:

$$\frac{\partial H^2}{\partial r} = \frac{\partial}{\partial r}(H_0 - s)^2 = \frac{\partial}{\partial r}\left(H_0{}^2 - 2sH_0 + s^2\right) = -2H_0\frac{\partial}{\partial r}\left(s - \frac{s^2}{2H_0}\right) = -2H_0\frac{\partial s_c}{\partial r} \tag{8.4.9}$$

where

$$s_c = s - \frac{s^2}{2H_0} \tag{8.4.10}$$

is the corrected drawdown. Similarly, we have

$$\frac{\partial H^2}{\partial t} = -2H_0\frac{\partial s_c}{\partial t} \tag{8.4.11}$$

Substitution of equations (8.4.9) and (8.4.11) into equation (8.4.8) yields

$$\frac{1}{r}\frac{\partial}{\partial r}\left(r(-2H_0)\frac{\partial s_c}{\partial r}\right) = \frac{S_y}{K(H_0 - s)}(-2H_0)\frac{\partial s_c}{\partial t} \tag{8.4.12}$$

which can be expressed as

$$\frac{1}{r}\frac{\partial}{\partial r}\left(r\frac{\partial s_c}{\partial r}\right) = \frac{S_y}{T}\frac{\partial s_c}{\partial t} \tag{8.4.13}$$

where $T(r, t) = Kh(r, t)$. If we assume that $H_0 >> s$, then $T(r, t)$ is essentially constant. Equation (8.4.13) is identical to that for a confined flow. The similarity here implies that the solution to Equation (8.4.13) will conform to the Theis solution, since $H_0 >> s$, then $s_c = s - \frac{s^2}{2H_0} \approx s$.

From the preceding analysis, we learn that the drawdown at a fully screened observation well induced by pumping in unconfined aquifers should behave similarly to that in a confined aquifer. In other words, the Theis solution for the confined aquifer can be used to describe the well hydrograph of an unconfined aquifer, even though the mechanisms responsible for either release or storage of water in these two types of aquifers are different (see Chapter 5). The applicability of the Theis solution to our conceptualized unconfined aquifer is attributed to the horizontal flow assumption embedded in the Boussinesq equation. Again, we emphasize that the specific yield (S_y) is a depth-averaged property, which includes both elastic effects of the saturated portion of the aquifer and instantaneous drainage of the pore due to falling of the water table (see equation 5.9.30 in Chapter 5).

While the preceding analysis leads us to the conclusion that the Theis solution adequately describes the drawdown-time behavior of an unconfined aquifer due to

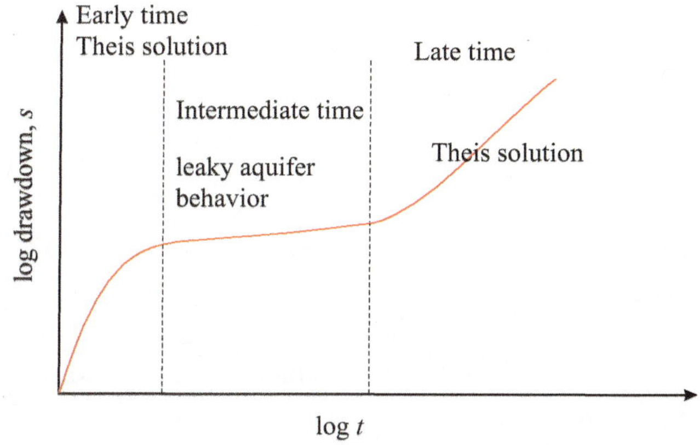

Figure 8.23. Conceptualized log drawdown versus log time plot for pumping in an unconfined aquifer to illustrate three characteristic time behaviors: a) Theis solution–like confined aquifer response at early time, b) recharge or leakage influenced behavior during intermediate time, c) a second Theis solution–like confined response at late time.

pumping, observed log drawdown-log time data often exhibit a distinct S-shaped curve (see Figure 8.23), deviating from the Theis solution.

That is, an observed well hydrograph induced by pumping in an unconfined aquifer commonly exhibits three distinct segments in time. At the early time, the drawdown in an unconfined aquifer behaves in a way similar to that in a confined aquifer. At an intermediate time, drawdown stabilizes – reaches a steady state, similar to that in a leaky aquifer, reflecting additional sources of water other than the storage of the saturated aquifer supplying the aquifer. Lastly, at late time, drawdown-time data again conform to the Theis solution–like confined aquifer response. Such a three-stage phenomenon has often been referred to as the delayed yield or delayed gravity response of the water table in an unconfined aquifer. The explanation and analysis of this response have been topics for debate and have evolved over the years. In the following section, we will examine the evolution of the conceptual models for the analysis of pumping tests in unconfined aquifers over the past decades.

8.4.3 Delayed Yield Response Conceptual Model

Three categories of conceptual mathematical models have been developed over the years to quantify and explain the delayed yield phenomenon in unconfined aquifers. The first category is the depth-averaged, radial flow model with a delayed drainage source term (Boulton, 1954, 1963). As stated by Boulton (1963), "the water-bearing material through which the water table has fallen during the early stages of pumping does not yield up its water immediately." As a result, the Boulton model includes an empirical delay coefficient to the specific yield term to represent the slow water-release process, and to account for the intermediate stage of the S-shaped well hydrograph (the log-log drawdown-time data plot). Mathematically, the "delayed yield response" concept is implemented in the following partial differential equation:

$$\frac{\partial^2 s}{\partial r^2} + \frac{1}{r}\frac{\partial s}{\partial r} = \frac{S}{T}\frac{\partial s}{\partial t} + \alpha \frac{S_y}{T}\int_0^t \frac{\partial s}{\partial t} e^{-\alpha(t-\gamma)} d\gamma \tag{8.4.14}$$

Equation (8.4.14) states that the amount of water derived from storage, due to an increment of drawdown Δs between t and $t + \Delta t$ after the onset of pumping, consists of two components: 1) a volume of water instantaneously released from storage per unit horizontal area of the aquifer, due to elasticity of the aquifer (the first term in the right-hand side of the equation) and 2) a delayed yield from storage, per unit horizontal area, at any time t ($t > \tau$) from the start of pumping, which is quantified by the second term on the right-hand side of the equation. This term represents an additional source (or flux) whose contribution decreases with time. The rate of decrease is determined by the delayed-yield index, $1/\alpha$ (empirical constant). According to the equation, the behavior of the drawdown conforms initially to the Theis solution, which depends on the parameters T and S. Then, the drawdown stabilizes, causing a deviation from the Theis solution due to additional source (or leakage from unsaturated zone). Finally, at the late time, the drawdown takes the form of the Theis solution that is controlled by the parameters T and ($S + S_y$).

The second group of models includes those based on a concept of instantaneous and complete drainage at the water table (e.g., Dagan, 1967; Brutsaert, 1970; Streltsova, 1972a, b; and Lakshminarayna and Rajagopalan, 1978). These are radial flow models that consider the 2-D flow field in the vertical plane along the radial distance, and the delayed-drawdown region of the time-drawdown curve is considered to be caused by the downward hydraulic-head gradient below the water table (i.e., delayed water table response). They neglect the influence of the unsaturated zone above the water table and assume instantaneous drainage of the initially saturated pores. Following is a snapshot of the analysis by Neuman (1972, 1973, 1975, and 1979).

8.4.4 Delayed Gravity Response of Water Table Model

Again, to avoid solving an unsaturated flow equation, the analytical analysis by Neuman (1972, 1973, 1975, and 1979) rested upon the following assumptions:

- Influence of the vadose zone can be omitted. The upper boundary of the saturated aquifer is the water table, which is dynamic and can be represented by a free surface boundary.

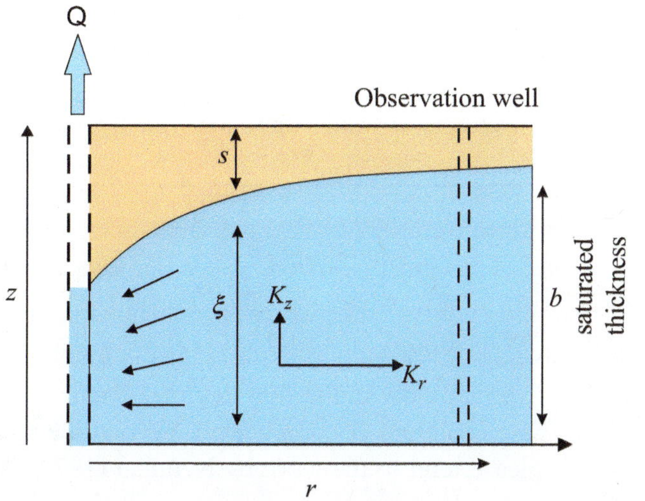

Figure 8.24. Schematic vertical cross section of an unconfined aquifer illustrating the cone of depression near a pumping well used in the analysis by Neuman (1972).

- Water release due to the lowering of the water table is instantaneous.
- The unconfined aquifer is of an infinite lateral extent.
- There is an impermeable horizontal layer (bottom).
- It is homogeneous, but could be isotropic or anisotropic.
- The principal hydraulic conductivity tensor is aligned with the r and z coordinate directions.
- There is a completely penetrating pumping well with a constant Q.

Since the aquifer is conceptualized as a homogeneous aquifer, the 3-D flow field can be visualized as an axisymmetric 2-D vertical cross-sectional flow field (Figure 8.24), which can be described by using the following partial differential equation

$$K_r \frac{\partial^2 s}{\partial r^2} + \frac{K_r}{r}\frac{\partial s}{\partial r} + K_z \frac{\partial^2 s}{\partial z^2} = S_s \frac{\partial s}{\partial t} \tag{8.4.15}$$

with initial condition $s(r, z, 0) = 0$, and the initial water table or the free surface $\xi(r, 0) = b$ (which is the initial thickness of the saturated zone) and the following boundary conditions

$$s(\infty, z, t) = 0, \ \frac{\partial s}{\partial z}(r, 0, t) = 0 \tag{8.4.16}$$

The free-surface boundary condition for the water table can be expressed as

$$K_r \frac{\partial s}{\partial r}\mathbf{n}_r + K_z \frac{\partial s}{\partial z}\mathbf{n}_z = \left(S_y \frac{\partial \xi}{\partial t} - I\right)\mathbf{n}_z \text{ at } (r, \xi, t) \tag{8.4.17}$$

where $\xi(r, t) = b - s(r, \xi, t)$, representing the water table position, S_y, is specific yield (a product of depth averaging as defined in Chapter 5 for a 2-D depth-average flow equation), which is assumed to be the porosity here, and I is the recharge from the land surface; $\mathbf{n}_r$ and $\mathbf{n}_z$ are vectors in r and z directions. Lastly, the boundary condition at the well is given by

$$\lim_{r\to 0}\int_0^{\xi} r\frac{\partial s}{\partial r}dr = \frac{Q}{2\pi Kr} \tag{8.4.18}$$

Notice that no depth averaging (or Dupuit assumption) is involved in this analysis, and the analysis allows vertical flow in the flow regime. The procedure to derive an analytical solution to these equations is complicated, beyond the scope of this book. For interested readers, please refer to the work by Neuman (1972). Nevertheless, we summarize some of the main findings here.

If the observation well is perforated through the entire saturated thickness of the aquifer, the drawdown is

$$\bar{s}(r,t) = \frac{1}{b}\int s(r,z,t)dz \tag{8.4.19}$$

which is the depth-averaged drawdown, and Neuman's solution gives

$$\bar{s}(r,t) = \frac{Q}{4\pi T}W(\mu_a,\mu_b,\eta) \tag{8.4.20}$$

where $W(\mu_a, \mu_b, \eta)$ is known as an unconfined well function, and

$$\eta = \frac{r^2K_z}{b^2K_r} \tag{8.4.21}$$

If $S_y >> S$ where S is the storativity, and $\bar{s} << b$, at early time, the depth-averaged drawdown is given by

$$\bar{s} = \frac{Q}{4\pi T}W(u_a,\eta) \tag{8.4.22}$$

which is similar to the $\frac{r}{B}$ solution for a leaky confined aquifer. The mechanism that allows water to be released from the aquifer is from the expansion of water as well as the compaction of a porous medium. The flow is predominantly horizontal flow (at a very early period) such that the drawdown behavior conforms to the Theis curve at very early time values.

During the intermediate time period, vertical flow dominates, and drawdown behavior is similar to that in a leaky confined aquifer, and drawdown tends to stabilize as it reaches a steady state. At later time, the drawdown can be depicted by

$$\bar{s} = \frac{Q}{4\pi T} W(u_b, \eta) \text{ and } u_b = \frac{r^2 S_y}{4Tt} \quad (8.4.23)$$

At this stage of the pumping test, the water table starts to lower noticeably as a result of gravity drainage (dewatering). Vertical flow is limited to an area near the water table, and horizontal flow dominates in the rest of the aquifer. Thus, the drawdown again behaves as the Theis curve. Neuman (1975) also presented procedures using the analytical solution to estimate transmissivity, storage coefficient, specific yield, and vertical hydraulic conductivity for such an idealized aquifer.

8.4.5 Transition of Water Release Mechanisms (Mao et al., 2011)

While the two conceptual models discussed earlier have been taken for granted for the past decades, applications of both models to time-drawdown data collected from field pumping tests generally yield specific yield (more precisely, porosity) values that are substantially smaller than those that would be expected on the basis of other methods of measurement (see Nwankwor et al., 1984; Endres et al., 2006).

As we advocated in Chapter 5, the governing equation for flow through variably saturated media is the most general model for subsurface hydrology without considering movement of air. As such, it is suitable for the analysis of pumping tests in unconfined aquifers. Again, the governing equation can be written as

$$\nabla \cdot [K(h, \mathbf{x}) \nabla (h + z)] + Q(\mathbf{x}_p) = \omega S_s(\mathbf{x}) \frac{\partial h}{\partial t} + \frac{\partial \theta}{\partial t} = (\omega S_s(\mathbf{x}) + C(h, \mathbf{x})) \frac{\partial h}{\partial t} \quad (8.4.24)$$

where ∇ is the spatial gradient, t is time, θ represents the volumetric moisture content, z is the elevation, which is positive upward, and h is the pressure head, and it is positive when the medium is fully saturated and negative when unsaturated. The saturation index ω is equal to 1 if the medium is saturated and 0 when the medium is unsaturated. The term $S_s(\mathbf{x})$ represents the specific storage; $C(h, \mathbf{x})$ is the soil moisture capacity, which can be derived from the moisture/pressure constitutive relationship; and $K(h, \mathbf{x})$ is the hydraulic conductivity-pressure constitutive function. $Q(\mathbf{x}_p)$ is the pumping rate at location $\mathbf{x}_p$.

The following section will present the results of numerical experiments of pumping in an unconfined aquifer by Mao et al. (2011), which are conducted by solving equation (8.4.24). The dimensions of the aquifer are 200 m × 200 m in the horizontal plane and 9 m in the vertical direction. A no-flux boundary condition is assigned to the top (no infiltration or evaporation) and bottom of the aquifer; no-flux boundaries are imposed on the other four lateral sides of the aquifer. The initial pressure head distribution in the aquifer is set to be hydrostatic with the water table at z = 6.7 m, representing a static condition. A well is represented by a line source of length 4.0 m from the bottom center of the aquifer (x = 100 m, and y = 100 m) with a constant rate of discharge (0.06 m^3/min), and the borehole storage is assumed negligible.

The hydraulic conductivity-pressure head and moisture-pressure head constitutive relationship of the aquifer are described by an exponential model (Gardner, 1958):

$$K(h) = K_s \exp(\alpha h) \tag{8.4.25}$$

and

$$\theta(h) = \theta_r + (\theta_s - \theta_r)\exp(\alpha h), \tag{8.4.26}$$

respectively. Parameter α is the pore-size distribution parameter for the unsaturated hydraulic conductivity/pressure head function and the moisture/pressure head relationship; θ_s and θ_r denote saturated and residual water content, respectively. Note that (θ_s − θ_r) is equivalent to the specific yield in previous studies of the aquifer tests. The aquifer was assumed homogeneous; values of the parameters in equations (8.4.25) and (8.4.26) were taken from the study by Akindunni and Gillham (1992): K_s = 0.00396 m/min, S_s = 0.000325/m, θ_s = 0.37, and θ_r = 0.07, and α = 4.0/m, respectively. The pumping test is simulated for 2000 minutes.

8.4.5.1 Drawdown-Time Curves

Simulated drawdowns in the unconfined aquifer as a function of time at five elevations (1.5 m, 3.0 m, 6.0 m, 7.1 m, and 7.5 m) at radial distance 5.0 m and 30.0 m from the pumping well are plotted using log scales in Figures 8.25a and 8.25b, respectively. Notice that elevations 7.1 m and 7.5 m are above the water table, and drawdowns at these elevations denote pressure change in the vadose zone due to pumping in the saturated zone. These two figures show that drawdowns in the saturated zone exhibit the characteristic S shape, which is more pronounced at the bottom of the aquifer. The flat part of the S curve at z = 6.0 m occurs earlier than at the bottom of the aquifer (i.e., z = 1.5 m and z = 3.0 m). On the other hand, the drawdown-time curves at elevations 7.1 m and 7.5 m in the unsaturated zone do not exhibit the S shape. These simulated results are consistent with field observations (e.g., Bevan et al., 2005).

Figures 8.25a and 8.25b also show drawdown-time curves at different elevations (z = 1.5 m, 3.0 m, and 6.0 m) in a confined aquifer that has an infinite lateral extent, with the same saturated thickness (6.7 m) and the saturated hydraulic properties as the unconfined aquifer. These curves are obtained by using the solution of Hantush (1961a, b), which considers the partial perforation of the pumping well in the aquifer. According to Figure 8.25a, drawdowns in the confined aquifer at r = 5 m, at higher elevation z = 6.0 m are smaller than those at z = 1.5 m and z = 3.0 m, indicative of vertical flow. At r = 30.0 m, far from the pumping well, identical drawdowns at the three

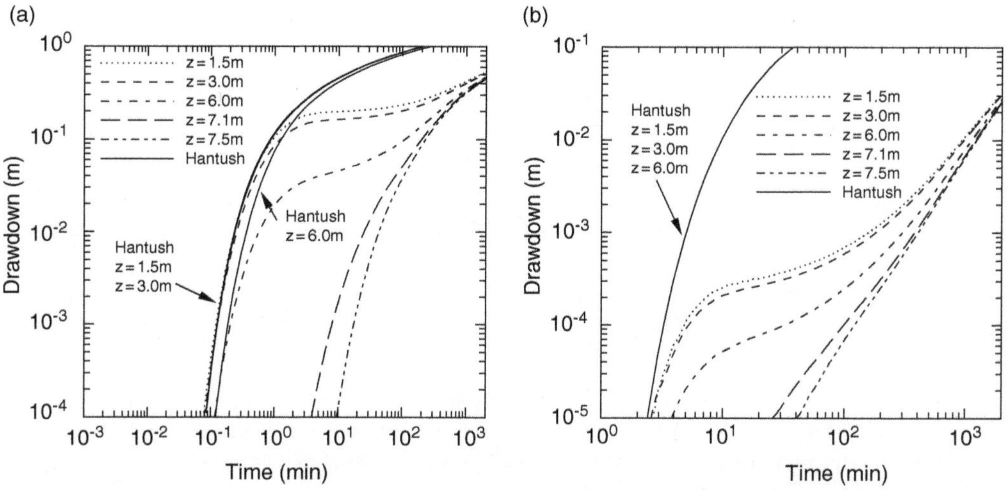

Figure 8.25. a) Log-log simulated drawdown-time curves at five different elevations (z = 1.5, 3.0, 6.0, 7.1, 7.5 m) at r = 5 m during pumping in an unconfined aquifer. b) The same at r = 30 m. Note that the elevation of the bottom of the aquifer is 0 (z = 0.0 m). Solid lines denote the drawdowns in a confined aquifer based on the solution for a partially screen pumping well by Hantush at different elevations in the saturated zone. The difference between the drawdowns at z = 1.5 m and z = 3.0 m exists but is very small at r = 5 m in a). The drawdowns at three elevations (1.5 m, 3.0 m, and 6.0 m) at r = 30 m are virtually the same, indicating flow is purely horizontal. (Mao et al., 2011)

elevations suggest that the effect of the partial perforation of the pumping well (i.e., vertical flow) is negligible (Figure 8.25b). As expected, no S-shaped curve is observed.

Notice that in both figures the drawdown-time curves in the unconfined aquifer depart from the initial steep drawdown portion and deviate from the corresponding drawdowns for a confined aquifer in a descending elevation order (i.e., z = 6.0 m, z = 3.0 m, and z = 1.5 m). This sequential departure (i.e., smaller drawdowns than the confined aquifer) implies that an apparent "additional" source of water is flowing from the upper to the lower part of the saturated zone of the unconfined aquifer as recharge, in comparison with the horizontal flow process in the confined aquifer. As will be explained later, such an "additional" source of water is mainly attributed to a different water release mechanism at this stage of the flow process in the unconfined aquifer.

8.4.5.2 Rate of Change in Storage

In order to explain the aforementioned departures in terms of different water release mechanisms, we examine the evolution of the "rate of change in storage" per volume of the porous medium (Ω) at various distances and depths during a pumping test in an unconfined aquifer. Mathematically, this rate of change in storage is the right-hand

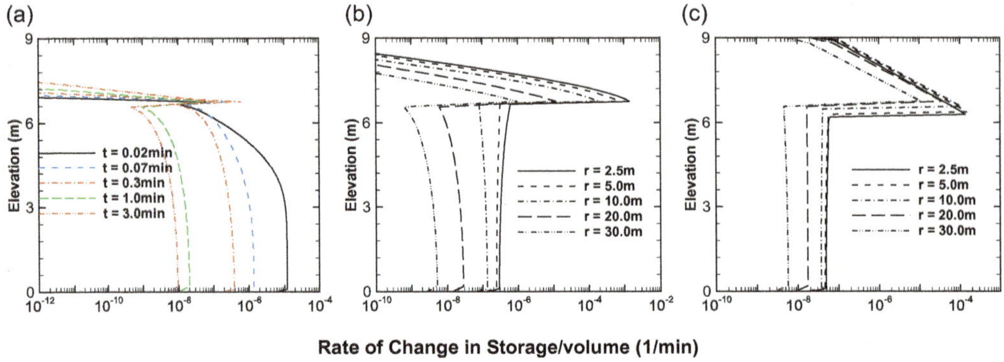

Figure 8.26. Ω as a function of elevation a) at five early times (t = 0.02, 0.07, 0.3, 1.0, and 3.0 min) corresponding to five different radii (r = 2.5, 5, 10, 20, and 30 m); b) at an intermediate stage (10 minutes after pumping); and c) a late stage (1000 minutes after pumping) at the five radii. The water table is located at the sharp discontinuity. Note that the elevation of the bottom of the aquifer is 0 (z = 0.0 m). (Mao et al., 2011)

side of equation (8.4.24), which represents the net increase in the mass flux in a unit volume of the porous medium

$$\Omega = \omega S_s(\mathbf{x})\frac{\partial h}{\partial t} + C(h, \mathbf{x})\frac{\partial h}{\partial t} \tag{8.4.27}$$

where the first term on the right-hand side of the equation denotes the change of storage due to elastic properties of the aquifer and water, while the second term represents filling up or draining of the pore.

Early times. In Figure 8.26a, the quantity Ω is plotted as a function of elevation at different radial distances from the pumping well, at various times of the early stage of the S-shaped drawdown-time curve. Specifically, these distances and times are $r = 2.5$ m at $t = 0.02$ min; $r = 5.0$ m at $t = 0.07$ min; $r = 10$ m at $t = 0.3$ min; $r = 20$ m at t = 1.0 min, r = 30 m at t = 3.0 min. We use different times for different radii to illustrate the behavior of Ω. At this early stage, the Ω value at $r = 2.5$ m is the largest at the bottom of the aquifer over the elevation from 0 m to 4 m where the screen for the pumping well is located. This value then decreases with elevation and reaches a minimum value at the water table; it then increases sharply at the water table and subsequently decreases with elevation in the unsaturated zone. Overall, the Ω values below the water table exhibit a strong nonlinear pattern with greater Ω values at the bottom of the aquifer than those near the water table. This pattern is consistent with the fact that the greatest change in pressure is at the bottom of the aquifer near the pumping well location.

Figure 8.26a shows that the Ω value in the unsaturated zone at early times is virtually zero except very close to the water table. Similar behaviors are also observed at the other radial distances. As pointed out previously, this behavior suggests that water flowing into the pumping well is mainly from the expansion of water and compaction of the aquifer material in the saturated zone, which is characterized by the specific storage of the aquifer. This period corresponds to the early time segment in the log-log drawdown-time plot (e.g. see Figure 8.23).

Intermediate times. During the intermediate stage (the flat portion) of the log-log drawdown-time curve (t = 10 min after pumping), the behavior of Ω as a function of elevation at r = 2.5 m, 5.0 m, 10.0 m, 20.0 m, and 30.0 m from the pumping well is shown in Figure 8.26b. In comparison with Figure 8.26a, the Ω values in the vadose zone above the water table (near the sharp break in the curve around z = 6.5 m) at all radii begin to increase. The Ω values in the saturated region below the water table decrease at all radial distances, compared with those in the early time. The closer to the pumping well the location, the greater reduction in Ω. Furthermore, the Ω values at locations close to the pumping well are almost constant with elevation. All these behaviors indicate that the flow field near the well is close to the steady state (quasi-steady state) in which the head field changes at the same rate at all elevations.

Figure 8.26b also suggests that the pumping has begun to induce a significant amount of water release from the vadose zone in order to supply the well discharge, whereas the elastic storage of the aquifer near the pumping location is almost exhausted. In other words, the drainage of water from pores in the unsaturated zone increasingly becomes the primary source of water for the well discharge. It yields more water to the saturated region below than does the elastic storage of the region itself and represents an "additional" source of water, compared with the similar flow in the confined aquifer. As a result, the drawdown-time curve deviates from that based on the Hantush solution (1961a, b). The transition of the water-release mechanism from compaction of the porous medium and expansion of water to dewatering of pores thus acts as an "additional" source of water and, in turn, the flat portion of the intermediate stage of the S-shaped drawdown-time curve. This is consistent with the importance of the vadose zone proposed by Nwankwor et al. (1992), Akindunni and Gillham (1992), and Narsimhan and Zhu (1993).

Late times. The behavior of Ω as a function of elevation at the five radii at t = 1000 min after pumping (i.e., the late stage of the S curve) is plotted in Figure 8.26c. At this stage, the water table at all radii starts to fall noticeably; the Ω values in the unsaturated zone are much greater at high elevations at this late stage than at the early or intermediate stages. The Ω values in the saturated region at all radii are very small and do not vary with the elevation. These behaviors of the Ω values suggest that flow in this region apparently approaches steady-state conditions; water released from the unsaturated zone and from previously saturated pores due to falling of the water table is equal to the amount of water withdrawn by the pumping well. Thus, little change in storage occurs in the saturated region of the aquifer. Note that although there is still some water released from the aquifer as a result of compaction of the aquifer and expansion of water, its amount is much smaller than the amount of water released from the falling water table and the initially unsaturated zone.

The preceding results show that the initial unsaturated zone above the water table is important, and the drainage of the initially saturated pores is continuous in space and time as the water table falls. The widely accepted terms "delayed yield," which indicates a lag in water release from the initially unsaturated zone, and the concept of "delayed water table response" do not elucidate the transition of the two water release mechanisms and are misleading. Although both delayed yield and delayed water table response models are either implicitly or explicitly related to the S-shaped behavior to transition in water release mechanisms, they do not capture the water release mechanism in the initially unsaturated zone above the water table and the initially saturated zone after falling of the water table. Specifically, the free-surface boundary with the specific yield term (equation (8.4.17)) misrepresents the pore drainage process, which is regulated by the moisture capacity term $C(h, \mathbf{x})$ in equation (8.4.27) and the capillary head before and during the falling of the water table. Under the same observed change in the water table position, the instantaneous release (or rapid drainage) model therefore has to force the specific yield or porosity value to be very small in order to match the drainage that is governed by the moisture capacity term. Therefore, application of the instantaneous model to drawdown-time data collected from field pumping tests generally yields specific yield (more correctly, porosity) values that are substantially below those that would be expected on the basis of other methods of measurement (see Nwankwor et al., 1984; Endres et al., 2007)

8.4.6 Information Content of Drawdown

Analysis of pumping test data from an unconfined aquifer based on an equivalent homogeneous conceptual model faces the same paradox (i.e., comparing apples and oranges) as do the classical analyses of pumping tests in a confined aquifer (see Section 7.4 in Chapter 7). In order to discuss this issue again, we will discuss the temporal and spatial cross-correlation between the observed head and hydraulic properties in the unconfined aquifer (see Section 9.3 in Chapter 9 for the mathematical analysis) or information content of drawdown about aquifer properties. The cross-correlation analysis for this case is more complex than those presented in Chapter 9, but the principle is the same. Details of the analysis are given in Mao et al. (2013b). We will briefly discuss results of the analysis in the following section.

8.4.6.1 Homogeneous Media

The temporal evolutions of the cross-correlation between the effective parameters, K_s, S_s, α, and θ_s, of the homogeneous medium and the heads at a radial distance ($r = 6$ m) and elevations (z = 1 m and z = 4 m) in the saturated zone are illustrated in Figures 8.27a and 8.27b. The cross-correlations between these parameters and the head in the vadose zone at $r = 1$ m and $z = 7$ m are shown in Figure 8.27c as a function of time. Besides the cross-correlations, the drawdowns with time are plotted as a blue solid curve in the figure corresponding to each observation location.

Generally speaking, the temporal evolutions of the cross-correlation between the heads at the two elevations in the saturated zone with the parameters are similar, but the timings of the cross-correlations at the higher elevation are delayed slightly. As illustrated in Figures 8.27a and 8.27b, the drawdown-log time curves at these two elevations exhibit the typical S-shaped response of an unconfined aquifer due to pumping. That is, at early times, drawdown rises rapidly, then it slows at intermediate times, and at late times it rises quickly again. The exact times of the three periods would vary with many factors. We therefore discuss the behaviors of the cross-correlations on the basis of the early, intermediate, and late times of the two S-shaped drawdown-time curves in the following sections.

At early times, flow to the pumping well is mainly from release of water from the storage in the saturated region within radial distances between the pumping well and the observation locations. Therefore, if the head at the observation location is high (i.e., the hydraulic gradient is steep toward the pumping well), the K_s of the aquifer therefore is likely low, and vice versa. Note that the qualifier "likely" is used to emphasize the influence of uncertainty in observed heads due to measurement errors, as well as spatial variability. The observed heads in the saturated zone are thus negatively correlated with K_s.

Aquifer compaction and water expansion are the major mechanisms responsible for the water release from the storage in the region at this period, and they are represented by the parameter S_s. The larger the S_s, the more water can be released per unit drawdown, resulting in the smaller drawdown (higher head) at the observation well for a given amount of release. The head in the saturated zone thus is positively correlated with S_s. This cross-correlation reaches the maximum around this time and decreases afterward. As expected, effects of parameter α and θ_s on the heads in the saturated zone are small. Their effects are greater on the head observed close to the vadose zone (see Figure 8.27a). Overall, there is no pressure response at the observation location in the vadose zone at this time.

At the intermediate times, the rate of drawdown at the observation locations starts to decrease, and the flat portion of the S-shaped log drawdown-log time curve in the saturated zone is formed. Water flowing to the pumping well is now mainly from drainage of the vadose zone and drainage of pores due to lowering of the water table at locations above the elevations of the two head measurement locations in the saturated zone. The zero cross-correlation between the heads and S_S at these two locations at this period further confirms this fact.

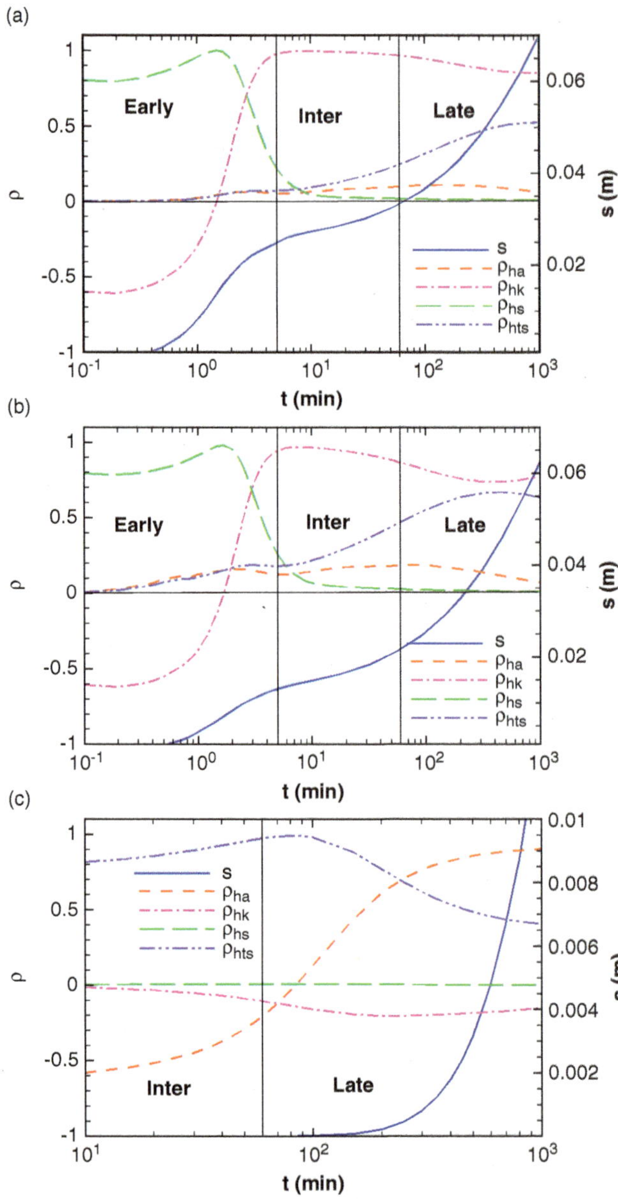

Figure 8.27. a) Cross-correlation as a function of t between K_S and observed heads (h) at r = 6 m at z = 1 m is denoted by ρ_{hk} (pink dash-dotted line); the relationship between S_S and h is represented by ρ_{hs} (green dashed line); ρ_{ha} (red dashed line) denotes the relationship between h and α; ρ_{hts} (the purple dash-dot-dot line) represents the relationship between h and θ_s. b) the same as a) but at z = 4 m. Both a) and b) are in the saturated zone. c) the same as a) and b) but at z^* = 7 m, which is in the vadose zone. The blue solid lines represent the drawdown, s. Two vertical lines separate the curves into early, intermediate (inter), and late periods. (Modified from Mao et al., 2013b)

Because flow is from locations above the observation location, if the observed head is higher (or smaller drawdown) than the one simulated with the mean parameter value, then the hydraulic gradient behind the observation location is likely smaller than the simulated, and thus the aquifer likely has a larger value of K_s than the mean value. The cross-correlation between the head in the saturated zone and K_s thus is positive.

However, the cross-correlation between the head in the vadose zone and K_s is negative and small. Such a negative correlation implies that if the head in the vadose zone is less negative (near saturation), the K_s value is likely to be small such that little water is drained. Pressure head responses at the measurement point in the vadose zone are also insignificant (Figure 8.27c). This implies that the head measurement at this time and location may carry little information about the K_s value of the aquifer.

The influence of aquifer parameters α and θ_s on the heads in the saturated zone increases during this period as shown in Figures 8.27a and 8.27.b. The cross-correlation between heads and θ_s is higher than that of α although both values are small.

At late times ($t > 60$ min), the rate of drawdown change at this period at the two observation locations in the saturated zone begins to increase to form the second rising limb of the S-shaped drawdown-time curves. Drawdown observed at the sampled location in the vadose zone increases significantly. The positive correlation between K_s and the heads at the two locations in the saturated zone decreases slightly but remains high. The head measurement in the vadose zone becomes more negatively correlated with K_s (Figure 8.27c). On the other hand, S_s does not correlate with any of these three heads, reflecting the fact that flow near these observation locations approaches a steady-state condition.

At this time, the positive relationship between the head at the two measurement points in the saturated zone and θ_s is increasing greatly. This positive correlation means that the larger the value of θ_s, the more water can be released from vadose zone drainage and pore drainage in the initially saturated zone as a result of lowering of the water table, and the higher heads in the saturated zone. Meanwhile, this correlation with the head in the vadose zone with θ_s reaches a maximum at the early part of this period and then decreases since the pore-size distribution (i.e., α) starts to take effect.

The relationship between α and the heads in the saturated zone increases slightly. Its cross-correlation with the head in the vadose zone, however, becomes significantly positive. This positive correlation implies that a larger α (i.e., a coarser material) will induce release of more water from the dewatering process near the water table and thus a larger (less negative) observed head can be sustained.

Overall, the temporal evolution of the cross-correlation between the head at different locations and the parameters (K_s, S_s, α, and θ_s) suggests a temporal data sampling strategy for better estimating the parameters of a homogeneous aquifer. Specifically, drawdown in the saturated zone at an early time where the maximum cross-correlation takes place will be optimal for estimating the S_s parameter of the aquifer. For the estimation of K_s, using the head measurement from the intermediate to the late periods is the most effective. To derive better estimates of α and θ_s parameters, using head measurements in the vadose zone at a late time would be the most desirable.

In addition, the analysis demonstrates that both α and θ_s control the drainage process in the vadose zone as well as that in the saturated zone resulting from the falling of the water table during the pumping test in the unconfined aquifer. This result corroborates the previous conclusion that the models assuming instantaneous release of water from drainage of pores (Section 8.4.4) misrepresent capillary effects on drainage of water from pores and could lead to significantly small porosity (specific yield incorrectly used in those models) values.

8.4.6.2 Heterogeneous Media

Similar to what we did in Section 7.4 in Chapter 7, here, we examine the influence of the head at an observation location due to the parameter at different parts of the unconfined aquifer. Discussion of the cross-correlation analysis for a heterogeneous medium follows the early-, intermediate-, and late-time classification used for the homogeneous case. Figure 8.28 shows at selected early and late periods the spatial distributions of the cross-correlations between the heads at a selected measurement location and each of the parameters (K_s, S_s, α, and θ_s), the mean drawdown distribution,

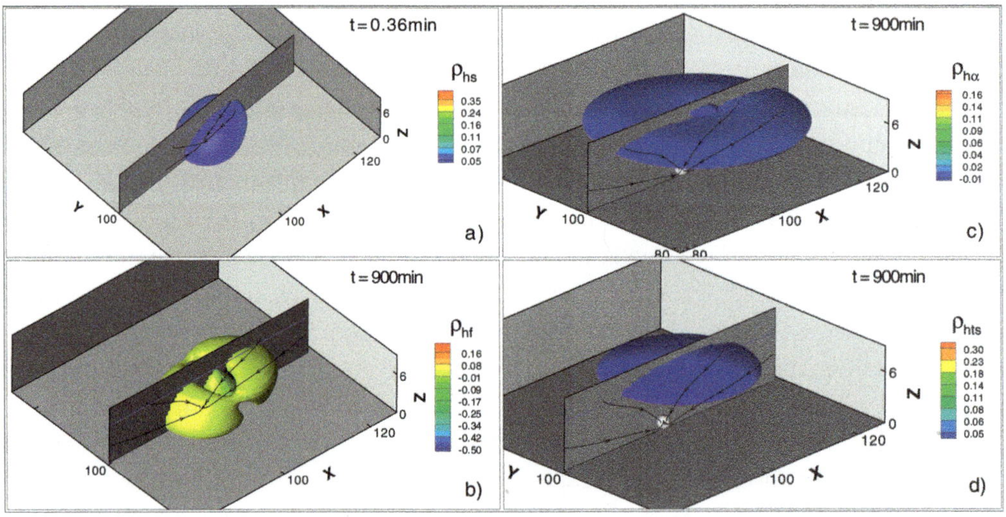

Figure 8.28. The 3D isosurfaces contour for the cross-correlation between the observed head at a location in the saturated zone (the dark blue dot) and parameters a) S_S at an early time, b) K_S, c) α, and d) θ_S at a late time. The darker cross section is the selected cross section that goes through the pumping well. The blue solid line is the locations of the water table, and the four black solid lines with arrows are the streamlines (or the flowlines). (Modified from Mao et al., 2013b)

as well as the mean streamlines during a pumping test in the unconfined aquifer. Physical meanings and implications of results in these figures are summarized in the following and a detailed discussion of these figures is provided in Mao et al. (2013b).

According to these figures, the cross-correlation analysis of observed heads and hydraulic properties during a pumping test in homogeneous and heterogeneous unconfined aquifers reveals that the head observed in the saturated zone at late times carries the greatest information about K_S heterogeneity distribution. In general, it contains nonsymmetrically weighted information about heterogeneity within the cone of depression. More weight is given to those K_S heterogeneity upstream of the observation location and the region between the observation location and the pumping location along the streamlines passing through the observation and pumping locations. Similar weights are also given to those K_S upstream of the pumping well along the streamlines on the opposite side. While these results are similar to those for the 2D depth-averaged confined aquifers, they reveal complex 3D spatial distributions of these cross-correlations. The information of K_S from head measured at the location in the vadose zone follows similar patterns but is less comprehensive. Identically to confined aquifers, the measured head in the saturated zone in an unconfined aquifer at the early time carries the most information about S_S heterogeneity in a narrow region along the streamline between the observation and pumping locations.

At intermediate times and late times, the heads measured in the saturated zone largely reflect the heterogeneity of α in the thin unsaturated region near the water table above the pumping and observation locations. The head measured in the vadose zone at late times has the most relevant information about α around the observation point in the vadose zone.

Likewise, θ_S has impacts on the head measurements in the saturated zone and the vadose zone during the intermediate and late times. Again, heterogeneity of θ_S in the vadose zone directly above the pumping and monitoring locations has the greatest impact on the head measurement in the saturated zone. Heads measured in the vadose zone correlate well with heterogeneity of θ_S in the area around the measurement location.

Implications of the cross-correlation analysis of heterogeneous aquifers are significant for the conventional analysis of unconfined aquifer tests. Conventional aquifer analyses rely on drawdown-time data collected at an observation well due to pumping at another well. They also assume aquifer homogeneity for mathematical simplicity (e.g., Tartakovsky and Neuman, 2007; Moench, 2008; Mishra and Neuman, 2010). The cross-correlation analysis shows that the influence of K_S at different regions of the aquifer on the head at an observation location changes with time. As a result, these conventional aquifer analyses will likely lead to ambiguous estimates of aquifer properties, as discussed in Section 7.4 of Chapter 7 for confined aquifers. In addition, the unique pattern of the cross-correlation between head and hydraulic parameters will likely lead to a scenario-dependent estimate of the aquifer properties that vary with the location of pumping well, as reported by Huang et al. (2011) and Sun et al. (2013).

While these complex 3D cross-correlation patterns between head and parameters may not be representative for all possible scenarios, they reinforce the need for hydraulic tomography to characterize unconfined aquifers (Zhu and Yeh, 2008; Cardiff and Barrash, 2011; Zhu et al., 2011), or joint interpretation of sequential pumping tests for unconfined aquifers (Mao et al., 2013b), which will be discussed in Chapter 9.

Lasty, the analysis by Mao et al. (2011; 2013b) reveals that the S-shaped drawdown-time curve in the saturated region during a pumping test in unconfined aquifers is sensitive to the spatial variability of hydraulic conductivity, specific storage, and saturated moisture content, and insensitive to the variability of other parameters for unsaturated hydraulic properties. This result implies that analysis of pumping tests in an unconfined aquifer for hydraulic conductivity and specific storage may assume homogeneity of the unsaturated zone. In order to identify the unsaturated hydraulic properties, measurements of negative pressure heads and moisture contents above the water table are necessary.

8.5 Summary

- Geologic boundaries and confined, leaky, and unconfined aquifers are merely our conceptual models of heterogeneity of aquifers.
- An impermeable boundary reduces the supply to the pumping well and increases the drawdown.
- A recharge boundary provides additional sources of water to the storage of the aquifer; it reduces drawdown and may lead to a steady flow field.
- Behaviors of leaky aquifers during a pumping test are similar to aquifers connected to a recharge boundary (or heterogeneity) that supplies additional water to the aquifers. Interpretation of aquifer characteristics based on limited observation wells requires caution.
- The S-shaped drawdown-time curves observed in pumping tests in unconfined aquifers are a manifestation of the transition of a water release mechanism from the elastic effect to drainage of pores, rather than the traditional concepts of the delayed yield or delayed water table response.
- The model for variably saturated flow in a heterogeneous model is most appropriate to analyze the pumping test in an unconfined aquifer. Hydraulic tomographic survey is a new generation of aquifer characterization methods.

8.6 Exercises

1. A nonleaky artesian aquifer is bounded on one side by an impermeable boundary, and the transmissivity and storage coefficient are determined to be 3.85×10^5 gpd/ft and $3.4 \times$

Table 8.1.

Distance from the Production Well ft	Drawdown ft
100	8.42
1,000	5.65
10,000	2.84

Table 8.2.

Time after Pumping Started, min	Drawdown, ft
2	0.801
3	0.912
4	1.061
5	1.172
6	1.231
7	1.323
8	1.376
9	1.432
10	1.481
20	1.882
30	2.112
40	2.341
50	2.522
60	2.702
70	2.831
80	3.003
90	3.171
100	3.301
200	4.212
300	4.431

10^{-4}, respectively. A fully penetrating production well has been discharging at a constant rate of 1485 gpm. The effect of the impermeable boundary is 0.61 foot for a pumping period of 100 minutes. This effect on drawdown is observed in an observation well 300 feet from the pumping well. Compute the distance from the observation well to the discharging image well associated with the barrier boundary.

2. A production well fully penetrating a nonleaky isotropic artesian aquifer with infinite areal extents was continuously pumped at a constant rate of 500 gpm for a period of 200 days. Table 8.1 shows the drawdowns observed at various distances from the production well in fully penetrating observation wells at the end of the 200-day pumping period. Compute the transmissivity and storage coefficient of the aquifer.
3. A production well fully penetrating a nonleaky isotropic artesian aquifer was continuously pumped at a constant rate of 1485 gpm for a period of 5 hours. This aquifer is delimited by two impermeable boundaries. Table 8.2 lists the drawdowns observed at a distance of 300 feet in a fully penetrating observation well. Compute the transmissivity and storage

coefficient of the aquifer and the distances to the image wells associated with the impermeable boundaries.

4. Derive the law of times for determining the time when the drawdown reaches the impermeable boundary. Discuss application of the law of times to detect the location of a recharge boundary.
5. Discuss the effects of the partial penetrating river on the drawdown-time curve.

9

Stochastic Approaches

9.1 Introduction

This chapter synthesizes the materials from all previous chapters to present a unified approach for analyzing groundwater flow through heterogeneous geologic media. The first part examines necessary conditions for mathematical (or numerical) models to be well defined such that a mathematically unique solution can be obtained. These mathematical models include those for prediction of groundwater flow – forward problems – and those for estimation of hydraulic properties of geologic media using their responses – inverse problems.

While the necessary conditions for real-world problems to be mathematically well defined can be collected and specified, the collected or specified conditions always involve uncertainty. Because of this uncertainty, the true solution to the problem is unknown (i.e., there are many possible solutions). As a result, stochastic methods to the forward and inverse problems are introduced in this chapter for deriving the statistically most likely solution, and quantifying its uncertainty. These stochastic techniques are grouped into two categories: unconditional and conditional approaches. The unconditional approach, which does not necessarily honor spatial or temporal observations at the sampled locations, includes the unconditional effective parameter and the Monte Carlo simulation methods. Both of them aim to provide unbiased estimates of the model parameters and predictions for flow field but at low resolutions. On the other hand, the conditional approach honors the observations and provides statistically unbiased and high resolution estimates of model parameters and predictions for the flow field. This approach is comprised of conditional effective parameter and conditional Monte Carlo simulation methods. In addition to discussion of these methods, a first-order analysis is presented for analyzing uncertainty in predicted heads and for studying the relationship between the observed head and parameters of the flow model. This relationship is the basis of the conditional methods, and it is used to explain the shortcomings of the unconditional effective parameter methods.

Conditional stochastic estimators (i.e., kriging and cokriging, successive linear estimator), which fuse the same or different types of information, are discussed next. At the end of this chapter, we discuss methods (i.e., hydraulic tomography) that collect nonredundant data sets in a cost-effective manner for reducing aquifer characterization uncertainties.

9.2 Well-Defined and Ill-Defined Forward and Inverse Problems

A forward problem here entails solving the governing flow equations for head in time and space over a solution domain (i.e., a basin or an arbitrarily defined field site), with given model input parameters, and boundary and initial conditions. For example, consider the following partial differential equation based on a heterogeneous geologic conceptual model. It describes the movement of water in a multidimensional, fully saturated geologic medium (see Chapter 5).

$$\nabla\cdot[K(\mathbf{x})\nabla H] + Q(\mathbf{x},t) = S_s(\mathbf{x})\frac{\partial H}{\partial t} \tag{9.2.1}$$

with boundary conditions

$$H(\mathbf{x},t) = H^*(\mathbf{x},t) \text{ at } \mathbf{x} = \Gamma_1 \text{ and } q(\mathbf{x},t) = -[K(\mathbf{x})\nabla H]\cdot\mathbf{n} \text{ and } \mathbf{x} = \Gamma_2 \tag{9.2.2}$$

where **n** is the normal vector, and an initial condition

$$H(\mathbf{x},0) = H^*(\mathbf{x},0) \tag{9.2.3}$$

where $\nabla = [\partial/\partial x, \partial/\partial y, \partial/\partial z]$, **x** is the position vector, t is time, H is the total head, $K(\mathbf{x})$ is hydraulic conductivity, $Q(\mathbf{x}, t)$ is a sink or source, $S_s(\mathbf{x})$ is the specific storage, $q(\mathbf{x}, t)$ is the boundary flux, and **n** is a unit vector normal to the boundary Γ_2. Here, K and S_s are defined as hydraulic properties or parameters of a geologic medium, and H is a state variable or system response.

If the properties (K and S_s), sink and source terms, $Q(\mathbf{x}, t)$, and boundary and initial conditions (i.e., necessary conditions) of a forward problem are fully specified, regardless of their accuracy, a particular (unique) solution, $H(\mathbf{x}$, t), to equation (9.2.1) exists, and this forward problem is then a **well-defined** problem. Otherwise, the forward problem is ill defined and it cannot be solved.

A well-defined problem, however, may not yield a "true" solution, because of the uncertainty or inaccuracy involved in the specified properties, sink and/or source terms, and boundary and initial conditions. That is, the solution to the equation is uncertain (i.e., the true solution is unknown), and therefore an infinite number of possible solutions exist.

The forward model, equation (9.2.1), can be rearranged to take the following form:

$$\mathrm{A}\nabla K(\mathbf{x}) + \mathrm{B}K(\mathbf{x}) + Q(\mathbf{x},t) = \mathrm{D}S_s(\mathbf{x}) \tag{9.2.4}$$

where $\mathrm{A} = \nabla H(\mathbf{x}, t)$, $\mathrm{B} = \nabla^2 H(\mathbf{x}, t)$, and $\mathrm{D} = \partial H(\mathbf{x}, t)/\partial t$. If $K(\mathbf{x})$ and $S_s(\mathbf{x})$ are unknowns to be estimated, equation (9.2.4) is then the first-order partial differential equation for an inverse problem for flow through a heterogeneous porous medium. We call this problem an inverse problem because the problem attempts to use the responses of an aquifer (i.e., head) to identify the hydraulic properties of the aquifer.

In order to solve equation (9.2.4), coefficients A, B. and D (i.e., head) at every **x**, at least at two times in the solution domain, and $Q(\mathbf{x}, t)$ must be specified (Yeh et al., 2011; Mao et al., 2013a). Since equation (9.2.4) is a first-order partial differential equation in terms of $K(\mathbf{x})$, specification of the boundary conditions for equation (9.2.4) in terms of $K(\mathbf{x})$ is a must for obtaining a particular solution. If the preceding necessary conditions are provided, a unique solution exists, and the inverse problem is **well defined.** Otherwise, it is **ill defined** and it cannot be solved mathematically, the same as an ill-defined forward problem.

In inverse problems that aim to identify aquifer properties, we will denote the term "primary variables" for properties, $K(\mathbf{x})$ and $S_s(\mathbf{x})$, that are being sought, and the term **primary information** for the available measurements of these properties. Variables such as heads and $Q(\mathbf{x}, t)$ are defined as secondary variables, and the data for these variables are called **secondary information,** since they are not the properties to be estimated.

To elucidate further the necessary conditions for an inverse problem for either a homogeneous or a heterogeneous aquifer conceptual model (see Chapter 5) to be well defined, we will consider some simple 1-D cases next.

9.2.1 Homogeneous Conceptual Model

Steady-state saturated flow. Consider one-dimensional, steady-state, saturated flow through a vertical homogeneous soil column with a length of L. If the specific discharge (a flux boundary condition) and the hydraulic head distribution or gradient along the column are known, then the hydraulic conductivity can be determined uniquely using Darcy's law: $K_s = -q/[d(h + z)/dz]$, with $z = L$. Otherwise, the inverse problem is ill defined and there is no solution for K_s. That is to say, in order to have a unique inverse solution to a steady-state saturated flow problem, the spatial hydraulic head distribution, from which the gradient can be determined, and a flux boundary condition must be specified beforehand. These are the necessary conditions for this inverse problem to be well defined.

Transient saturated flow. The inverse problem is more complicated if the flow is transient. Consider the governing flow equation for a one-dimensional transient saturated flow in the vertical direction:

$$K_s \frac{\partial^2 (h+z)}{\partial z^2} = K_s \frac{\partial^2 H}{\partial z^2} = S_s \frac{\partial H}{\partial t} \tag{9.2.5}$$

where $H = h + z$. Notice that equation (9.2.5) can be rewritten as a diffusion equation where the diffusion coefficient is the ratio of K_s to S_s. Thus, it is impossible to identify K_s and S_s uniquely solely on the basis of hydraulic head measurements. To overcome this problem, we integrate the preceding equation from the bottom ($z = 0$) to the top ($z = L$) of the soil column to obtain

$$-K_s \frac{\partial H}{\partial z}\bigg|_{z=0,\,t} = S_s \frac{\partial}{\partial t}\left[\int_{z=0}^{z=L} H(z,t)dz\right] - K_s \frac{\partial H}{\partial z}\bigg|_{z=L,\,t} \tag{9.2.6}$$

The first term on the right-hand side of equation (9.2.6) represents the change in storage along the column at time t, while the second term denotes the specific discharge at $z = L$ at t. The sum of these two terms results in the specific discharge at the end of the column, $q(0, t)$:

$$-K_s \frac{\partial H}{\partial z}\bigg|_{z=0,t} = q(0,t) \tag{9.2.7}$$

Provided that the specific discharge, $q(0, t)$ and the gradient, $\partial H/\partial z$ at $z = 0$ at time t, are known, the hydraulic conductivity can then be determined.

Once K_s is known, according to equation (9.2.6), knowledge of the spatial distribution of the hydraulic head along the column, at least at two times, yields the flux at top and bottom of the column, the change in head over the column, and, in turn, the value of the parameter S_s. Therefore, the necessary conditions for a transient 1-D saturated flow inverse problem to be a well-defined problem are knowledge of 1) the spatial distribution of the hydraulic head at least at two times and 2) the flux at one boundary. The requirement of the knowledge of the spatial head distribution, however, can be relaxed when a predetermined mathematical model is used to fit the observed data (Yeh et al., 2011).

Steady-state unsaturated flow. The unsaturated hydraulic conductivity, being a function of pressure head or moisture content, generally exacerbates the difficulty of inverse modeling of flow through an unsaturated porous medium. The necessary conditions for the saturated flow problem (see the steady-state saturated flow discussion) still apply for the inverse problems of this category. For example, consider steady flow in the previous soil column, but now under an unsaturated condition. Assuming that Darcy's law is still applicable for such flow, the following 1-D equation (see Chapter 3) applies:

$$q = -K(h)[dh/dz + 1] \tag{9.2.8}$$

where h is the negative pressure head. Note that specific discharge (q) is constant throughout the soil column as a result of 1-D steady flow. If the pressure gradient at a location along the column is known along with the specific discharge, then the $K(h)$ value at that location is determined. To relate the conductivity to the pressure head, the pressure head must be determined at the location where the gradient is evaluated.

On the basis of this principle, a unit gradient approach ($dh/dz = 0$) has been a popular method for determining the unsaturated hydraulic conductivity. Under a unit gradient condition, the specific discharge (q) equals $K(h)$ at a given h. Since the conductivity is a function of the pressure head (h), many steady unit gradient experiments with different flow rates must be conducted. The number of steady flow experiments can be reduced if a constitutive relationship between K and h is prescribed. For example, if we assume that $K(h) = K_s \exp(\beta h)$, where K_s is saturated hydraulic conductivity and β is a pore-size distribution parameter (i.e., the Gardner model; see Chapter 3, equation 3.5.1), we have

$$q = -K_s \exp(\beta h) \tag{9.2.9}$$

In this case, two independent equations are required to determine the value of the two unknowns uniquely (i.e., K_s and β parameters). Therefore, at least two steady unit-gradient flow experiments with different pressure heads must be conducted. If the Mualem and Van-Genuchten (MVG) model (see Chapter 3, equation 3.5.2), which has three model parameters (K_s, α, and n), is used, then three unit gradient experiments with different flow rates must be conducted.

When the unit gradient condition does not exist, at least two (for the Gardner model) or three (for the MVG model) different pressure heads and their gradients must be known, in addition to the known specific discharge. With this information, two or three independent equations can be established and then the inverse problem becomes well defined. However, nonlinearity in these equations can make a mathematical solution difficult to obtain. Again, for such inverse problems, the pressure profiles (many pressure heads) along the flow path and boundary flux are the necessary conditions for a well-defined inverse problem.

Transient unsaturated flow. For transient unsaturated flow, we will integrate the Richards' equation (Chapter 5, equation 5.5.4) over the length of the soil column (from z = 0, to z = L, the top of the column) to derive the following equation:

$$-K(h)\left[\frac{\partial h}{\partial z}+1\right]\Bigg|_{z=0,\,t} = \frac{\partial}{\partial t}\left[\int_{z=0}^{z=L} \theta(z,t)dz\right] - K(h)\left[\frac{\partial h}{\partial z}+1\right]\Bigg|_{z=L,\,t} \qquad (9.2.10)$$

The term on the left-hand side of equation (9.2.10) is the outflow at z = 0. The first term on the right-hand side of equation (9.2.10) denotes change in moisture storage, while the second term on the right-hand side is the specific discharge at z = L.

According to this equation, knowledge of the pressure head and moisture content along the column at any two times will allow evaluation of the pressure head gradients at z = 0 and z = L, as well as the change in moisture content along the column over the time interval (the first term on the right-hand side of the equation). Equation (9.2.10), however, remains an ill-defined inverse problem, since $K(h)$ at z = 0 and z = L are unknown, and the number of unknowns is greater than the number of independent equations. Nonetheless, the sum of the two terms on the right-hand side of equation (9.2.10) yields the specific discharge at the end of the column, $q(0, t)$.

$$-K(h)\left[\frac{\partial h}{\partial z}+1\right]\Bigg|_{z=0} = q(0,t) \qquad (9.2.11)$$

Accordingly, an additional measurement of discharge at z = 0 at the two given times will allow estimation of K for the observed pressure head. The minimum number of $K(h)$ estimates again depends on the form of the conductivity model used. This scenario actually forms the basis of the well-known instantaneous profile method (see Chapter 3, Section 3.9.2).

Notice that if the measurements of both the pressure head and the moisture content at different times at a point in a homogeneous soil column are available, the moisture retention curve $\theta(h)$ of the medium can be constructed. Furthermore, if the parameters of the retention curve are equivalent to those of the relative conductivity curve, the relative conductivity versus pressure head relationship can then be derived without inverting the flow model.

Consequently, the necessary conditions for the inverse problem to be well defined are knowledge of the head and moisture content distributions at different times and of one of the boundary fluxes in equation (9.2.10) at different times (assuming the magnitudes are nonzero). The aforementioned conditions are also valid for multidimensional flow (Yeh et al., 2011; Mao et al., 2013a).

9.2.2 Heterogeneous Conceptual Model

Spatially varying hydraulic properties complicate the inverse problems for flow through heterogeneous conceptual models. However, principles discussed previously for the homogeneous conceptual model still apply to both steady and transient flows. For example, consider a steady-state saturated flow through a soil column, consisting of 10 material blocks (zones); all zones have the same length but different hydraulic conductivity, as shown in Figure 9.1. Suppose the spatial distribution of head along the column is known. The gradient along the column can thus be calculated. As the gradient is determined, the spatial pattern of the relative magnitude of hydraulic conductivity along the column is known. Since the flow is steady and one-dimensional, the specific discharge must be constant throughout the flow path. Consequently, knowledge of the boundary flux or the boundary hydraulic conductivity allows the determination of K for each zone exactly, using Darcy's law: that is, a mathematically unique solution. Accordingly, the head distribution along the flow path and the boundary flux constitute the necessary conditions for this inverse problem to be well defined. Furthermore, these conditions for a 1-D problem hold for each streamline for any multidimensional flow problem.

Finally, we emphasize that the necessary conditions are a must to make a forward or an inverse mathematical problem well defined such that it has a unique solution. When, on the contrary, these necessary conditions are not specified, the mathematical

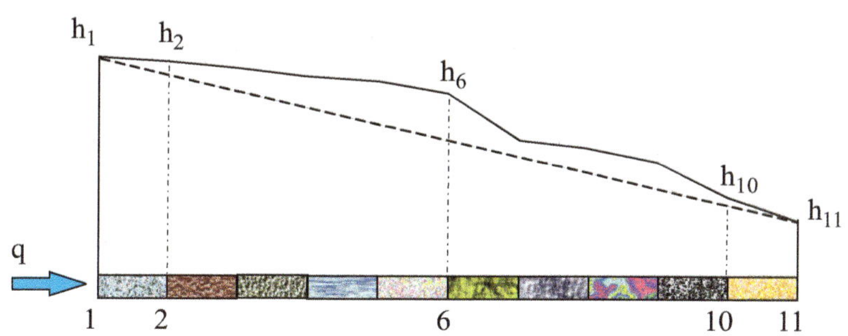

Figure 9.1. Distribution of the steady-state head values along a heterogeneous soil column.

problem is ill defined, and the mathematical problem cannot be solved, unless the definition of the problem is altered, as discussed below.

9.3 Stochastic Conceptualization of Forward and Inverse Problems

To shed some light on solutions to an ill-defined problem that has insufficient information about the necessary conditions for a given problem, let us revisit the steady flow problem illustrated in Figure 9.1. Suppose the specific discharge, q; pressure heads, h_1 and h_{11}; and the size of each block (L) are the only information available to us. Using Darcy's law, we can formulate the following relationships:

$$\begin{aligned}&-K_1(h_2-h_1)=-K_2(h_3-h_2)=-K_3(h_4-h_3)=-K_4(h_5-h_4)=-K_5(h_6-h_5)=\\&-K_6(h_7-h_6)=-K_7(h_8-h_7)=-K_8(h_9-h_8)=-K_9(h_{10}-h_9)=-K_{10}(h_{11}-h_{10})=qL,\end{aligned} \tag{9.3.1}$$

According to equation (9.3.1), there are 19 unknowns (i.e., 10 Ks and 9 heads) and only 10 equations (all 10 equations are contained in equation (9.3.1)). Therefore, neither determining the exact head distribution along the column (a forward problem), nor identifying the K for each of the 10 blocks along the column (an inverse problem) is well defined. That is, there is no unique solution to either problem. Or, one may say that there are an infinite number of possible head and K distributions that can yield the same discharge with the given two boundary head values (h_1 and h_{11}; see Figure 9.2) and specific discharge q.

These possible heads are nevertheless bounded by the two limiting head distributions, indicated by dashed lines, one at the top (blue dashed line) and one at the bottom (black dashed line), because of constraints imposed by Darcy's law (a physical principle) (see Figure 9.2). Likewise, a large number of K values will satisfy the set of equations in equation (9.3.1), but all of these are bounded by a maximum value, $K_{\max}=10Lq/(h_1-h_{10})$, if we consider the upper bound heads, assuming that h_{10} is 0.999 of h_1), or $K_{\max}=10Lq/(h_2-h_{11})$, if we consider the lower bound heads, assuming that h_{11} is 0.999 of h_2. They are also bounded by a minimum value

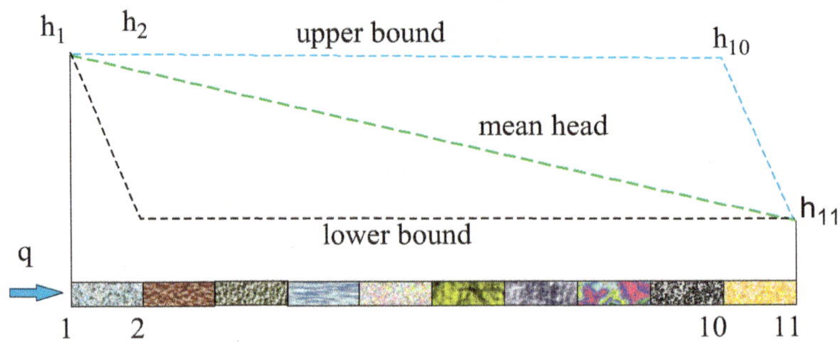

Figure 9.2. Upper and lower bounds of all possible head distributions along the soil column in Figure 9.1, with 11 nodes and 10 blocks of different hydraulic conductivities under a steady flow condition, with given q and boundary heads, h_1 and h_{11}.

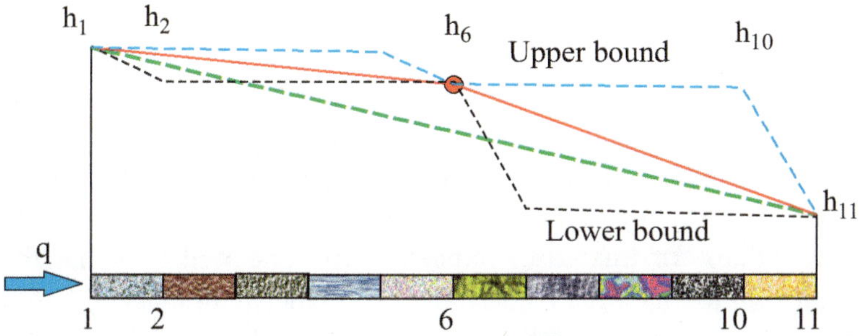

Figure 9.3. Possible head distributions along the heterogeneous soil column in Figure 9.2 with h_6 value specified.

$K_{\min} = Lq/(h_{10} - h_{11})$ based on the upper bound heads and $K_{\min} = Lq/(h_1 - h_2)$ based on the lower bound. Notice that these are approximations, since neither h_2 nor h_{10} is known. We guess what these bounds are according to the fact that the minimum head h_2 must be slightly greater than h_{10}, so that a nonzero gradient exists, and the given q is maintained.

While an infinite number of heads and Ks exist for the illustrated case, one can define an unconditional ensemble average hydraulic head distribution and the mean gradient, given the boundary heads (h_1 and h_{11}) (see Chapter 6, also Figure 9.2). This distribution and gradient are the most likely head distribution and gradient, without considering any observation inside the domain (green dashed line in Figure 9.2). Once the mean gradient is known, with the given specific discharge, an effective hydraulic conductivity of an equivalent homogeneous conceptual model can be uniquely determined (see Section 9.2.1). Notice that this effective K is not the true K for each block; it is merely the hydraulic conductivity for an equivalent homogeneous medium that reproduces the specific discharge and yields the most probable head distribution with the given information. As shown in Chapter 6, the effective hydraulic conductivity for this case is the harmonic mean of the unknown Ks. Furthermore, the effective K, in essence, defines the approximate mean K along the column.

In addition to the information given in Figure 9.2, suppose that an observed head h_6 is specified along the column, shown as a red dot above the unconditional ensemble mean head (green dashed line) in Figure 9.3. With this new information, all possible head distributions must agree with the observed head at the observation location (i.e., head conditioning). Then, following the previous reasoning, one can deduce the upper and the lower bound for the heads, which are indicated by the blue and the black dashed line, respectively. The red line, which passes through the observed head h_6 and the two boundary heads, then represents the conditional ensemble mean head, given the observed head h_6. Comparison of the upper and the lower bound in this figure with those in Figure 9.2 reveals that the overall range of possible head distributions is smaller after specification of h_6. In addition, the range of possible heads is narrow on the upstream side (from h_1 to h_6), where the mean gradient of the conditional mean head is small. However, the range is large on the downstream side (from h_6 to h_{11}), where the mean gradient is large. These relationships therefore suggest that the uncertainty in head is proportional to the mean gradient, even without any knowledge of the variability of the K along the column (see equation 6.2.12 for the head variance in Chapter 6).

Following the analysis for Figure 9.2, one can also derive the possible maximum and minimal K values for this case. Subsequently, one will find that the range of possible K values on the upstream side must be smaller than that on the downstream side. In addition, the range of K values for the entire column is smaller than those in Figure 9.2, where possible heads are not conditioned by any observed head.

Furthermore, the conditional mean head along the column permits determination of the conditional mean gradient. With additional information about q, one subsequently can uniquely identify the effective K of the upstream side and that of the downstream side. These effective Ks represent the conditional effective Ks, given the observed head.

If the q is not specified, the relative magnitude of the conditional effective Ks becomes apparent once an observed head is specified. For example, if the observed head is higher than the unconditional mean head (green dashed line as in Figure 9.3), the mean head gradient on the upstream side must be smaller than that on the downstream side. Thus, one must conclude that the conditional effective K of the upstream side ought to be larger than the unconditional effective K of the equivalent homogeneous media, and the conditional effective K on the downstream side must be smaller. That is, while the exact K for each block is unknown, this observed head already has revealed that the mean of the K value of all blocks upstream from the observation point must be greater than that of all blocks downstream from the same observation point. On the other hand, if the observed head is lower than the unconditional mean head, the relative magnitudes of the effective K upstream and downstream will be reversed. Accordingly, we can conclude that in a statistical sense, the observed head is positively correlated with the effective K upstream and negatively correlated with that downstream of the observation location.

The preceding example demonstrates the fact that if a forward or an inverse problem is ill defined, the objective or the definition of the problem can be changed to make the problem well defined. That is, instead of identifying all hydraulic conductivity values and the corresponding head distribution, we reduce the resolution of the model (see Chapter 6) to determine the effective K and to derive the most probable (mean) head distribution (conditional mean head). Consequently, we have unique solutions to both forward and inverse problems.

However, these unique solutions may not satisfy our desired resolutions. At the level of our resolutions, there are many possible solutions. Besides, modeling of real-world problems faces other issues, such as representativeness of the governing flow equations, effects of multiscale heterogeneity of the geologic media, scale discrepancies between our observation and our model based on the CV or REV concept (see

Chapters 2, 5, and 6), noise/measurement errors, and others. Because of these uncertainties, there are many possible solutions.

Accordingly, any forward or any inverse problem associated with real-world scenarios is always ill posed, in terms of the definition by Hadamard (1902), which states that a well-posed problem of physical phenomena should have the properties that

1. A solution exists.
2. The solution is unique.
3. The solution's behavior changes continuously with the initial conditions.

At this end, we iterate the fact that if the mathematical forward or inverse problem is well defined, it always has a unique solution. Because of uncertainty, the problem can be ill posed. A probabilistic approach is an appropriate solution to any ill-posed problem. Stochastic conceptualization of the problem becomes necessary.

9.3.1 Unconditional Approaches

Unconditional and conditional stochastic approaches have been developed over the past decades to address the aforementioned nonunique (i.e., uncertain) solutions of forward and inverse hydrological problems (Dagan, 1982). They are probabilistic approaches, which are slightly different from the illustrative example presented earlier. These stochastic approaches first conceptualize the spatially varying, local-scale hydraulic properties as random fields, which have an infinite number of possible realizations, and they are characterized by the mean, variance, and correlation scales of each property (see Chapter 4).

The goal of the unconditional approach is to provide the best unbiased flow field predictions, and to quantify their uncertainty without honoring point measurements of properties or observed hydrological behaviors in time or space (similar to the approach discussed for Figure 9.2). This objective can be accomplished by using either 1) effective parameter methodology or 2) Monte Carlo simulation (Yeh, 1992).

9.3.1.1 The Effective Parameter Approach

The effective parameter approach adopts the equivalent homogeneous conceptual model. It defines the model's effective parameters in terms of the statistics that describe measurement errors or uncertainty about the parameter, or the spatial variability of hydraulic properties at a small-scale CV (see Chapter 6). The effective parameters derived from the approach can be used in an equivalent homogeneous forward model to simulate flow fields under different stresses (e.g., Gelhar and Axness, 1983; Yeh et al., 1985a, b, c; Yeh et al., 2005). These approaches also can be used to estimate the uncertainty of the flow field predictions based on the effective parameters. In the following, we illustrate how a first-order analysis can be applied to analyze uncertainty in the flow field predictions arising from two cases: Case 1, uncertainty in the parameters of the homogeneous conceptual model, and Case 2, spatial variability of the parameters in the heterogeneous conceptual model.

Uncertainty of the parameters of homogeneous conceptual models. In this case, we employ the probabilistic approach to analyze the uncertainty of the prediction of transient hydraulic head response to a pumping test in an equivalent homogeneous aquifer with uncertain effective T and S parameters (Sun et al., 2013). The uncertainty in those parameters, in essence, is due to either the measurement errors or a lack of their measurements. This approach first assumes that the natural log of T ($\ln T$) and that of S ($\ln S$) are random variables, and $\ln T = Y + y$ and $\ln S = Z + z$, where Y and Z are their means, and y and z denote their perturbations from the mean. Note that logs of the parameters are dimensionless (but their magnitudes depend on the parameter units). Likewise, the uncertainty in the head is represented by $H = \overline{H} + h$, where $\overline{H}$ is the mean, and h is the perturbation about the mean. Expanding the transient hydraulic head in equation (9.2.5), using a Taylor series about the mean values of parameters, and neglecting second- and higher-order terms, the head perturbation at location $\mathbf{x}$ at a given time t can be expressed as

$$h(\mathbf{x},t) \approx \left[\left. \frac{\partial H(\mathbf{x},t)}{\partial \ln T} \right|_{Y,Z} \right] y + \left[\left. \frac{\partial H(\mathbf{x},t)}{\partial \ln S} \right|_{Y,Z} \right] z = J_{\mathrm{hy}}(\mathbf{x},t)y + J_{\mathrm{hz}}(x,t)z \qquad (9.3.2)$$

where J_{hy} $[L]$ and J_{hz} $[L]$ are scalars and are the sensitivity of h at location $\mathbf{x}$ at a given time t with respect to change in $\ln T$ and $\ln S$ of the homogeneous aquifer, respectively. Equation (9.3.2) states that the head perturbation at location $\mathbf{x}$ and time t is approximately a weighted sum of the perturbations y and z. The weights are the sensitivity of h to y and of h to z at ($\mathbf{x}$, t).

If the perturbations are cast in a probabilistic framework, and if the perturbation of $\ln T$ and that of $\ln S$ are mutually independent of each other, the cross-covariance of h and y and that of h and z can be given as

$$\sigma_{hy}^2(\mathbf{x},t) = J_{hy}(\mathbf{x},t)\sigma_y^2 \text{ and } \sigma_{hz}^2(\mathbf{x},t) = J_{hz}(\mathbf{x},t)\sigma_z^2 \qquad (9.3.3)$$

where σ_y^2 and σ_z^2 are the variances of $\ln T$ and $\ln S$, which represent the uncertainty associated with $\ln T$ and $\ln S$ of the equivalent homogeneous aquifer. The corresponding head variance at ($\mathbf{x}$, t) based on equation (9.3.3) is given as

$$\sigma_h^2(\mathbf{x},t) = (J_{hy}(\mathbf{x},t))^2\sigma_y^2 + (J_{hz}(\mathbf{x},t))^2\sigma_z^2 \qquad (9.3.4)$$

Equation (9.3.4) manifests that the uncertainty of the head prediction is not just a function of the sensitivity of the head to the parameter, but also the uncertainty of the parameters. That is, while the head may not be very sensitive to the change in a parameter, large uncertainty in this parameter may significantly contribute to the overall uncertainty of the head prediction.

The first-order analysis can also be employed to investigate the relative importance of each parameter uncertainty to overall uncertainty in predictions (i.e., cross-correlation). If the cross-covariance, $\sigma_{hy}^2(\mathbf{x},t)$ [L], and $\sigma_{hz}^2(\mathbf{x},t)$ [L] (equation 9.3.3) are normalized by the square root of the product of the variances of $h(\mathbf{x}, t)$ and ln$T$ (i.e., $\sigma_h^2(\mathbf{x},t)$ and σ_y^2, respectively), or those of h ($\mathbf{x}$, t) and ln$S$ (i.e., $\sigma_h^2(\mathbf{x},t)$ and σ_z^2, respectively) to obtain their corresponding cross-correlations ρ_{hy} and ρ_{hz} at location $\mathbf{x}$ at time t, one obtains

$$\rho_{hy}(\mathbf{x},t) = \frac{\sigma_{hy}^2(\mathbf{x},t)}{\sqrt{\sigma_h^2(\mathbf{x},t)\sigma_y^2}} = \frac{J_{hy}(\mathbf{x},t)\sigma_y^2}{\sqrt{\left((J_{hy}(\mathbf{x},t))^2\sigma_y^2 + (J_{hz}(\mathbf{x},t))^2\sigma_z^2\right)\sigma_y^2}} \tag{9.3.5}$$

$$\rho_{hz}(\mathbf{x},t) = \frac{\sigma_{hz}^2(\mathbf{x},t)}{\sqrt{\sigma_h^2(\mathbf{x},t)\sigma_z^2}} = \frac{J_{hz}(\mathbf{x},t)\sigma_z^2}{\sqrt{\left[(J_{hy}(\mathbf{x},t))^2\sigma_y^2 + (J_{hz}(\mathbf{x},t))^2\sigma_z^2\right]\sigma_z^2}} \tag{9.3.6}$$

These cross-correlations are dimensionless and range from −1 to +1. Further, they represent the relationship between the uncertainty of a given parameter of the homogeneous aquifer and the uncertainty in the head at a given location and time caused by uncertainty of all parameters. In effect, the cross-correlation between the head and a parameter is a product of a weighted sensitivity and uncertainty of the parameter.

To illustrate the cross-correlation between the hydraulic head at a given location and T and S parameters in an unbounded homogeneous and isotropic aquifer, we use the Theis solution (see Chapter 7). That is, the drawdown, $s(r, t)$, at time t in an observation well at a radial distance r from a pumping well, which discharges at a constant rate Q, is given by

$$s(r,t) = \frac{Q}{4\pi T}W(u),\ W(u) = \int_u^\infty \frac{e^{-u}}{u}du \text{ and } u = \frac{r^2S}{4Tt} \tag{9.3.7}$$

Equation (9.3.7) leads to the following dimensionless sensitivities

$$\frac{T\partial H(r,t)}{Q\partial \ln T} = -\frac{T\partial s(r,t)}{Q\partial \ln T} = \frac{1}{4\pi}\left[w(u) - e^{-u}\right] \tag{9.3.8}$$

$$\frac{T\partial H(r,t)}{Q\partial \ln S} = -\frac{T\partial s(r,t)}{Q\partial \ln S} = \frac{1}{4\pi}e^{-u} \tag{9.3.9}$$

Subsequently, the head variance can be expressed as

$$\sigma_h^2(r,t) = \frac{Q^2\sigma_y^2}{16\pi^2T^2}\left[[w(u)]^2 - 2w(u)e^{-u} + (1+1/b)e^{-2u}\right] \tag{9.3.10}$$

where $\sigma_y^2 = b\sigma_z^2$. Furthermore, the cross-correlation between h and ln$T$ can be expressed as

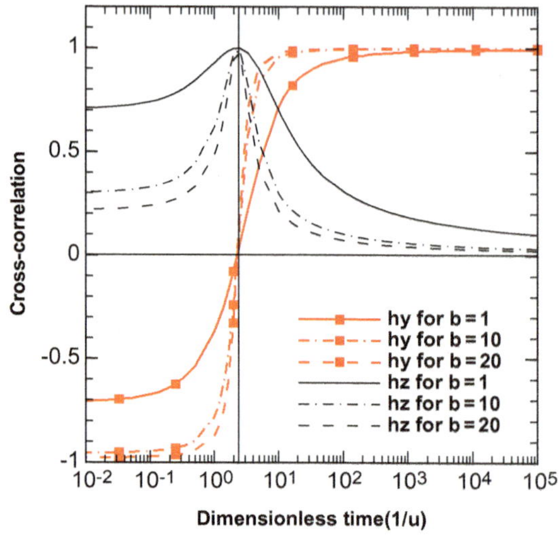

Figure 9.4. Cross-correlation between the head at an observation well and aquifer properties ln$T$ (ρ_{hy}) and ln$S$ (ρ_{hz}) as a function of dimensionless time for a homogeneous aquifer with three different ratios of the variance of ln$T$ to that of ln$S$ (i.e., $b = 1$, 10, and 20). (Modified from Sun et al., 2013)

$$\rho_{hy}=\frac{\left.\frac{\partial H(r,t)}{\partial \ln T}\right|_{Y,Z}\sigma_y^2}{\sqrt{\left[\left(\left.\frac{\partial H(r,t)}{\partial \ln T}\right|_{Y,Z}\right)^2\sigma_y^2+\left(\left.\frac{\partial H(r,t)}{\partial \ln S}\right|_{Y,Z}\right)^2\sigma_z^2\right]\sigma_y^2}}=\frac{w(u)-e^{-u}}{\sqrt{[w(u)]^2-2w(u)e^{-u}+(1+1/b)e^{-2u}}}$$

(9.3.11)

The cross-correlation between h and ln$S$ can be easily derived as well; it is

$$\rho_{hz}=\frac{\left.\frac{\partial H(r,t)}{\partial \ln S}\right|_{Y,Z}\sigma_z^2}{\sqrt{\left[\left(\left.\frac{\partial H(r,t)}{\partial \ln T}\right|_{Y,Z}\right)^2\sigma_y^2+\left(\left.\frac{\partial H(r,t)}{\partial \ln S}\right|_{Y,Z}\right)^2\sigma_z^2\right]\sigma_z^2}}=\frac{e^{-u}}{\sqrt{b[w(u)]^2-2bw(u)e^{-u}+(b+1)e^{-2u}}}$$

(9.3.12)

Note that in equations (9.3.11) and (9.3.12), we assume that $\sigma_y^2 = b\sigma_z^2$, where b is a constant.

The cross-correlations between h and ln$T$ and that between h and ln$S$ (i.e., equations (9.3.11) and (9.3.12)) are plotted as a function of dimensionless time t^*, which equals $1/u$ in three homogeneous aquifers for three different bs in Figure 9.4.

As illustrated by the solid black and red lines (for an unbounded aquifer with an infinite lateral extent) in Figure 9.4, at early times, the head at the observation well is positively correlated with ln$S$ and negatively correlated with ln$T$. Physically, this means that if the observed head is high (or drawdown is small), then S is likely to be large and T is likely to be small, and vice versa. At the intermediate time ($t^* = 2.2$) in the figure, when the correlation of h and ln$T$ is zero, the cross-correlation of the h and ln$S$ reaches maximum. This dimensionless time can be converted to real time as

$$t_m = \frac{2.2r^2S}{4T} = \frac{r^2S}{1.8T} \tag{9.3.13}$$

Furthermore, the time t_m is approximately equal to the intercept t_0 on the time axis of a log time-drawdown plot of a well hydrograph at any r, when the straight-line portion of the plot is extrapolated to the point where the drawdown is zero according to Cooper-Jacob's approach (see Chapter 7). That is,

$$t_0 = \frac{r^2S}{2.25T} \tag{9.3.14}$$

Afterward, ρ_{hy} continuously grows and approaches an asymptotic value (close to +1) at large times, whereas ρ_{hz} decreases rapidly and approaches some value (zero if a steady-state condition is met). These results are consistent with the fact that at early times, the head is controlled by both T and S, and it is only influenced by T near steady state. Therefore, if the head in an observation well is high, then T of the aquifer must also be high near the steady state.

The larger the b value (ratio of σ_y^2 to σ_z^2), ρ_{hy} starts at a more negative value, and the earlier it reaches 1. On the other hand, the larger the b value is, the smaller the ρ_{hz} value during early time, and the faster it rises, the sharper it drops. The value of b does not alter the time to reach maximum ρ_{hz}, however. At this moment, one should ask what the physical meaning is for the flow field as ρ_{hy} approaches 1 in an unbounded aquifer. Is it the time when the drawdown reaches a steady-state shape regime (or a quasi-steady state), in which T is the major controlling factor before a true steady-state flow field is achieved?

Spatial variability of heterogeneous conceptual models. In this case, we show how the first-order analysis can be used to analyze uncertainty in predicted head (i.e., how the true head deviates from the predicted head, on the basis of the effective parameter) due to spatial variability of the parameters T and S. In order to do so, we assume that ln$T$ and ln$S$ at every location of the aquifer are random fields (Chapter 4). Again, $\ln T = Y + y$ and $\ln S = Z + z$, where Y and Z are mean values and y and z denote perturbations, which represent spatial variability or uncertainty due to the absence of measurements of these parameters at each location. Likewise, the head is represented by $H = \overline{H} + h$, where $\overline{H}$ is the mean and h is the perturbation about the mean value caused by the spatial variability or uncertainty of the parameters T and S. The first-order approximation of the head perturbation at location $\mathbf{x}_i$ at a given time t is then given as

$$h(\mathbf{x}_i, t) \approx \left[\frac{\partial H(\mathbf{x}_i, t)}{\partial \ln T(\mathbf{x}_j)}\bigg|_{Y,Z}\right] y(\mathbf{x}_j) + \left[\frac{\partial H(\mathbf{x}_i, t)}{\partial \ln S(\mathbf{x}_j)}\bigg|_{Y,Z}\right] z(\mathbf{x}_j)$$
$$= J_{\mathrm{hy}}(\mathbf{x}_i, \mathbf{x}_j, t) y(\mathbf{x}_j) + J_{\mathrm{hz}}(\mathbf{x}_i, \mathbf{x}_j, t) z(\mathbf{x}_j) \qquad (9.3.15)$$

where $y(\mathbf{x}_j)$ and $z(\mathbf{x}_j)$ are perturbation of $\ln T$ and $\ln S$ at location $\mathbf{x}_j$ for $j = 1, \ldots, N$, where N is the total number of parameters in the aquifer (or a number of elements in a finite element domain); $J_{\mathrm{hy}}(\mathbf{x}_i, \mathbf{x}_j, t)$ and $J_{\mathrm{hz}}(\mathbf{x}_i, \mathbf{x}_j, t)$ are the sensitivity of h at location $\mathbf{x}_i$ at a given time t, with respect to $\ln T$ and $\ln S$ perturbations at location $\mathbf{x}_j$. Notice the difference between these sensitivities and those in equation (9.3.2). Here, Einstein's summation convention over the repeated suffix is used. In other words, the head perturbation at $(\mathbf{x}_i, t)$ is a weighted sum of perturbation of parameters $\ln T$ and $\ln S$ everywhere in the aquifer. The weights are the corresponding sensitivity values. Accordingly, the head may not be very sensitive to a unit perturbation of a parameter at a far distance, but a large magnitude of the perturbation (e.g., impermeable boundary) may induce a significant perturbation in head. For example, a drawdown-time curve observed at a well during a pumping test may manifest the effect of an impermeable boundary at far distance.

Suppose that $\ln T$ and $\ln S$ are mutually independent of each other. The cross-covariance matrices between h and y and between h and z are then

$$\begin{aligned} C_{hy}(\mathbf{x}_i, \mathbf{x}_j, t) &= J_{hy}(\mathbf{x}_i, \mathbf{x}_j, t) C_{yy}(\mathbf{x}_i, \mathbf{x}_j) \\ C_{hz}(\mathbf{x}_i, \mathbf{x}_j, t) &= J_{hz}(\mathbf{x}_i, \mathbf{x}_j, t) C_{zz}(\mathbf{x}_i, \mathbf{x}_j) \end{aligned} \quad \text{for } i, = 1, \ldots, N \qquad (9.3.16)$$

$C_{yy}(\mathbf{x}_i, \mathbf{x}_j)$ and $C_{zz}(\mathbf{x}_i, \mathbf{x}_j)$ are covariance matrices or functions of perturbation of $\ln T$ and $\ln S$, which represent the spatial variability and structure (spatial fabrics) of the parameters (see Chapter 4). Different from the variance expression for the parameters in equation (9.3.3), equation (9.3.16) uses covariance matrices since we aim to consider spatial variability of $\ln T$ and $\ln S$ at every location of the aquifer. The corresponding head covariance matrix based on equation (9.3.15) is given as

$$\begin{aligned} C_{hh}(\mathbf{x}_i, \mathbf{x}_j, t) = {} & J_{hy}(\mathbf{x}_i, \mathbf{x}_j, t) C_{yy}(\mathbf{x}_i, \mathbf{x}_j) J_{hy}^T(\mathbf{x}_i, \mathbf{x}_j, t) + \\ & J_{hz}(\mathbf{x}_i, \mathbf{x}_j, t) C_{zz}(\mathbf{x}_i, \mathbf{x}_j) J_{hz}^T(\mathbf{x}_i, \mathbf{x}_j, t) \end{aligned} \qquad (9.3.17)$$

The superscript T denotes transpose. The components of $C_{hh}(\mathbf{x}_i, \mathbf{x}_j, t)$ at $\mathbf{x}_i = \mathbf{x}_j$ are the head variances $\sigma_{hh}^2(\mathbf{x}_i, t)$, which represent the uncertainty in head at location $\mathbf{x}_i$ at a given time t due to the unknown heterogeneity in the aquifer. Generally speaking, the head variance is proportional to the variance of the parameters and correlation scales, in addition to the mean hydraulic gradient (see equation 6.2.12). For example, the head variance is large near the pumping well, where the mean gradient is high.

The cross-covariance, C_{hy} and C_{hz}, can be normalized by the square root of the product of the variances of h at ($\mathbf{x}_i$, t) and ln$T$ or those of h at ($\mathbf{x}_i$, t) and ln$S$ to obtain their corresponding cross-correlation ρ_{hy}, ρ_{hz} at location i and j at time t. These can be expressed as

$$\begin{aligned}\rho_{hy}(\mathbf{x}_i,\mathbf{x}_j,t) &= \frac{J_{hy}(\mathbf{x}_i,\mathbf{x}_j,t)C_{yy}(\mathbf{x}_i,\mathbf{x}_j)}{\sqrt{\sigma_h^2(\mathbf{x}_i,t)\sigma_y^2}} = \frac{J_{hy}(\mathbf{x}_i,\mathbf{x}_j,t)C_{yy}(\mathbf{x}_i,\mathbf{x}_j)}{\sqrt{C_{hh}(\mathbf{x}_i,\mathbf{x}_i,t)C_{yy}(\mathbf{x}_j,\mathbf{x}_j)}} \\ \rho_{hz}(\mathbf{x}_i,\mathbf{x}_j,t) &= \frac{J_{hz}(\mathbf{x}_i,\mathbf{x}_j,t)C_{zz}(\mathbf{x}_i,\mathbf{x}_j)}{\sqrt{\sigma_h^2(\mathbf{x}_i,t)\sigma_z^2}} = \frac{J_{hz}(\mathbf{x}_i,\mathbf{x}_j,t)C_{zz}(\mathbf{x}_i,\mathbf{x}_j)}{\sqrt{C_{hh}(\mathbf{x}_i,\mathbf{x}_i,t)C_{zz}(\mathbf{x}_j,\mathbf{x}_j)}}\end{aligned} \tag{9.3.18}$$

where $\sigma_h^2(\mathbf{x}_i, t)$ is the head variance at $\mathbf{x}_i$ and t; σ_y^2 and σ_z^2 are the variances of ln$T$ and ln$S$, respectively. Furthermore, the cross-correlations ρ_{hy} and ρ_{hz} (dimensionless) represent how the head perturbation at the location $\mathbf{x}_i$ at a given time t is influenced by the ln$T$ or ln$S$ perturbation at location $\mathbf{x}_j$ in a statistical sense.

As shown in Chapter 7, the cross-correlation analysis of heterogeneous aquifers can be used to understand the physical meaning of the effective T and S of an equivalent homogeneous aquifer, estimated from a cross-hole pumping test and Theis or Jacob-Cooper analysis. Moreover, the cross-correlation analysis is the foundation of the cokriging approach, nonlinear geostatistical inverse approaches, and joint inverse modeling approaches, which will be discussed in Sections 9.4, 9.5, and 9.6.

9.3.1.2 The Monte Carlo Simulation

Unlike the aforementioned effective parameter approach, the Monte Carlo simulation approach adopts the heterogeneous conceptual model and generates many possible realizations of random fields of hydraulic properties with given spatial statistics. It then solves the governing equations with these possible fields to yield corresponding possible flow fields for any given stresses. Statistical analysis of these possible flow fields then yields the most likely flow field and its deviations from the true field (e.g., Smith and Freeze, 1979a, b). Monte Carlo simulation is thus a straightforward but brute-force methodology, in comparison with the effective parameter approach. Both unconditional approaches generally yield predictions with large uncertainty or at low resolutions.

9.3.2 Conditional Approaches

The conditional approaches can be divided into the conditional effective parameters approach and the conditional Monte Carlo simulation approach, as were the unconditional approaches. Here, the conditional Monte Carlo simulation approach will be discussed first.

9.3.2.1 The Conditional Monte Carlo Simulation Approach

Rather than generating the unconditional realizations of fields of hydraulic properties, the conditional Monte Carlo approach creates realizations that preserve measurements of the primary information (K and/or S_s) at the sampling locations. It then uses these realizations to simulate corresponding flow fields and discards those K and S_s realizations that do not reproduce secondary information (e.g., head measurements) at sampling locations. The final remaining realizations of K and S_s fields are conditional realizations, which are the equally likely solutions to the parameter identification or inverse problem. If the necessary conditions are used to condition the realizations, only one realization remains, which is the true solution. Otherwise, these conditional realizations can then be used to conduct forward simulations to derive possible flow fields under different stresses to evaluate the most likely predicted flow field and the associated uncertainties (e.g., Gómez-Hernández et al., 1997; Capilla et al., 1997; Hanna and Yeh, 1998; Hendricks et al., 2008).

9.3.2.2 The Conditional Effective Parameter Approach

As in the example depicted in Figure 9.3, rather than seeking all possible conditional realizations, the conditional effective parameter approach looks for a set of K and S_s fields that 1) agrees with the measured parameters (if any) at sample locations and 2) yields the head field that honors observed heads at sample locations, and that is the most likely unbiased head predictions at locations where no samples are available (e.g., Hoeksema and Kitanidis, 1984; Graham and McLaughlin, 1989a, b; Kitanidis, 1995; Yeh et al., 1995; Harter and Yeh, 1996a, b; Yeh et al., 1996; Yeh and Zhang, 1996; Zhang and Yeh, 1997; Li and Yeh, 1999; Hughson and Yeh, 2000). Again, as the number of observations becomes larger than or equal to the number of parameters to be estimated and other necessary conditions (Section 9.2) are met, only one conditional realization exists. This realization is the conditional effective parameter field and is the true field. Conversely, if the necessary conditions are not met, there is a common and unique conditional effective field, even though a large number of conditional realizations exist. This conditional effective field is not the true field, but it is the best and unbiased estimate that honors all observations. Deviations from the true field (uncertainty) are then quantified by the residual covariance. The amount of reduction in uncertainty rests upon the number of spatial data and specifications of other necessary conditions. The uncertainty associated with the prediction using the conditional approaches is generally smaller than the uncertainty associated with the predictions using the unconditional approaches. As discussed in Section 6.7 of Chapter 6, the conditional approach, in essence, adopts the heterogeneous conceptual model. As more data sets are used to enhance the information about the heterogeneous field, the predicted flow field progressively gets closer to the reality. We will introduce some techniques used in the conditional stochastic approach in Sections 9.4, 9.5, and 9.6.

9.4 Conditional Linear Estimator

Geostatistics is a conditional mean approach, which consists of two parts: spatial structure identification (i.e., variogram analysis, Chapter 4) and spatial estimation (i.e., spatial interpolation or extrapolation) of the parameters of interest. The estimator is a stochastic linear estimator that takes advantage of measured or known values of variables at different locations and prior knowledge of the spatial statistical structure of the variables, to derive the best unbiased estimate of a geologic variable, and it quantifies uncertainties associated with the estimate.

To discuss geostatisitcs, we will consider a hydrogeology estimation problem in which a hydraulic property (such as hydraulic conductivity, porosity, or specific storage) of a groundwater basin at a given location $\mathbf{x}$ is represented by the variable $W(\mathbf{x})$. In this section, we will consider hydraulic conductivity as the variable. Therefore, this problem aims to estimate the hydraulic conductivity $W(\mathbf{x}_0)$ at locations $\mathbf{x}_0$, using hydraulic conductivity or head values, $W(\mathbf{x}_1)$, $W(\mathbf{x}_2)$, ... , and $W(\mathbf{x}_m)$ measured at locations $\mathbf{x}_1$, $\mathbf{x}_2$, ... , and $\mathbf{x}_m$. Notice that the subscript is a location index and m is the total number of sampled locations. A mathematical statement for this problem is

$$\hat{W}(\mathbf{x}_0) = \mathcal{F}[W(\mathbf{x}_1), W(\mathbf{x}_2), \ldots, W(\mathbf{x}_m)] \tag{9.4.1}$$

where $\hat{W}(\mathbf{x}_0)$ denotes the estimate, which is a function of the measured values at the sampled locations, $\mathbf{x}_1$, $\mathbf{x}_2$, ... , and $\mathbf{x}_m$. The question then arises, "What should the function, $\mathcal{F}$, be?" In fact, this mathematical problem and its solutions are nothing new to hydrogeologists. Hydrogeologists often manually construct a contour map of hydraulic conductivity, based on knowledge of well-log data at several adjacent wells. By manual contouring, hydrogeologists essentially define a function $\mathcal{F}$ for the problem, based on their intuition and experience. This function is, however, defined subjectively and qualitatively, and one therefore cannot assess the uncertainty associated with the contour map or the probability that the contoured hydraulic conductivity field is correct or incorrect. Geostatistics, in contrast, is a mathematical tool that defines this function in a quantitative manner and provides a way to quantify the uncertainty associated with the estimate.

Before we develop the function, we have to ask ourselves the following question: What is the objective of the function, or what do we expect the function to do? Since we do not anticipate that the function will provide an exact estimate of the hydraulic conductivity value, the only hope is to develop a function that gives us a "best" estimate at the point where there is no sample. Then, what does the "best estimate" mean? To define it, one needs some criteria by which one can measure the accuracy of $\hat{W}(\mathbf{x}_0)$ as a predictor of $W(\mathbf{x}_0)$. The following are the criteria used in geostatistics:

1. Bias

$$B = E[\hat{W}(\mathbf{x}_0)] - W(\mathbf{x}_0) \tag{9.4.2}$$

In other words, if we estimate $\hat{W}(\mathbf{x}_0)$ at a location $\mathbf{x}_0$ many times, the estimation bias is then defined as a measure of the difference between the average of the estimates and the true value of the random variable, $W(\mathbf{x}_0)$, at the location. A good estimator should possess small bias. If the estimator yields zero difference, it is called an unbiased estimator.

2. Variance of an estimator

In addition to the minimal bias, one would like to have an estimator that has a small variance. That is, we would like to have an estimator that produces a very small variation between estimates. Mathematically,

$$\mathrm{var}[\hat{W}(\mathbf{x}_0)] = E\left[(\hat{W}(\mathbf{x}_0) - E[\hat{W}(\mathbf{x}_0)])^2\right] \tag{9.4.3}$$

Mean Square Error (MSE)

Mean square error (MSE) is mathematically defined as the expected value of the square of the difference between our estimate and the true value:

$$\begin{aligned}
E\left[(\hat{W}(\mathbf{x}_0) - W(\mathbf{x}_0))^2\right] &= E\left[\{[\hat{W}(\mathbf{x}_0) - E(\hat{W}(\mathbf{x}_0))] + [E(\hat{W}(\mathbf{x}_0)) - W(\mathbf{x}_0)]\}^2\right] \\
&= E\left[\left(\hat{W}(\mathbf{x}_0) - E(\hat{W}(\mathbf{x}_0))\right)^2\right] + E\left[\left(E(\hat{W}(\mathbf{x}_0)) - W(\mathbf{x}_0)\right)^2\right] \\
&\quad + 2E\left[[\hat{W}(\mathbf{x}_0) - E(\hat{W}(\mathbf{x}_0))](E(\hat{W}(\mathbf{x}_0)) - W(\mathbf{x}_0))\right] \\
&= \mathrm{var}[\hat{W}(\mathbf{x}_0)] + (E[\hat{W}(\mathbf{x}_0)] - W(\mathbf{x}_0))^2 = \mathrm{var}[\hat{W}(\mathbf{x}_0)] + B^2
\end{aligned} \tag{9.4.4}$$

It is essentially the sum of the variance of our estimate and the square of the estimation bias of our estimator. As stated previously, we want an estimator to meet the minimal bias and variance criteria. Now, we simply state that the estimator must have a minimal mean square error, i.e.,

$$E[(\hat{W}(\mathbf{x}_0) - W(\mathbf{x}_0))^2] = \text{minimum} \tag{9.4.5}$$

The objective for the function, $\mathcal{F}$, thus has been defined. However, in addition to the minimal mean square error requirement, the estimation function must use information that we already have. That is,

$$\hat{W}(\mathbf{x}_0) = \mathcal{F}[W(\mathbf{x}_1), W(\mathbf{x}_2), \ldots, W(\mathbf{x}_m)] \tag{9.4.6}$$

These two requirements, that is, equations (9.4.5) and (9.4.6), lead us to the conditional expectation operator in statistics, which is to be discussed next.

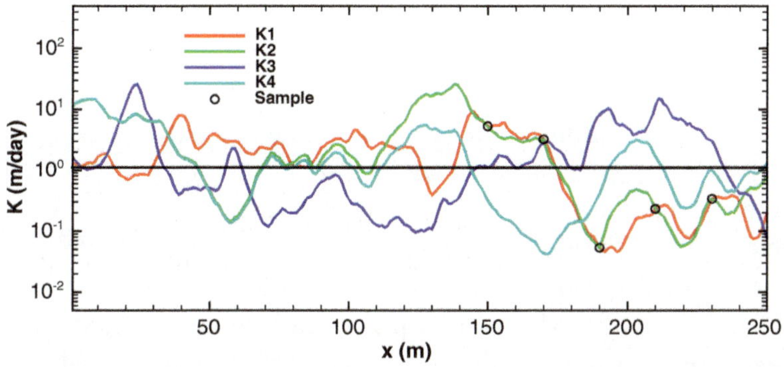

Figure 9.5. A schematic illustration of four possible realizations of K ($\boldsymbol{x}$) along a transect line (i.e., $K1$, $K2$, $K3$, and $K4$). Black circles indicate sample values and locations. $K1$ and $K2$ are the realizations of the stochastic process, which honor sample values at the sampling locations: conditional realizations. Lines $K3$ and $K4$ represent unconditional realizations, which do not agree with the sample values at the sampling locations. The black solid horizontal line is the unconditional mean value.

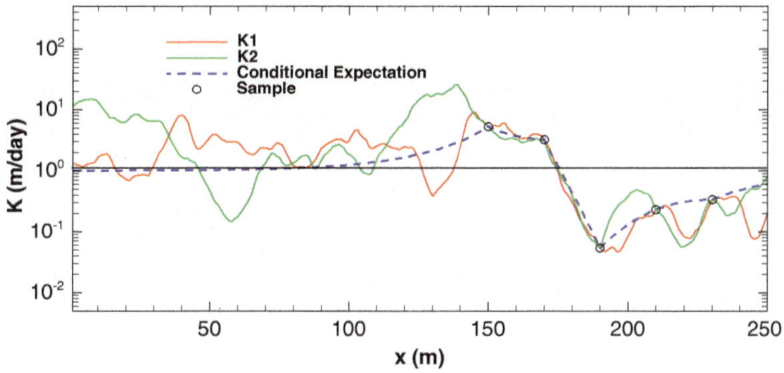

Figure 9.6. A schematic illustration of conditional realizations (i.e., $K1$ and $K2$ lines) and conditional expectation (the dashed blue line).

Conditional Expectation

Conditional expectation can be written in a mathematical format such as

$$E[W(\mathbf{x}_0)|W(\mathbf{x}_1), W(\mathbf{x}_2), \ \dots \ W(\mathbf{x}_m)] \tag{9.4.7}$$

This is the expected value of W, given the fact that we already know $W(\mathbf{x}_1)$, $W(\mathbf{x}_2)$, . . . , and $W(\mathbf{x}_m)$.

In order to explain the physical meaning of the conditional expectation, we will use Figures 9.5 and 9.6. As shown in these figures, there are measurements of hydraulic conductivity, K, only at five locations along a transect line (the black circles in Figures 9.5 and 9.6). Our objective is to estimate the hydraulic conductivity values at the rest locations. Suppose we conceptualize the hydraulic conductivity along the transect line as a stochastic process, characterized by a joint probability distribution with a mean and a covariance function (see Chapter 4). We essentially imagine an infinite number of possible realizations of hydraulic conductivity values (i.e., ensemble) along the transect line with the same spatial statistics. We expect that within this ensemble, many realizations (a subset of the ensemble) will have the conductivity values that agree with the measured values at the sampled locations (for example, red and green lines, or realizations $K1$ and $K2$ in Figure 9.5). Meanwhile, many realizations (a subset of the ensemble) will have values at the sample locations that do not agree with the measured values (for example, light and dark blue lines or realizations $K3$ and $K4$ in Figure 9.5). If we discard the latter subset, we are then left with a subset of the stochastic process, which has the same values as the measurements at the sampled locations (red and green lines in Figure 9.6). This subset is then called the conditional realizations.

The conditional expectation mentioned earlier simply means that we take the ensemble average of this remaining subset (the conditional realizations). This ensemble average, as a consequence, yields a conductivity field that agrees with measurements at sample locations, and it represents the average of the possible conductivity values at the unsampled locations of the subset (dashed blue line in Figure 9.6). In comparison with the conditional realizations, the conditional expectation is smoother, since it is the statistically most likely outcome that agrees with the observations.

The question one may ask next is, "How can this concept of the conditional expectation be translated into a mathematical operator?"

If K is a multivariate normal (i.e., it has a jointly normal distribution), the conditional expectation is a linear sum of all the weighted sampled values (see Priestley, 1981, p. 728). That is,

$$\begin{aligned}\hat{K}(\mathbf{x}_0) &= E[K(\mathbf{x}_0)|K(\mathbf{x}_1), K(\mathbf{x}_2), \ldots K(\mathbf{x}_m)] = \sum_{i=1}^{m} \lambda_{0i} K(\mathbf{x}_i) \\ &= \lambda_{01}K(\mathbf{x}_1) + \lambda_{02}K(\mathbf{x}_2) + \ldots + \lambda_{0m}K(\mathbf{x}_m)\end{aligned} \tag{9.4.8}$$

In other words, if the stochastic process has a normal joint probability distribution, the conditional expectation is a linear sum of weighted measurements of conductivity at the sampled locations. This statement basically defines the function that we are looking for, which is often referred to as the linear predictor.

Now, we have the mathematical function (a linear predictor) and the values of K at the sampled locations. The values of the weights, λ_i, of the linear estimator, however, still have to be determined, so that they can be used to estimate the unknown K at the unsampled locations. Determining the weights and the estimate then becomes the task of kriging. In the next section, we will focus only on the simple kriging method, which is the fundamental and the most general one.

9.4.1 Simple Kriging (Linear Estimator Using the Same Types of Information)

Let us assume that $\ln K(\mathbf{x})$ (i.e., the natural logarithm of the hydraulic conductivity) is a stochastic process (see Chapter 4). It has a spatially constant mean, $E[\ln K(\mathbf{x})] = F$, and a spatial covariance:

$$\mathrm{Cov}(\mathbf{x}_i, \mathbf{x}_j) = E[(\ln K(\mathbf{x}_i) - F)(\ln K(\mathbf{x}_j) - F)] = C(\mathbf{x}_i, \mathbf{x}_j) = C(\boldsymbol{\tau})$$

where

$$\begin{aligned} C(\mathbf{x}_i, \mathbf{x}_j) &= E[\ln K(\mathbf{x}_i) \ln K(\mathbf{x}_j)] - F\,E[\ln K(\mathbf{x}_i)] - F\,E[\ln K(\mathbf{x}_j)] + F^2 \\ &= E[\ln K(\mathbf{x}_i) \ln K(\mathbf{x}_j)] - F^2 \end{aligned}$$

Now, we consider the perturbation of the process, which has a mean zero:

$$f(\mathbf{x}) = \ln K(\mathbf{x}) - F \text{ where } E[f(\mathbf{x})] = 0 \qquad (9.4.9)$$

Then, the estimate of f at the point x_o (if $f_i = f(\mathbf{x}_i)$, $i = 1 \ldots n$ at locations $\mathbf{x}_i'$s are known) becomes

$$\hat{f}_0 = \hat{f}_0(\mathbf{x}_0) = \lambda_{01} f_1 + \lambda_{02} f_2 + \ldots \lambda_{0n} f_n$$

or

$$\hat{f}_0 = \sum_{i=1}^{n} \lambda_{0i} f_i \qquad (9.4.10)$$

where $\hat{f}_0$ is the estimate of f at location $\mathbf{x}_0$, f_i represents the observation of f at the sampled locations $\mathbf{x}_i'$s, and λ_{0i} is the weight associated with f_0 and f_i. The weight represents the contribution of the observed f value at location $\mathbf{x}_i$ to the estimate f at location $\mathbf{x}_0$. As discussed before, an optimal estimator requires that the error of estimation be minimal. That is,

$$E[(\hat{f}_0 - f_0)^2] = \text{minimum} \qquad (9.4.11)$$

where $\hat{f}_o$ and f_o are random variables. Before finding the optimal estimator, we will evaluate the variance:

$$\begin{aligned} E[(\hat{f}_0 - f_0)^2] &= E\left[\left(\sum_{i=1}^{n} \lambda_{0i} f_i - f_0\right)^2\right] \\ &= E\left[\left(\sum_{i=1}^{n} \lambda_{0i} f_i\right)\left(\sum_{j=1}^{n} \lambda_{0j} f_j\right)\right] - 2E\left[\sum_{i=1}^{n} \lambda_{0i} f_i f_0\right] + E\left[(f_0)^2\right] \qquad (9.4.12) \\ &= \sum_{i=1}^{n}\sum_{j=1}^{n} \lambda_{0i}\lambda_{0j} E[f_i f_j] - 2\sum_{i=1}^{n} \lambda_{0i} E[f_i f_0] + E(f_0{}^2) \end{aligned}$$

The following is an example showing how the product of two sums can be written in the double sums form, as used in the preceding equation.

$$
\begin{aligned}
&E[(\lambda_{01}f_1+\lambda_{02}f_2)(\lambda_{01}f_1+\lambda_{02}f_2)]\\
&=\lambda_{01}\lambda_{01}E[f_1f_1]+\lambda_{01}\lambda_{02}E[f_1f_2]+\lambda_{02}\lambda_{01}E[f_2f_1]+\lambda_{02}\lambda_{02}E[f_2f_2]\\
&=\sum_{i=1}^{2}\sum_{j=1}^{2}\lambda_{0i}\lambda_{0j}E[f_if_j]
\end{aligned}
\tag{9.4.13}
$$

Using the fact that

$$E(f_if_j)=C(\mathbf{x}_i,\mathbf{x}_j) \text{ and } E(f_if_i)=C(\mathbf{x}_i,\mathbf{x}_i)=\sigma_f^2$$

one can express the variance or equation (9.4.12) as

$$E[(\hat{f}-f_0)^2]=\sum_{i=1}^{n}\sum_{j=1}^{n}\lambda_{0i}\lambda_{0j}C(\mathbf{x}_i,\mathbf{x}_j)-2\sum_{i=1}^{n}\lambda_{0i}C(\mathbf{x}_i,\mathbf{x}_0)+C(\mathbf{x}_i,\mathbf{x}_i) \tag{9.4.14}$$

Now, we will find the values of λ_{0i} that minimize the variance equation (9.4.14). In order to do so, we differentiate the variance with respect to λ_{0i} and set the resultant equation equal to zero. That is,

$$\frac{\partial}{\partial\lambda_{0i}}E[(\hat{f}_0-f_0)^2]=0 \tag{9.4.15}$$

The result of the differentiation is

$$\sum_{j}\lambda_{0j}C(\mathbf{x}_i,\mathbf{x}_j)=C(\mathbf{x}_i,\mathbf{x}_0) \tag{9.4.16}$$

for a given subscript i. Since i = *1,. . .,* n, we have a system of n equations in the following matrix form:

$$
\begin{bmatrix}
c(\mathbf{x}_1,\mathbf{x}_1) & c(\mathbf{x}_1,\mathbf{x}_2) & \dots & c(\mathbf{x}_1,\mathbf{x}_n)\\
\cdot & \cdot & & \\
\cdot & \cdot & & \\
c(\mathbf{x}_n,\mathbf{x}_1) & \cdot & &
\end{bmatrix}
\begin{bmatrix}
\lambda_{01}\\ \lambda_{02}\\ \dots\\ \lambda_{0n}
\end{bmatrix}
=
\begin{Bmatrix}
c(\mathbf{x}_1,\mathbf{x}_0)\\ c(\mathbf{x}_2,\mathbf{x}_0)\\ \dots\\ c(\mathbf{x}_n,\mathbf{x}_0)
\end{Bmatrix}
\tag{9.4.17}
$$

We then solve for λ_{0i}, where $i = 1, \dots, n$, which are the weights to be used in the linear predictor. Notice that the value of λ_{0i} depends on the covariance function – more precisely, the correlation function and the variance do not affect the weights. Our estimate can subsequently be obtained by using

$$\hat{f}_0=\sum_{i=1}^{n}\lambda_{0i}f_i \tag{9.4.18}$$

Now, we will check whether the estimate thus derived satisfies the minimum bias and variance criterion we specified previously:

Bias:

$$\begin{aligned} E(\hat{f}_0 - f_0) &= E[\hat{f}_0] - E[f_0] \\ &= E\left[\sum \lambda_{0i} f_i\right] - E[f_0] \\ &= \sum \lambda_{0i} E[f_i] - E[f_0] = 0 \end{aligned} \tag{9.4.19}$$

Therefore, we can say our estimator is an unbiased estimator. We also can check the variance.

Kriging variance:

$$\begin{aligned} \mathrm{var}(\hat{f}_0 - f_0) &= E\left[(\hat{f}_0 - f_0)^2\right] - \left[E(\hat{f}_0 - f_0)\right]^2 \\ &= E\left[(\hat{f}_0 - f_0)^2\right] - 0 \end{aligned} \tag{9.4.20}$$

from equation (9.4.16), we know that

$$\sum_i \sum_j \lambda_{0i} \lambda_{0j} C(\mathbf{x}_i, \mathbf{x}_j) = \sum_i \lambda_{0i} C(\mathbf{x}_i, \mathbf{x}_0) \tag{9.4.21}$$

using equation (9.4.14), the kriging variance then becomes

$$\mathrm{var}(\hat{f}_0 - f_0) = E\left[(\hat{f}_0 - f_0)^2\right] = C(0) - \sum_i \lambda_{0i} C(\mathbf{x}_i, \mathbf{x}_0) \tag{9.4.22}$$

This kriging variance (or conditional variance) is a measure of the likely deviation of the estimate from the true field or the uncertainty of the estimate. According to equation (9.4.22), the kriging variance is smaller than the variance of the process itself, because of the measurements at $\mathbf{x}_i$. Also, at the measurement locations, the kriging variance is zero (if there is no nugget effect; see Chapter 4). That is, this linear estimator preserves the data at the sampling locations.

Now, we can change $\hat{f}_o$ back to ln$K$. Recall that $f(\mathbf{x}) = \ln K(\mathbf{x}) - F$, then

$$\ln \hat{K}_0 = F + \sum_i \lambda_{0i} (\ln K_i - F) \tag{9.4.23}$$

Equation (9.4.23) is our best, linear, and unbiased estimate of ln$K$ at $\mathbf{x}_0$. However, the estimate is simply a conditional expectation, which yields the most likely value of ln$K$ at the location. It is still an estimate, and therefore it does not necessarily represent an exact or true value of ln$K$ at the location.

Lastly, as is the inverse-distance weighted estimation technique, kriging is an interpolation and extrapolation algorithm, but it takes into account our prior knowledge of the spatial statistic structure (the correlation function). This prior knowledge is useful. For instance, in a statistically perfectly stratified geologic formation (the horizontal correlation scale is much longer than the vertical correlation scale, Chapter 4), the sampled attribute at a point in a layer is expected to be highly correlated with that attribute at other locations within the same layer. Since kriging weights are derived from the spatial correlation structure, the contribution of the sampled locations will be weighted heavily for the estimate at any point in the same layer, in spite of the distances between the sampled location and the estimated location. On the other hand, the inverse distance approach assigns the influence of the sampled locations solely on the basis of the distance between the sampled and the estimated location, thus ignoring the fact that the formation is perfectly stratified (our prior statistical information about the geology). The estimate by kriging, therefore, can yield a more correct estimate (geologically speaking) than the inverse distance approach. However, we must emphasize here again that the kriging estimate is not the true property, but the most likely one in a statistical sense.

9.4.2 Cokriging (Linear Estimator Using Different Types of Information)

Cokriging is a stochastic linear interpolator, which is similar to the simple kriging estimator, but it uses data sets of two or more different types of variables and their spatial autocorrelation and cross-correlation relationships. For example, if some transmissivity (T) data of an aquifer are available at certain locations and water level data (h) at others, then cokriging can be employed to estimate the water level and/or T at any unsampled location by utilizing measured water level and T data sets, information on geologic structure (spatial correlation of Ts), and flow processes (cross-correlation between T and h, as discussed in Section 9.3). Furthermore, cokriging provides the best linear unbiased estimates, as well as estimation variance that represents uncertainty associated with the estimates.

Formulation of the cokriging equations. For convenience, we will develop the cokriging equations for two variables, transmissivity and head, only. Consider that the transmissivity, $T(\mathbf{x})$, and the hydraulic head under steady-state flow conditions, $\phi(\mathbf{x})$, are two correlated random fields and assume that

$$\ln T(\mathbf{x}) = F(\mathbf{x}) + f(\mathbf{x}) \text{ and } E[\ln T(\mathbf{x})] = F(\mathbf{x}) \tag{9.4.24}$$

$$\phi(\mathbf{x}) = H(\mathbf{x}) + h(\mathbf{x}) \text{ and } E[\phi] = H(\mathbf{x}) \tag{9.4.25}$$

where f and h are perturbations of $\ln T$ and $\phi(\mathbf{x})$ about their respective mean values, $F(\mathbf{x})$ and $H(\mathbf{x})$, respectively.

The estimation of f using the cokriging method can be obtained as follows:

$$\hat{f}(\mathbf{x}_m) = \sum_{i=1}^{N_f} \lambda_{mi} f(\mathbf{x}_i) + \sum_{j=1}^{N_h} \beta_{mj}[\phi(\mathbf{x}_j) - H(\mathbf{x}_j)]$$
$$= \sum_{i=1}^{N_f} \lambda_{mi} f(\mathbf{x}_i) + \sum_{j=1}^{N_h} \beta_{mj} h(\mathbf{x}_j) \quad (9.4.26)$$

where $\hat{f}(\mathbf{x}_m)$ is the estimate of f (the perturbation of ln$T$) at location $\mathbf{x}_m$; N_f and N_h are the total number of f and h measurements, respectively; $f(\mathbf{x}_j)$, $j = 1...N_f$ are the measured values of f at locations $\mathbf{x}_1, \mathbf{x}_2, ... \mathbf{x}_{N_f}$; $h(\mathbf{x}_j)$, $j = 1, ...N_h$ are the measured values of h at locations $\mathbf{x}_1, \mathbf{x}_2, ... \mathbf{x}_{N_h}$. Note that the sampled locations of f and h can be different. Furthermore, the cokriging weights, λ_{mi} and β_{mj}, assign contributions of $f(\mathbf{x}_j)$ and $h(\mathbf{x}_j)$ measurements, respectively, to the f estimate at $\mathbf{x}_m$.

As in the kriging method, the estimate based on the cokriging method satisfies the criterion of minimal mean square error. That is,

$$E\left[\left(\hat{f}(\mathbf{x}_m) - f(\mathbf{x}_m)\right)^2\right] = E\left[(\hat{f}_m - f_m)^2\right] = \text{minimal} \quad (9.4.27)$$

Substituting equation (9.4.26) into equation (9.4.27), we have

$$E\left[(\hat{f}_m - f_m)^2\right] = E\left[\left(\sum_{i=1}^{N_f} \lambda_{mi} f_i + \sum_{j=1}^{N_h} \beta_{mj} h_j - f_m\right)\left(\sum_{k=1}^{N_f} \lambda_{mk} f_k + \sum_{p=1}^{N_h} \beta_{mp} h_p - f_m\right)\right]$$
$$= \sum_{i=1}^{N_f}\sum_{k=1}^{N_f} \lambda_{mi}\lambda_{mk} C_{ff}(\mathbf{x}_i, \mathbf{x}_k) + \sum_{i=1}^{N_f}\sum_{p=1}^{N_h} \lambda_{mi}\beta_{mp} C_{fh}(\mathbf{x}_i, \mathbf{x}_p) - \sum_{i=1}^{N_f} \lambda_{mi} C_{ff}(\mathbf{x}_i, \mathbf{x}_m) +$$
$$\sum_{k=1}^{N_f}\sum_{j=1}^{N_h} \beta_{mj}\lambda_{mk} C_{hf}(\mathbf{x}_j, \mathbf{x}_k) + \sum_{j=1}^{N_h}\sum_{p=1}^{N_h} \beta_{mj}\beta_{mp} C_{hh}(\mathbf{x}_j, \mathbf{x}_p) - \sum_{j=1}^{N_h} \beta_{mj} C_{hf}(\mathbf{x}_j, \mathbf{x}_m) -$$
$$\sum_{k=1}^{N_f} \lambda_{mk} C_{ff}(\mathbf{x}_m, \mathbf{x}_k) - \sum_{p=1}^{N_h} \beta_{mp} C_{fh}(\mathbf{x}_m, \mathbf{x}_p) + C_{ff}(\mathbf{x}_m, \mathbf{x}_m) \quad (9.4.28)$$

Again, $C_{ff}(\mathbf{x}_i, \mathbf{x}_k)$ denotes the covariance between f at location $\mathbf{x}_i$ and f at location $\mathbf{x}_k$, while $C_{fh}(\mathbf{x}_i, \mathbf{x}_k)$ represents the cross-covariance between f at location $\mathbf{x}_i$ and h at location $\mathbf{x}_{k;}$. λ_{mi} is the weight of the measurement of f at location $\mathbf{x}_i$ to the estimate of f at location $\mathbf{x}_m$. Likewise, β_{mp} represents the weight of the observation of h at location $\mathbf{x}_p$ to the estimate of f at location $\mathbf{x}_m$.

The same as we did in developing the kriging equation, we will minimize the mean square error by differentiating equation (9.4.28) with respect to the cokriging coefficients, λ_{mk} and β_{mk}, and set the resultant equations to zero. Subsequently, the cokriging system of equations can be expressed as follows:

$$\sum_{i=1}^{N_f} \lambda_{mi} C_{ff}(\mathbf{x}_i, \mathbf{x}_k) + \sum_{j=1}^{N_h} \beta_{mj} C_{hf}(\mathbf{x}_j, \mathbf{x}_k) = C_{ff}(\mathbf{x}_m, \mathbf{x}_k) \text{ for } k = 1, \ldots, N_f \quad (9.4.29)$$

$$\sum_{i=1}^{N_f} \lambda_{mi} C_{fh}(\mathbf{x}_i, \mathbf{x}_p) + \sum_{j=1}^{N_h} \beta_{mj} C_{hh}(\mathbf{x}_j, \mathbf{x}_p) = C_{fh}(\mathbf{x}_m, \mathbf{x}_p) \text{ for } p = 1, \ldots, N_h \quad (9.4.30)$$

Solutions to the system of equations are the cokriging coefficients, which can be used in equation (9.4.26) to estimate $f(\mathbf{x}_m)$. The covariance functions and cross-covariance function in equations (9.4.29) and (9.4.30) can then be evaluated from either field data or theories.

Afterward, one can calculate the uncertainty associated with the estimate (cokriging variance) at location $\mathbf{x}_\mathrm{m}$ with a simplified form of equation (9.4.28).

$$E\left[\left(\hat{f}_m - f_m\right)^2\right] = C_{ff}(\mathbf{x}_m, \mathbf{x}_m) - \sum_{k=1}^{N_f} \lambda_{mk} C_{ff}(\mathbf{x}_m, \mathbf{x}_k) - \sum_{p=1}^{N_h} \beta_{mp} C_{fh}(\mathbf{x}_m, \mathbf{x}_p) \quad (9.4.31)$$

Notice that for estimating f at a different location, only the right-hand sides of equation (9.4.29) and equation (9.4.30) have to be changed. Afterward, these equations are solved again for the new weights, which then are used in equation (9.4.26) to obtain the estimate and used in equation (9.4.31) to derive the variance at the new location. This procedure is repeated for as many locations as needed. Notice that the preceding theory can also be extended to estimate the spatial distribution of head along with T.

If no h measurements are to be used, the system of equations is reduced to equation (9.4.29) and the term $\sum_{j=1}^{N_h} \beta_{mj} C_{hf}(\mathbf{x}_j, \mathbf{x}_k)$ in the equation, representing the cross-correlation between h and T, will be discarded. The rest of the equation is the same as the simple kriging equation, and the interpolation relies on the spatial correlation of Ts alone.

In cases where h measurements are the only data available, the linear estimator can also be used to estimate the T field. In this case, one needs only equation (9.4.30) and this equation can be simplified to

$$\sum_{j=1}^{N_h} \beta_{mj} C_{hh}(\mathbf{x}_j, \mathbf{x}_p) = C_{fh}(\mathbf{x}_m, \mathbf{x}_p) \quad (9.4.32)$$

The weight, β_{mj}, depends on both the head covariance and the cross-covariance of head measurement at $\mathbf{x}_\mathrm{p}$ and T at location $\mathbf{x}_\mathrm{m}$, where it is to be estimated. This weight, in essence, represents the contribution of the head measurement at $\mathbf{x}_\mathrm{p}$ to the T value at location $\mathbf{x}_\mathrm{m}$, or the normalized cross-correlation between h and T (see the discussion in Section 9.3). Subsequently, the linear estimator can be expressed by using a simplified form of equation (9.4.26) as

$$\hat{f}(\mathbf{x}_m) = \sum_{j=1}^{N_h} \beta_{mj} h(\mathbf{x}_j) \tag{9.4.33}$$

Equation (9.4.33) is then used for the estimation of $\hat{f}(\mathbf{x}_m)$, which will be added to the unconditional mean $F(\mathbf{x}_m)$ to obtain ln $T(\mathbf{x})$. Afterward, one can calculate the uncertainty (or cokriging variance) associated with the estimate at location $\mathbf{x}_m$ with a simplified form of equation (9.4.31) as

$$E\left[\left(\hat{f}_m - f_m\right)^2\right] = C_{ff}(\mathbf{x}_m, \mathbf{x}_m) - \sum_{p=1}^{N_h} \beta_{mp} C_{fh}(\mathbf{x}_m, \mathbf{x}_p) \tag{9.4.34}$$

The cokriging technique has been applied to flow and transport in saturated and unsaturated media by Hoeksema and Kitanidis (1984), Harter and Yeh (1996a, b), Yeh and Zhang (1996), and Li and Yeh (1999). As discussed previously, cokriging is a linear estimator. It yields the best estimate if the relationship between the estimate and observed data is linear. The relationship between the hydraulic properties and the response of aquifers or geologic formations, however, is nonlinear. That is, Darcy's law tells us that for a given flux, the hydraulic conductivity is inversely proportional to the hydraulic gradient or the head. Cokriging, thus, does not maximize the information content in head measurements to determine hydraulic conductivity, or vice versa.

9.5 Successive Linear Estimator

The successive linear estimator (SLE) (Yeh et al., 1996, Zhang and Yeh, 1997; Hughson and Yeh, 2000) is an extension of the spatial stochastic linear estimator (kriging or cokriging), which incorporates the nonlinear relationship between the system parameter (e.g., T) and the system response (e.g., h) during estimation of T using the head data. Specifically, SLE employs the governing equation for flow to calculate the h field that corresponds to the T field obtained from the first linear estimation (i.e., kriging or cokriging). The difference between observed and simulated h at each observation location is then determined. The difference at each observation location is subsequently employed in the linear estimator to improve the first T estimates. These newly improved T estimates are then used to simulate the new head field. The differences between the observed and improved heads are then evaluated, and the differences are again employed in the linear estimator to improve the T estimates further. This procedure is used iteratively until the improvement becomes negligible. The mathematics of the algorithm for steady-state flow problems is briefly presented next.

Consider the case where the hydraulic heads during a steady flow condition are used to estimate the transmissivity. For this scenario, the successive linear estimator takes the following form:

$$\begin{aligned}\hat{f}^{(r+1)}(\mathbf{x}_m) &= \hat{f}^{(r)}(\mathbf{x}_m) + \sum_{j=1}^{n_h} \omega_{mj}^{(r)} \left(\phi_j^*(\mathbf{x}_j) - \phi_j^{(r)}(\mathbf{x}_j) \right) \\ &= \hat{f}^{(r)}(\mathbf{x}_m) + \sum_{j=1}^{n_h} \omega_{mj}^{(r)} \left[\left(h_j^*(\mathbf{x}_j) + H(\mathbf{x}_j) \right) - \left(h_j^{(r)}(\mathbf{x}_j) + H(\mathbf{x}_j) \right) \right] \\ &= \hat{f}^{(r)}(\mathbf{x}_m) + \sum_{j=1}^{n_h} \omega_{mj}^{(r)} \left(h_j^*(\mathbf{x}_j) - h_j^{(r)}(\mathbf{x}_j) \right)\end{aligned} \tag{9.5.1}$$

in equation (9.5.1), $\hat{f}^{(r)}(\mathbf{x}_m)$ is the estimate of f at location $\mathbf{x}_m$ at iteration r. The superscript r is the iteration index; when $r = 1$, the estimate is from kriging or cokriging. The variable, $\phi_j^*(\mathbf{x}_j)$, stands for the observed head at observation locations, $\mathbf{x}_j$, and $\phi_j^{(r)}(\mathbf{x}_j)$ the simulated head, which is derived from the simulation using the previous estimated $\hat{f}^{(r)}(\mathbf{x})$. On the other hand, $H(\mathbf{x}_j)$ is the simulated mean head based on the unconditional mean T. $h_j^*(\mathbf{x}_j)$ and $h_j^{(r)}(\mathbf{x}_j)$ are the observed and simulated head perturbations, respectively. The term $\omega_{mj}^{(r)}$ represents the contribution (or the weight) of the discrepancy between the observed and the simulated head at location $\mathbf{x}_j$ to the estimate of f at location $\mathbf{x}_m$ at the rth iteration. In the following analysis, we will use f_m to denote $f(\mathbf{x}_m)$, $\hat{f}_m$ to denote $\hat{f}(\mathbf{x}_m)$, h_j^* to denote $h^*(\mathbf{x}_j)$, and $h_j^{(r)}$ to denote $h^{(r)}(\mathbf{x}_j)$ for convenience.

This linear estimator has to satisfy the minimum mean square error criterion, that is,

$$E\left[\left(f_m - \hat{f}_m^{(r+1)} \right)^2 \right] = \text{minimum} \tag{9.5.2}$$

substituting equation (9.5.1) into equation (9.5.2), we have

$$E\left[\left(f_m - \hat{f}_m^{(r+1)} \right)^2 \right] = E\left\{ \left[\left(f_m - \hat{f}_m^{(r)} \right) - \sum_{j=1}^{n_h} \omega_{mj}^{(1)} \left(h_j^* - h_j^{(r)} \right) \right]^2 \right\} \tag{9.5.3}$$

Equation (9.5.3) can be expanded and rewritten as

$$\begin{aligned}E\left[(f_m - \hat{f}_m^{(r+1)}))^2 \right] &= \varepsilon_{ff}^{(r)}(\mathbf{x}_m, \mathbf{x}_m) - \sum_{j=1}^{n_h} \omega_{mj}^{(r)} \varepsilon_{fh}^{(r)}(\mathbf{x}_m, \mathbf{x}_j) \\ &\quad - \sum_{j=1}^{n_h} \omega_{mj}^{(r)} \varepsilon_{hf}^{(r)}(\mathbf{x}_j, \mathbf{x}_m) + \sum_{j=1}^{n_h} \sum_{k=1}^{n_h} \omega_{mj}^{(r)} \omega_{mk}^{(r)} \varepsilon_{hh}^{(r)}(\mathbf{x}_j, \mathbf{x}_k)\end{aligned} \tag{9.5.4}$$

where

$$\varepsilon_{ff}^{(r)}(\mathbf{x}_m,\mathbf{x}_m)=E\left[\left(f_m-\hat{f}_m^{(r)}\right)\right]^2,\ \varepsilon_{fh}^{(r)}(\mathbf{x}_m,\mathbf{x}_j)=E\left[\left(f_m-\hat{f}_m^{(r)}\right)\left(h_j^*-h_j^{(r)}\right)\right],$$
$$\text{and } \varepsilon_{hh}^{(r)}(\mathbf{x}_j,\mathbf{x}_k)=E\left[\left(h_j^*-h_j^{(r)}\right)\left(h_k^*-h_k^{(r)}\right)\right] \tag{9.5.5}$$

To find the weights that minimize equation (9.5.4), we differentiate equation (9.5.4) with respect to ω_{mj} and then set the resultant equation to zero. We have

$$-\varepsilon_{fh}^{(r)}(\mathbf{x}_m,\mathbf{x}_j)-\varepsilon_{hf}^{(r)}(\mathbf{x}_j,\mathbf{x}_m)+2\sum_{k=1}^{n_h}\omega_{mk}^{(r)}\varepsilon_{hh}^{(r)}(\mathbf{x}_j,\mathbf{x}_k)=0 \text{ where } j=1,N_h. \tag{9.5.6}$$

Since $\varepsilon_{fh}^{(r)}(\mathbf{x}_m,\mathbf{x}_j)=\varepsilon_{hf}^{(r)}(\mathbf{x}_j,\mathbf{x}_m)$, equation (9.5.6) can be reduced to

$$\sum_{k=1}^{n_h}\omega_{mk}^{(r)}\varepsilon_{hh}^{(r)}(\mathbf{x}_j,\mathbf{x}_k)=\varepsilon_{fh}^{(r)}(\mathbf{x}_m,\mathbf{x}_j) \text{ where } j=1,N_h. \tag{9.5.7}$$

If $\varepsilon_{hh}^{(r)}(\mathbf{x}_j,\mathbf{x}_k)$ and $\varepsilon_{fh}^{(r)}(\mathbf{x}_m,\mathbf{x}_j)$ are known, equation (9.5.7) can be solved for the weights, $\omega_{mj}^{(r)}$, which can be used in equation (9.5.1) to obtain a new estimate of $\hat{f}^{(r+1)}(\mathbf{x}_m)$. Similarly, the weights for other locations, $\omega_{\ell j}^{(r)}$, can be obtained and in turn the entire $\hat{f}^{(r+1)}(\mathbf{x})$ field can be estimated.

Subsequently, a new $h^{(r+1)}(x)$ field is simulated using the conditional mean flow equation, which is the same as the deterministic equations in Chapter 6, for instance, a steady-state mean flow equation

$$\nabla[T_c^{eff}\nabla\phi_c]=0 \tag{9.5.8}$$

with specified boundary conditions. The term T_c^{eff} in equation (9.5.8) is the antilog of $\ln\hat{T}_c^{(r+1)}(\mathbf{x})=F(\mathbf{x})+\hat{f}^{(r+1)}(\mathbf{x})$ field and the subscript, c, represents the conditioning, and the superscript *eff* indicates the effective parameter. The procedures from implementation of equations (9.5.1) through (9.5.8) are repeated until the following two criteria are met:

1) $\Delta h_j^{(r+1)}=\left(h_j^*-h_j^{(r+1)}\right)$ at all j's are smaller than a given tolerance, or they stabilize.
2) The variance of $\hat{f}^{(r+1)}(x)$ field stabilizes.

In the preceding procedures, the conditional (or residual) covariance functions, $\varepsilon_{hh}^{(r)}(\mathbf{x}_j,\mathbf{x}_k)$ and $\varepsilon_{fh}^{(r)}(\mathbf{x}_0,\mathbf{x}_j)$, are approximated using the first-order Taylor series expansion of the head as shown in the following:

$$\phi(\mathbf{x})=H(\mathbf{x})+h(\mathbf{x})=G[F(\mathbf{x})+f(\mathbf{x})]\simeq G[F(\mathbf{x})]+\left.\frac{\partial G}{\partial \ln T}\right|_{F(\mathbf{x})}f(\mathbf{x}) \tag{9.5.9}$$

where G denotes the governing flow equation. Thus, the first-order head perturbation becomes

$$\mathbf{h}^{(r)} \simeq \frac{\partial \mathbf{G}}{\partial \ln \mathrm{T}}\bigg|_{F(\mathbf{x})^{(r)}} \mathbf{f}^{(r)} = \mathrm{J}^{(r)}\mathrm{f}^{(r)} \tag{9.5.10}$$

where **J** is the sensitivity matrix of head with respect to lnT evaluated at **F**. The residual covariances thus are

$$\begin{aligned} \varepsilon_{hh}^{(r)}(\mathbf{x}_i, \mathbf{x}_j) &= \mathrm{J}^{(r)}\varepsilon_{ff}^{(r)}(\mathbf{x}_i, \mathbf{x}_j)\mathrm{J}^{(r)T} \\ \varepsilon_{fh}^{(r)}(\mathbf{x}_i, \mathbf{x}_j) &= \mathrm{J}^{(r)}\varepsilon_{ff}^{(r)}(\mathbf{x}_i, \mathbf{x}_j) \end{aligned} \tag{9.5.11}$$

To evaluate equation (9.5.11), $\varepsilon_{ff}^{(r)}(\mathbf{x}_i, \mathbf{x}_j)$ must be determined first. For $r = 0$, $\varepsilon_{ff}^{(0)}(\mathbf{x}_i, \mathbf{x}_j)$ is the cokriging covariance function. For $r > 0$, it is determined as follows:

$$\varepsilon_{ff}^{(r+1)}(\mathbf{x}_m, \mathbf{x}_n) = \varepsilon_{ff}^{(r)}(\mathbf{x}_m, \mathbf{x}_n) - \sum_{j=1}^{n_h} \omega_{mj}^{(r)}\varepsilon_{fh}^{(r)}(\mathbf{x}_n, \mathbf{x}_j) \tag{9.5.12}$$

which is the conditional covariance function for the estimated transmissivity after iteration r. Furthermore, this represents an approximated uncertainty of the estimated ln$T$ field.

For more details, refer to the work by Yeh et al. (1996). The main idea behind SLE is that it is an iterative stochastic estimator, which seeks the best unbiased conditional fields of the effective parameters. Additionally, the iteration scheme is to utilize the information content about heterogeneity contained in an observed head successively to its maximum for estimating the T field. Therefore, the residual covariance has to be updated iteratively, since the successive estimate becomes closer to the true value at each iteration.

9.6 Joint Interpretation of Multiple Pumping Tests

Uncertainty in the estimated property fields and in the predicted flow behavior can be sufficiently reduced only if we have enough nonredundant data to condition the outcome. Cost-effective collection of nonredundant data sets thus becomes an important issue. Noninvasive geophysical surveys, such as seismic, geoelectrical, or electromagnetic surveys, are often thought to be the tools that are able to address this issue. For example, noninvasive geophysical surveys using surface electrode or geophone arrays can cover areas where no wells are available to monitor water and contaminant movements, or to map the geologic structures (Yeh et al., 2006, 2007). It has been recognized, however, that geophysical surveys detect geophysical anomalies, which may not be the same as hydrological anomalies. They must be translated to hydrologic heterogeneities using some petrophysical relationship (such as Archie's law). This relationship inherently varies spatially (see Yeh et al., 2002), as do hydrologic properties. Consequently, geophysical surveys that fail to consider spatial variability of petrophysical relationships will likely add further uncertainty to hydrologic property characterization. That is to say, in spite of the utility of geophysical surveys, hydraulic tests remain the method of choice for characterizing hydrologic properties.

Since the work by Yeh and Liu (2000) and Liu et al. (2002), hydraulic tomography (HT) (a new aquifer test method – sequential pumping tests) has become an active research topic for detailed aquifer characterization, and it has evolved to a mature technology. In principle, hydraulic tomography is a joint interpretation of sequential pumping tests or multiwell interference tests to map hydraulic property distributions in aquifers. Specifically, one disturbs an aquifer at one location by injecting or withdrawing water to create a pressure excitation or disturbance. This disturbance is modified by hydraulic heterogeneity in the aquifer as it propagates away from the excitation location. One then collects the disturbance at several monitoring locations in the aquifer to obtain one excitation-response data set. Afterward, another disturbance is created at a different location and signals are collected at the monitoring locations. By repeating the excitation at different locations, one acquires many sets of excitation and response data. Each set of data carries nonredundant information about aquifer heterogeneity. Subsequently, assimilation of these data sets leads to a detailed map of spatial distribution of aquifer heterogeneity.

In essence, HT facilitates collection of nonredundant information using a limited number of wells. Yeh et al. (2014) used 1-D examples and probabilistic reasoning based on Darcy's law to elucidate the information content embedded in a single head measurement about aquifer heterogeneity during a pumping test in a confined aquifer. On the basis of this information content associated with a head measurement, they explained why HT collects nonredundant information and why it can depict aquifer heterogeneity. To illustrate the intuitive nature of HT further, we will examine some drawdown data collected during sequential pumping tests conducted by Berg and Illman (2011) using nine wells at the North Campus Research Site (NCRS) at the University of Waterloo, Ontario, Canada.

Figures 9.7 through 9.9 show the dimensionless drawdown hydrographs observed at different ports of different wells during three pumping events. Dimensionless drawdown is used to eliminate the effects of different pumping rates and is obtained by multiplying the drawdown with the mean hydraulic conductivity and thickness of the aquifer and dividing by the pumping rate. Each figure is subdivided into nine subfigures: The center one is the well field, and the others show hydrographs at various elevations of the wells that are not pumped. No drawdown data set was collected at the pumping well. The pumping location at the well is marked by a triangle. In each of the wells, a circle marks the elevation at which the maximum drawdown or fastest drawdown is observed. This circle is also linked to the corresponding drawdown curve by a solid line. The lines, circles, and triangles are color-coded by their elevation. A legend defining the observed drawdowns at different elevations at each well is shown in each subfigure. Ports of the same number are of a similar elevation. For each hydrograph, the solid lines are drawdown measured by

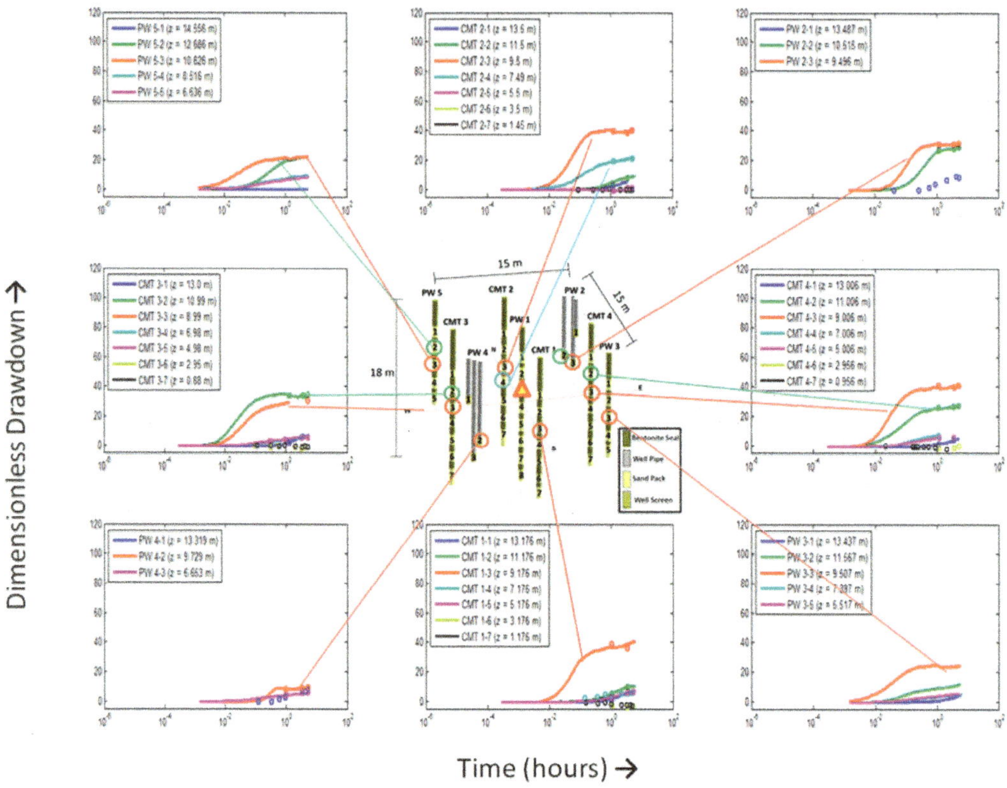

Figure 9.7. Drawdown responses of 44 observation ports for pumping at PW1–3. Circles indicate the location of highest or fastest drawdown response for each well.

using a pressure transducer, while the open circles are drawdown values that are manually recorded using a water level tape.

Figure 9.7 shows drawdown-time plots at different observation ports of different wells during pumping at locations PW1–3 (at elevation 10.24 m). According to this figure, observed drawdowns at elevation levels 2 and 3 of all eight wells are much greater than those at other levels (with the exception that drawdowns at CMT2–4 are greater than at CMT2–2) when pumping occurs at PW1–3. This suggests that the porous materials at levels 2 and 3 and CMT2–4 are well connected with that at the pumping location PW1–3. The greater and faster responses at elevation level 2 (CMT3–2, at elevation 10.99 m) than at level 3 (CMT3–3, at elevation 8.99 m) of CMT3 well suggest that the high permeable layer is located at around 10.24 m to 10.99 m. On the other hand, the rapid and large drawdowns occur at level 4 (CMT2–4, elevation 7.49 m) rather than level 2 (CMT2–2 at elevation 11.5 m) at well CMT2, indicating that this permeable zone at CMT2 is below 11.5 m and extends down to 7.49 m. The drawdown responses at CMT3–2 and CMT2–4 imply that the high permeable layer slightly dips or becomes thicker toward the CMT2 direction (i.e., to the north). Furthermore, all noticeable drawdowns tend to level off at late time,

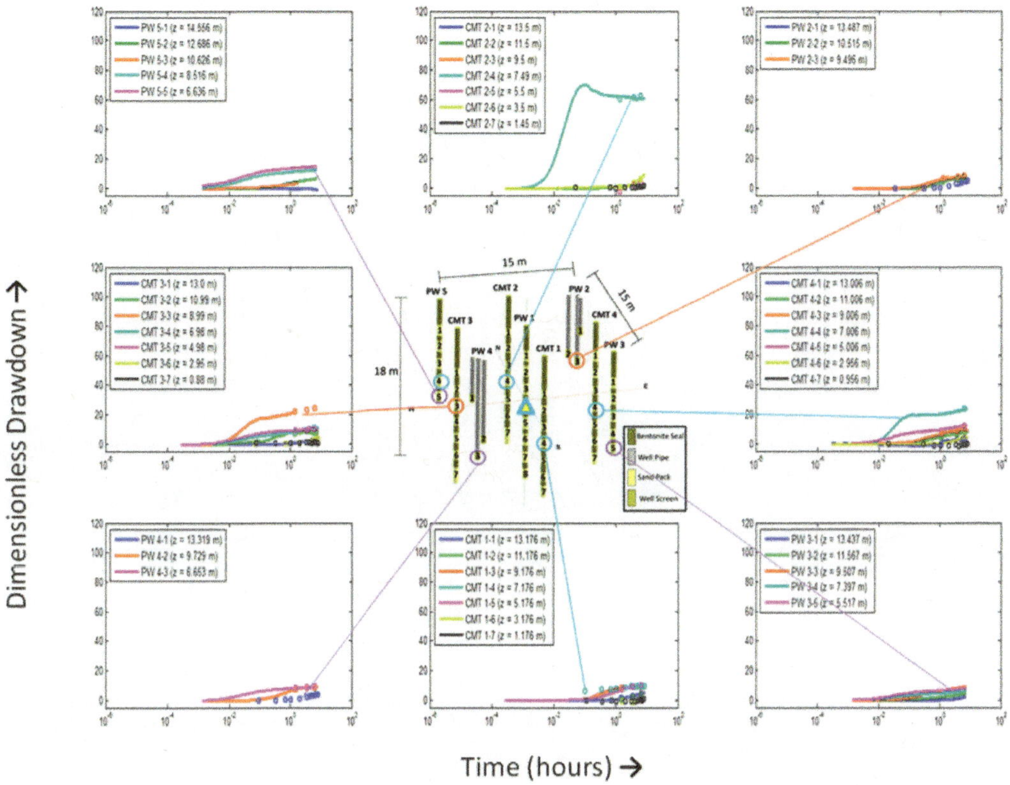

Figure 9.8. Drawdown responses of 44 observation ports for pumping at PW1–4. Circles indicate the location of highest or fastest drawdown response for each well.

approaching a steady state. This implies that that there are sources of water replenishing the pumping.

Figure 9.8 shows that when pumping is at port PW1–4 (2 m below port PW1–3), responses at most elevations in all wells (except at CMT2–4) are smaller than those induced by pumping at PW1–3. Drawdown at levels 4 and 5 of all wells are the most responsive except at well CMT3, where drawdown at level 3 is significantly greater. This again suggests that levels 4 and 5 and PW1–4 are within the same layer. This layer appears to be less permeable than the layer where levels 2 and 3 are located, since all drawdowns are smaller. Also, CMT2–4 records anomalously large drawdowns, while other ports of CMT2 measure negligible drawdowns. Notice that CMT3–3 (elevation 8.99 m), CMT2–4 (elevation 7.49 m), and CMT4–4 (elevation 7.01 m) are the only locations where dimensionless drawdown reaches 20. Consequently, we may postulate that (i) a conduit exists between PW1–4 and CMT2–4, and (ii) a more permeable region that spans most of level 3 exists and it is slightly tilted toward the east (i.e. CMT4).

When the pumping source is moved farther downward from PW1–4 to PW1–5 (Figure 9.9), the magnitudes of drawdowns at all levels are small. In many wells, late-

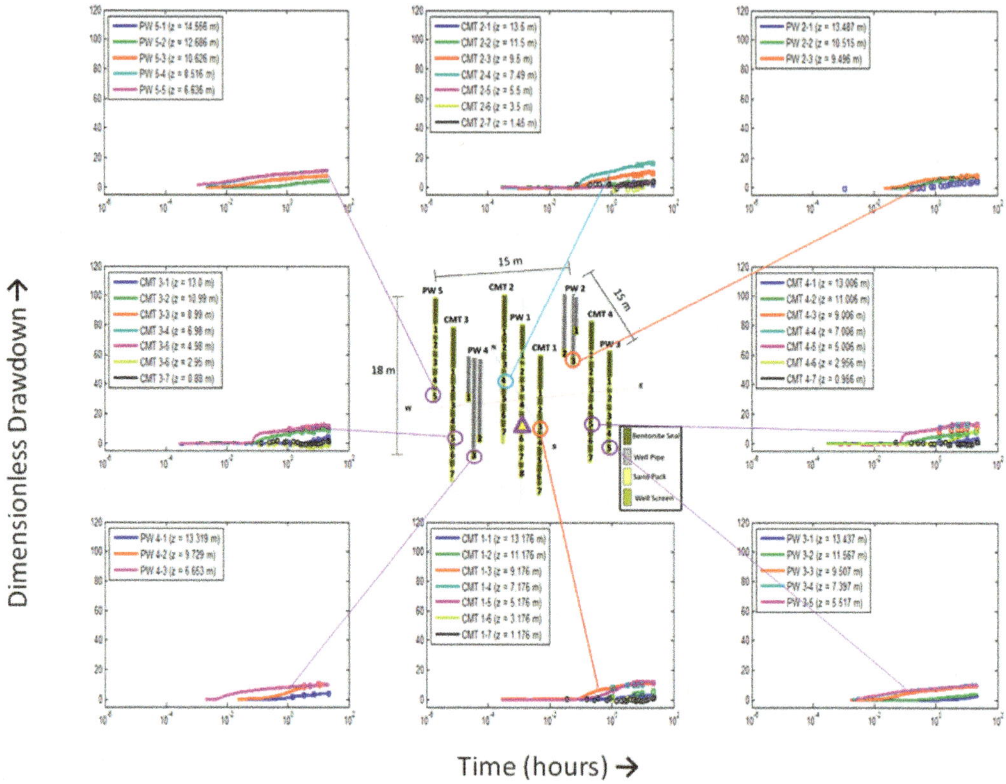

Figure 9.9. Drawdown responses of 44 observation ports for pumping at PW1–5. Circles indicate the location of highest or fastest drawdown response for each well.

time drawdowns at many ports are stacked together. We may thus conclude that the subsurface at and below CMT2–5 is highly impermeable.

As illustrated in Figures 9.7 through 9.9, the takeoff time, maximum drawdown, and time evolution of the drawdown hydrographs recorded at different locations during a given pumping test allow interpretation of the heterogeneity of the aquifer. Each figure is tantamount to a snapshot of aquifer heterogeneity as experienced by the water flow induced by pumping at a given location. Hence, sequential pumping tests at different locations are equivalent to taking many snapshots at different angles and perspectives. Using each snapshot to confirm or improve another is the fundamental principle of hydraulic tomography – letting the data tell the story.

The cross-correlation relationship between aquifer heterogeneity and observed head during a pumping test discussed in Section 7.4 (Chapter 7) also offers a theoretical explanation of the nonredundancy for the data collected by the hydraulic tomography surveys. Figures 7.18 and 7.19 show that the spatial patterns of ρ_{hy} (i.e., the cross-correlation between head perturbation, h, and natural log transmissivity perturbation, y) evolve from a symmetric pattern at early times, similar to those of ρ_{hz} (i.e., the cross-correlation between h and natural log storage coefficient perturbation, z), to a radially nonsymmetrical two-hump pattern at late times (see equation 9.3.18). According to these figures, the head at the observation well at the early time is essentially affected by T and S in a small circular area centered between the pumping well and the observation well. Therefore, if the location of the observation, or the pumping well, is different from that in the previous pumping test, and the heterogeneity in the new circular area is different from the previous one, the new head carries nonredundant information about T and S in the circular area.

At late times, the highest ρ_{hy} values are at the ridges of the two humps at the upstream of the observation and the pumping wells. This unique two-hump pattern suggests that if one of the two wells is moved to a new location, and if the heterogeneity at the hump associated with the new location is different from that at the previous location, the head at the observation well (regardless of whether its location is changed or not) will be different. That is, this new head data set carries new (nonredundant) information about aquifer heterogeneity. Furthermore, ρ_{hy} values over the entire area of the cone of depression are relatively high at late times in comparison with those at earlier times. Therefore, adding more head data from the new observation well at a late time or the head data from the same observation well due to pumping at new locations helps to decipher the spatial distribution of T anomalies. This is the reason that a joint interpretation of sequential pumping tests or multiwell pumping tests is superior to an interpretation based on one pumping test, as articulated by Huang et al. (2011) and as demonstrated in Figure 9.10.

Huang et al. (2011) used a synthetic heterogeneous aquifer to demonstrate the robustness of HT, since the true properties of the aquifer are known exactly. Figure 9.10a shows the synthetic heterogeneous T field of a 2-D plane aquifer, which has a well field consisting of ten wells, and no drawdown or constant head boundary conditions along all sides. Seven sequential steady-state pumping tests (a HT survey) in this aquifer were simulated, and drawdowns at the other ten wells were collected during each pumping test. Using the pumping rate, boundary conditions, and simulated drawdowns, the SLE (Section 9.5) was used to estimate the synthetic T field. Figure 9.10b illustrates the estimated T field, based on only one pumping test data set. The estimated T fields based on joint interpretation of two, three, four, five, six, and seven pumping test data sets are shown in Figure 9.10c, d, e, f, g, and h, respectively. After examination of these figures, it is apparent that joint interpretation of four or five pumping test data sets has led to an estimated T field that captures the general features of the synthetic T field. Further inclusion of other pumping test data sets has resulted in some refinements of the estimate only. These figures also demonstrate that HT can depict heterogeneity patterns not only within the well field but also away from the well field.

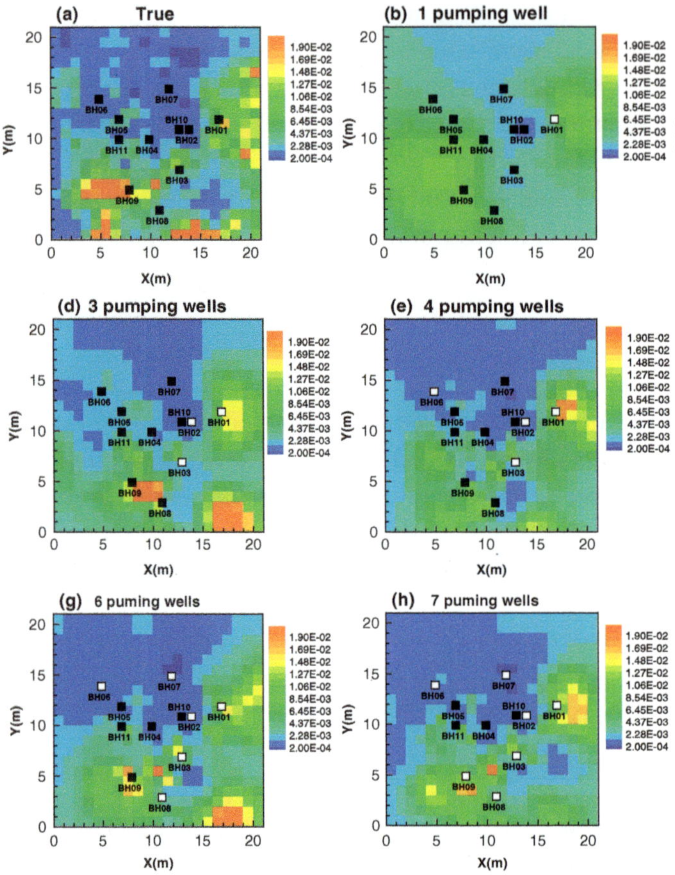

Figure 9.10. True T field of the synthetic aquifer (a). Estimated T field of the aquifer based on data sets from joint interpretation of different number of pumping tests. The white square represents the pumping well location. (Huang et al., 2011)

After decades of active theoretical, laboratory, and field studies, HT is no doubt a viable technology for characterizing the spatial distributions of hydraulic parameters at a higher level of detail than traditional methods (e.g., Vasco et al., 2000; Vasco and Karasaki, 2006; Yeh and Liu, 2000; Liu et al., 2002; Bohling et al., 2002, 2007; Brauchler et al., 2003, and 2011; Zhu and Yeh, 2005, 2006; He et al., 2006; Liu et al., 2007; Ni and Yeh, 2008; Straface et al., 2007; Xiang et al., 2009; Illman et al., 2007, 2008, 2009, 2010,d 2011; Hao et al., 2008; Kuhlman et al., 2008; Castagna and Bellin, 2009; Berg and Illman, 2011; Cardiff et al., 2009, 2012; Cardiff and Barrash, 2011; Huang et al., 2011; Li et al., 2008; Sun et al., 2013; Mao et al., 2013c). These studies have proved that the transient HT can identify not only the pattern of the heterogeneous K or T field, but also the variation of specific storage (S_s) or the storage coefficient (S) (see Zhu and Yeh, 2005; Liu et al., 2007; Xiang et al., 2009, in particular). More importantly, they have demonstrated that the hydraulic property fields estimated by HT can yield much more accurate predictions of flow and solute transport processes than can other conventional characterization approaches (e.g., Ni et al., 2009; Illman et al., 2011). In particular, Berg and Illman (2011, 2013, 2014) substantiated the robustness of HT for a highly heterogeneous geologic medium with a variance of log hydraulic conductivity of 5.4 and a vertical correlation scale of 0.15 m. Likewise, robust results were obtained when HT was applied to map fractures and faults in granite rocks over a scale of kilometers (Illman et al., 2009), as well as microcracks in laboratory rocks (Sharmeen et al., 2012).

In summary, these studies unequivocally support the call for a change in the way we collect and analyze data for characterizing aquifers, as advocated by Yeh and Lee (2007); it also promotes the concept of computed axial tomographic (CAT) scanning of a groundwater basin using natural stimuli, as proposed by Yeh et al. (2008).

9.7 Summary

- Applications of the forward or inverse models to real-world problems always involve uncertainty: many possible solutions or nonuniqueness or ill-posed problems. Therefore, stochastic modeling approaches, which conceptualize the problems in a probabilistic framework, are most appropriate.
- The stochastic approaches can be classified into unconditional and conditional approaches. Both approaches are used to derive the best unbiased estimates of parameters (inverse modeling) or system states (forward modeling) and to quantify their uncertainties. The conditional approach, however, honors observed data in space and/or time, and it can yield high resolution estimates and predictions with small uncertainty, as more data are used.
- Conditional approaches can be classified into the conditional effective parameter approach and the conditional simulation (or conditional realization, or conditional ensemble) approach. The conditional effective approach directly seeks the best unbiased estimates of conditional fields of effective properties, which provide the best unbiased prediction of groundwater flow processes. On the other hand, conditional simulation explores equally likely realizations of property fields that preserve data at observation locations. These realizations are then employed to make predictions of the possible corresponding flow processes from which the best unbiased flow prediction is derived and uncertainty in such predictions is estimated.
- Conditional approaches are built upon spatial and temporal autocorrelations between observations or cross-correlation relationships between observed head and hydraulic properties. Additionally, cross-correlation analysis identifies redundancy of data or information.
- Kriging and cokriging approaches are conditional effective (or mean) parameter approaches. Both are linear estimators. The former uses the linear sum of weighted observed attributes to estimate the same attribute in space, while the latter uses the

linear sum of weighted observed different types of attributes to estimate one type of attribute. These weights are related to their spatial correlation and cross-correlations with other attributes.

- The successive linear estimator (SLE) successively incorporates nonlinear relationships between hydraulic parameters and heads during the estimation process, such that information about hydraulic parameters in the observed head is fully utilized. Additionally, SLE is an effective methodology for fusion of the same or different types of information and for addressing uncertainty.
- Regardless of the effectiveness of any estimator, resolution and uncertainty depend on the amount of nonredundant information used to condition the outcome.
- The hydraulic tomographic survey is an effective way to collect nonredundant data sets in a cost-effective manner to enhance aquifer characterization.

9.8 Exercise

1. Consider the 1-D steady flow through a column setup in Figure 9.3. Suppose that the observed head has a value lower than the ensemble mean head (green line in Figure 9.3). Please plot the conditional mean head and its upper and lower bounds. In addition, discuss how this observation head reduces the range of possible K values along the column in comparison with the one in Figure 9.3.
2. On the basis of the results of Problem 1 and that of Figure 9.3, discuss how the value of the head observation in comparison with the ensemble mean head would affect the distribution of possible heads and K values.
3. Using the Theis solution, derive the cross-correlation between h and $\ln T$ and that between h and $\ln S$ for an impermeable line boundary at different distances from the observation well. Do the same for a recharge boundary. Plot results in a form similar to Figure 9.4 and discuss the results (see Sun et al., 2013).
4. Plot head variance as a function of distance from the pumping well, r, and time in unbounded aquifers (equation 9.3.10). Derive and plot the head variance as a function of u for an impermeable boundary and for a recharge line boundary, and then discuss the results.
5. Table 9.1 lists twenty synthetic hydraulic conductivity (K) values at 1 m intervals along a 1-D transect line. Suppose that we have only five measurements (the values marked with *) of the K distribution, and we know that the spatial covariance function is an exponential function with a correlation scale of 5 m. In addition, the mean and variance of $\ln K$ are 0.1 and 0.5, respectively. Reconstruct and plot the K values along the transect line with +1 and –1 standard deviation based on the kriging variance. Discuss the result in terms of the effects of conditioning, goodness of the estimates, and behaviors of the estimate at different distances in relation to the correlation scale.
6. Solve Problem 5 again using an incorrect correlation scale, variance, and mean. Discuss how errors of these inputs affect your estimates. Also examine the effects of the nugget of the correlation function on the estimates (Chapter 4, homework 2). Discuss the results.

Table 9.1. *The synthetic K values along a transect line*

Distance (m)	***K* (m/hr)**	
1	0.352259	
2	0.360366	*
3	0.427775	
4	0.705911	*
5	1.317432	
6	1.622882	
7	1.208899	
8	0.746709	
9	0.561846	*
10	0.595582	
11	0.736615	
12	1.331198	
13	2.521854	
14	2.637034	
15	2.790671	
16	2.623194	*
17	2.852272	
18	2.06019	
19	1.314095	
20	0.80969	*

7. Derive the cokriging equations for estimating head at a location using some known heads and hydraulic conductivity measurements at different locations.
8. On the basis of the cross-correlation analysis, can you answer the following question? Can long-term temporal groundwater data at few wells replace data from a dense spatial groundwater level network in terms of aquifer characterization?

Reference

Abramowitz, M., and I. A. Stegun. 1964. *Handbook of Mathematical Functions*. National Bureau of Standards: 231.

Akindunni, F. F., and R. W. Gillham. 1992. Unsaturated and saturated flow in response to pumping of an unconfined aquifer: Numerical investigation of delayed drainage. *Ground Water*, 30: 873–884.

Barenblatt, G. I., I. P. Zheltov, and I. N. Kochina. 1960. Basic concepts in the theory of seepage of homogeneous liquids in fissured rocks. *Journal of Applied Mathematics*, 24: 1286–1303.

Barrash, W., and T. Clemo. 2002. Hierarchical geostatistics and multifacies systems: Boise Hydrogeophysical Research Site, Boise, Idaho. *Water Resources Research*, 38, no. 10: 1196.

Bear, J. 1972. *Dynamics of Fluids in Porous Media*. New York: Elsevier (reprinted by Dover Publications, 1988) 784 p.

Bear, J. 1979. *Hydraulics of groundwater*. McGraw-Hill, New York, N.Y., 241 p.

Berg, S. J., and W. A. Illman. 2011. Three-dimensional transient hydraulic tomography in a highly heterogeneous glaciofluvial aquifer-aquitard system. *Water Resources Research*, 47, W10507, doi:10.1029/2011WR010616.

Berg, S. J., and W. A. Illman. 2013. Field study of subsurface heterogeneity with steady-state hydraulic tomography. *Ground Water* 51, no. 1: 29–40, doi:10.1111/j.1745–6584.2012.00914.x.

Berg, S. J., and W. A. Illman. 2014. Comparison of hydraulic tomography with traditional methods at a highly heterogeneous site. *Ground Water*, doi:10.1111/gwat.12159.

Bevan, M. J., A. L. Endres, D. L. Rudolph, and G. Parkin. 2005. A field scale study of pumping-induced drainage and recovery in an unconfined aquifer. *Journal of Hydrology*, 315: 52–70.

Boggs, J. M., S. C. Young, D. J. Benton, and Y. C. Chung. 1990. *Hydrogeologic characterization of the MADE site. Interim Report EN-6915, Electric Power Res. Inst.*, Palo Alto, CA.

Bohling, G. C., J. J. Butler Jr., X. Zhan, and M. D. Knoll. 2007. A field assessment of the value of steady-shape hydraulic tomography for characterization of aquifer heterogeneities. *Water Resources Research*, 43, W05430, doi:10.1029/2006WR004932.

Bohling, G. C., X. Zhan, J. J. Butler Jr., and L. Zheng. 2002. Steady shape analysis of tomographic pumping tests for characterization of aquifer heterogeneities. *Water Resources Research*, 38, no. 12: 1324.

Boulton, N. S. 1954. Unsteady radial flow to a pumped well allowing for delayed yield from storage. *International Association of Scientific Hydrology*, 2: 472–477.

Boulton, N. S. 1963. Analysis of data from nonequilibrium pumping tests allowing for delayed yield from storage. *Proceedings Institution of Civil Engineering*, 26: 469–482.

Bouwer, H. 1978. *Groundwater Hydrology*. McGraw-Hill College, New York: 480 p.

Brouwer, J., and H. Anderson. 2000. Water holding capacity of ironstone gravel in a typic plinthoxeralf in Southeast Australia. *Soil Science Society of America Journal*, 64: 1603–1608.

Brauchler R., R. Hu, P. Dietrich, and M. Sauter. 2011. A field assessment of high-resolution aquifer characterization based on hydraulic travel time and hydraulic attenuation tomography. *Water Resources Research*, 47, W03503, doi:10.1029/2010WR009635.

Brauchler, R., R. Liedl, and P. Dietrich. 2003. A travel time based hydraulic tomographic approach. *Water Resources Research*, 39, no. 12: 1370.

Brutsaert, W. 1970. *Immiscible multiphase flow in ground water hydrology: A computer analysis of the well flow problem*, PhD Thesis, Colorado State University, Fort Collins.

Buckingham, E. 1907. *Studies on the movement of soil moisture. Bulletin 38. USDA Bureau of Soils*, Washington, DC

Cacas, M. C., E. Ledoux, G. de Marsily, B. Tillie, A. Barbreau, E. Durand, B. Feuga, and P. Peaudecerf. 1990. Modeling fracture flow with a stochastic discrete fracture network: Calibration and validation. 1. The flow model. *Water Resources Research*, 26, no. 3: 479–489.

Capilla, J. E., J. J. Gmez-Hernández, and A. Sahuquillo. 1997. Stochastic simulation of transmissivity fields conditional to both transmissivity and piezometric Data. 2. Demonstration on a synthetic aquifer: *J. of Hydrology*, v. 203, no. 1, 175–188.

Cardiff, M., and W. Barrash. 2011. 3D transient hydraulic tomography in unconfined aquifers with fast drainage response. *Water Resources Research*, 47, W12518, doi:10.1029/2010WR010367.

Cardiff, M., W. Barrash, and P. K. Kitanidis. 2012. A field proof-of-concept of aquifer imaging using 3D transient hydraulic tomography with modular, temporarily-emplaced equipment. *Water Resources Research*, 48, W05531, doi:10.1029/2011WR011704.

Cardiff, M., W. Barrash, P. K. Kitanidis, B. Malama, A. Revil, S. Straface, and E. Rizzo. 2009. A potential-based inversion of unconfined steady-state hydraulic tomography. *Ground Water*, 47, no. 2: 259–270.

Carman, P. C. 1937. Fluid flow through granular beds. *Transaction of Institution of Chemical Engineering*, 15: 150–166.

Carsel, R. F., and R. S. Parrish. 1988. Developing joint probability distributions of soil water retention characteristics. *Water Resources Research*, 24: 755–769.

Castagna, M., and A. Bellin. 2009. A Bayesian approach for inversion of hydraulic tomographic data. *Water Resources Research*, 45, W04410, doi:10.1029/2008WR007078.

Chen, Z. X., Pang Z. Y., Jiang L. S., and Liu M. X. 1986, Exact solution for the problem of crossflow in a bounded two-aquifer system with an aquitard, *Water Resources Research*, 22, no. 8: 1225–1236.

Clayton, V. D., and Andre, G. 1992. *Gslb: Geostatistical Software Library and User's Guide*. Oxford University Press, Oxford: 340 p.

Coile, T. S. 1953. Moisture content of small stone in soil. *Soil Science*, 75: 203–207.

Conca, J. L., M. J. Apted, W. Zhou, R. C. Arthur, and J. Kessler. 1998. Flow barrier system for long-term high-level-waste isolation: Experimental results. *Nuclear Technology*, 124: 88–99.

Cooper, H. H. Jr., and C. E. Jacob. 1946. A generalized graphical method for evaluating formation constants and summarizing well field history, *Eos Transactions of the American Geophysical Union*, 27, 526–534, doi:10.1029/TR027i004p00526.

Corey, A. T. 1994. Mechanics of immiscible fluids in porous media, Water Resources Publications, 252 p.

Dagan, G. 1967. Method of determining permeability and effective porosity of unconfined anisotropic aquifers. *Water Resources Research*, 3, no. 4: 1059–1071.

Dagan, G. 1982. Stochastic modeling of groundwater flow by unconditional and conditional probabilities. 2. The solute transport. *Water Resources Research*, 18, 835–848.

De Marsily, G. 1986. *Quantitative Hydrogeology: Groundwater Hydrology for Engineers*. Edition 2, Academic Press, Orland, FL: 440 p.

Dershowitz, W. S., and H. H. Einstein. 1988. Characterizing rock joint geometry with joint system models. *Rock Mechanics and Rock Engineering*, 21, no. 1: 21–51.

Duguid, J. O., and P. C. Y. Lee. 1977. Flow in fractured porous media. *Water Resources Research*, 13, no. 3: 558–566.

Dverstorp, B., J. Andersson, and W. Nordqvist. 1992. Discrete fracture network interpretation of field tracer migration in sparsely fractured rock. *Water Resources Research*, 28: 2327–2343.

Dykhuizen, R. C. 1990. A new coupling term for dual porosity models. *Water Resources Research*, 26, no. 2: 351–356.

Endres, A. L., J. P. Jones, and E. A. Bertrand. 2007. Pumping-induced vadose zone drainage and storage in an unconfined aquifer: A comparison of analytical model predictions and field measurements. *Journal of Hydrology*, 335, no. 1–2: 207–218.

Flint, A. L., and S. Childs. 1984. Physical properties of rock fragments and their effect on available water in skeletal soils, in J. E. Box, Jr. (ed.) *Erosion and productivity of soils containing rock fragments*, pp. 91–103. Soil Science Society of America Spec. Publ. 13. SSSA, Handbook No. 436. Madison, WI.

Forchheimer, P. 1914–16. Teaching and the hydraulic manual, 5 volumes.

Freeze, R. A., and J. A. Cherry. 1979. *Groundwater*, Englewood Cliffs, NJ, Prentice-Hall, 604 p.

Gardner, W. R. 1958. Some steady-state solutions of unsaturated moisture flow equations with application to evaporation from a water table. *Soil Science* 85: 228–232.

Gee, G. W., and A. L. Ward. 2001. *Vadose Zone Transport Field Study: Status Report. PNNL-13679*, Pacific Northwest National Laboratory.

Gelhar, L. W. 1993. *Stochastic Subsurface Hydrology*. Prentice-Hall, Englewood Cliffs, N.J.: 390 p.

Gelhar, L. W., and C. L. Axness. 1983. Three-dimensional stochastic analysis of macrodispersion in aquifers, *Water Resources Research*, 19, no. 1: 161–180.

Gerke, H. H., and M. T. van Genuchten. 1993a. A dual-porosity model for simulating the preferential movement of water and solutes in structured porous media. *Water Resources Research*, 29, no. 2: 305–319.

Gerke, H. H., and M. T. van Genuchten. 1993b. Evaluation of a first-order water transfer term for variably saturated dual-porosity models. *Water Resources Research*, 29, no. 4: 1225–1238.

Glass R. J., J. R. Brainard, T.-C. J. Yeh. 2005. Infiltration in unsaturated layered fluvial deposits at Rio Bravo: Macroscopic anisotropy and heterogeneous transport Vadose Zone. *Journal*, 4, no. 1: 22–31.

Gómez-Hernández, J. J., A. Sahuquillo, and, J. E. Capilla. 1997. Stochastic simulation of transmissivity fields conditional to both transmissivity and piezometric data, 1. *Theory. J. of Hydrology*, 203, no. 1–4: 162–174.

Graham, W. D., and D. B. McLaughlin. 1989a. Stochastic Analysis of Non-Stationary Subsurface Solute Transport, 1. *Unconditional Moments, Water Resources Research*, 25, no. 2: 215–232.

Graham, W. D., and D. B. McLaughlin. 1989b. Stochastic Analysis of Non-Stationary Subsurface Solute Transport. 2. *Conditional Moments, Water Resources Research*, 25, no. 11: 2331–2355.

Gutjahr, A. L. 1989. *Fast Fourier transforms for random field generation,* project report. Dept. of Math., N. M. Tech., Socorro.
Hadamard, J. 1902. Sur les problèmes aux dérivés partielles et leur signification physique. *Princeton University Bulletin,* 13: 49–52.
Hanna, S., and T.-C. J. Yeh. 1998. Estimation of co-conditional moments of transmissivity, hydraulic head, and velocity fields. *Advances in Water Resources,* 22: 87–93.
Hantush, M. S. 1961a. Drawdown around a partially penetrating well. *J. of the Hyd. Div. Proc. of the Am. Soc. of Civil Eng.* 87, no. HY4: 83–98.
Hantush, M. S. 1961b. Aquifer tests on partially penetrating wells. *Jour. of the Hyd. Div., Proc. of the Am. Soc. of Civil Eng.* 87, no. HY5: 171–194.
Hantush, M. S. 1964. Hydraulics of wells, in Ven Te Chow (ed.), *Advances in Hydrosciences,* vol. 1, Academic Press Inc., New York.
Hantush, M. S., and C. E. Jacob. 1955. Non-steady radial flow in an infinite leaky aquifer. *Transactions of the American Geophysical Union* 36, no. 1: 95–100.
Hao, Y., T.-C. J. Yeh, J. Xiang, W. A. Illman, K. Ando, K.-C. Hsu et al. 2008. Hydraulic tomography for detecting fracture zone connectivity. *Ground Water,* 46: 183–92, doi:10.1111/j.1745–6584.2007.00388.x.
Harter, T., and T.-C. J. Yeh. 1996a. Stochastic analysis of solute transport in heterogeneous, variably saturated soils, *Water Resources Research,* 32, no. 6: 1585–1596.
Harter, T., and T.-C. J. Yeh. 1996b. Conditional stochastic analysis of solute transport in heterogeneous, variably saturated soils, *Water Resources Research,* 32, no. 6: 1597–1610.
He, Z., A. Datta-Gupta, and D. W. Vasco. 2006. Rapid inverse modeling of pressure interference tests using trajectory based travel time and amplitude sensitivities. *Water Resources Research,* 42, W03419, doi:10.1029/2004WR003783.
Hendricks Franssen, H. J., P. Brunner, P. Makobo, and W. Kinzelbach. 2008. Equally likely inverse solutions to a groundwater flow problem including pattern information from remote sensing images, *Water Resources Research,* 44, W01419, doi:10.1029/2007WR006097.
Hess, K. M. 1989. Use of a borehole flowmeter to determine spatial heterogeneity of hydraulic conductivity and macrodispersion in a sand and gravel aquifer, Cape Cod, Massachusetts, in F. J. Molz, J. G. Melville, and O. Gtiven (ed.), *Proceedings of the Conference on New Field Techniques for Quantifying the Physical and Chemical Properties of Heterogeneous Aquifers,* pp. 497–508. National Water Well Association, Dublin, Ohio.
Hoeksema, R. J., and P. K. Kitanidis. 1984. An application of the geostatistical approach to the inverse problem in two-dimensional groundwater modeling. *Water Resources Research,* 20, no. 7: 1003–1020.
Huang, S. Y., J. C. Wen, T.-C. J. Yeh, W. Lu, H. L. Juan, C. M. Tseng, J. H. Lee, and K. C. Chang. 2011. Robustness of joint interpretation of sequential pumping tests: Numerical and field experiments. *Water Resources Research,* 47, W10530, doi:10.1029/2011WR010698.
Hughson, D., and T.-C. J. Yeh. 2000. An inverse model for three-dimensional flow in variably saturated porous media. *Water Resources Research,* 36: 829–839.
Illman, W. A., S. J. Berg, and T.-C. J. Yeh. 2011. Comparison of approaches for predicting solute transport: Sandbox experiments. *Ground Water,* 50: 421–431, doi:10.1111/j.1745–6584.2011.00859.x.
Illman, W. A., A. J. Craig, and X. Liu. 2008. Practical issues in imaging hydraulic conductivity through hydraulic tomography. *Ground Water,* 46, no. 1: 120–132, doi:10.1111/j.1745–6584.2007.00374.x.

Illman, W. A., X. Liu, and A. Craig. 2007. Steady-state hydraulic tomography in a laboratory aquifer with deterministic heterogeneity: Multimethod and multiscale validation of K tomograms. *J. Hydrol.*, 341, no. 3–4: 222–234, doi:10.1016/j.jhydrol.2007.05.011.

Illman, W. A., X. Liu, S. Takeuchi, T.-C. J. Yeh, K. Ando, and H. Saegusa. 2009. Hydraulic tomography in fractured granite: Mizunami Underground Research Site, Japan. *Water Resources Research*, 45, W01406, doi:10.1029/2007WR006715.

Illman, W. A., J. Zhu, A. J. Craig, and D. Yin. 2010. Comparison of aquifer characterization approaches through steady state groundwater model validation: A controlled laboratory sandbox study. *Water Resources Research*, 46, W04502, doi:10.1029/2009WR007745.

Isaaks, E. H., and R. M. Srivastava. 1989. *Applied Geostatistics*. Oxford University Press, New York: 561 p.

Jacob, C. E. 1963a. Determining the permeability of water table aquifers. *U.S. Geological Survey Water-Supply*, 1536-I: 245–271.

Khaleel, R., and E. J. Freeman. 1995. Variability and scaling of hydraulic properties for 200 area soils, Hanford Site. WHC-EP-0883, Westinghouse Hanford Co., Richland, WA.

Khaleel, R., and P. R. Heller. 2003. On the hydraulic properties of coarse-textured sediments at intermediate water contents. *Water Resources Research*, 39, no. 9, doi:10.1029/2003WR002387.

Khaleel, R., and J. F. Relyea. 1997. Correcting laboratory-measured moisture retention data for gravels. *Water Resources Research*, 33: 1875–1878.

Khaleel, R., and J. F. Relyea. 2001. Variability of Gardner's α for coarse-textured sediments. *Water Resources Research*, 37: 1567–1575.

Khaleel, R., J. F. Relyea, and J. L. Conca. 1995. Evaluation of Van Genuchten–Mualem relationships to estimate unsaturated hydraulic conductivity at low water contents. *Water Resources Research*, 31, no. 11: 2659–2668.

Khaleel, R., T.-C. J. Yeh, and Z. Lu. 2002. Upscaled flow and transport properties for heterogeneous unsaturated media. *Water Resources Research*, 38, no. 5: 10.1029.

Kitanidis, P. K. 1995. Quasi-Linear geostatistical theory for Inversing. *Water Resources Research*, 31, no. 10: 2411–2419, doi:10.1029/95WR01945.

Klute, A., 1965. Laboratory measurement of hydraulic conductivity of saturated soil. in C. A. Black (ed.), *Methods of Soil Analysis*. American Society of Agronomy, Madison, WI: 210–221.

Kuhlman K. L., A. C. Hinnell, P. K. Mishra, and T.-C. J. Yeh. 2008. Basin-scale transmissivity and storativity estimation using hydraulic tomography. *Ground Water*, 46, no. 5: 706–715, doi:10.1111/j.1745-6584.2008.00455.x.

Lakshminarayana, V., and S. P. Rajagopalan. 1978. Type-curve analysis of time-drawdown data for partially penetrating wells in unconfined anisotropic aquifers. *Ground Water*, 16: 328–333.

Last G. V., and T. G. Caldwell. 2001. *Core Sampling in Support of the Vadose Zone Transport Field Study*. PNNL-13454, Pacific Northwest National Laboratory, Richland, WA.

Last G. V., T. G. Caldwell, and A. T. Owen. 2001. *Sampling of Borehole WL-3A through -12 in Support of the Vadose Zone Transport Field Study*. PNNL-13631, Pacific Northwest National Laboratory, Richland, WA.

Lenhard, R. J., and, J. C. Parker. 1987. Model for hysteretic constitutive relations governing multiphase flow. 2. Permeability-saturation relations. *Water Resources Research*, 23: 2197–2206.

Li, B., and T.-C. J. Yeh. 1999. Cokriging estimation of the conductivity field under variably saturated flow conditions, *Water Resources Research*, 35, no. 12: 3663–3674.

Li, W., A. Englert, O. A. Cirpka, and H. Vereecken. 2008. Three-dimensional geostatistical inversion of flowmeter and pumping test data. *Ground Water*, 46, no. 2: 193–201, doi:10.1111/j.1745–6584.2007.00419.x.

Liu, S. Y., T.-C. J. Yeh, and R. Gardiner. 2002. Effectiveness of hydraulic tomography: Sandbox experiments. *Water Resources Research*, 38, no. 4: Article Number: 1034, doi:10.1029/2001WR000338.

Liu, X., W. A. Illman, A. J. Craig, J. Zhu, and T.-C. J. Yeh. 2007. Laboratory sandbox validation of transient hydraulic tomography. *Water Resources Research*, 43, W05404, doi:10.1029/2006WR005144.

Long, J. C. S., J. S. Remer, C. R. Wilson, and P. A. Witherspoon. 1982. Porous media equivalents for networks of discontinuous fractures. *Water Resources Research*, 18, no. 3: 645–658.

Mao, D., L. Wan, T.-C. J. Yeh, C. H. Lee, K. C. Hsu, J. C. Wen, and W. Lu. 2011. A revisit of drawdown behavior during pumping in unconfined aquifers. *Water Resources Research*, 47, W05502, doi:10.1029/2010WR009326.

Mao, D., T.-C. J. Yeh, and L. Wan. 2013a. Necessary conditions for inverse modeling of flow through variably saturated porous media. *Advances in Water Resources*, 52: 50–61.

Mao, D., T.-C. J. Yeh, L. Wan, C.-H. Lee, K.-C. Hsu, J.-C. Wen, and W. Lu. 2013b. Cross-correlation analysis and information content of observed heads during pumping in unconfined aquifers, *Water Resources Research*, 49, no. 2: 713–731.

Mao, D., T.-C. J. Yeh, L. Wan, J.-C. Wen, W. Lu, C.-H. Lee, and K.-C. Hsu. 2013c. Joint interpretation of sequential pumping tests in unconfined aquifers. *Water Resources Research*, 49, no. 4: 1782–1796.

Mehuys, G. R., L. H. Stolzy, J. Letey, and L. V. Weeks. 1975. Effect of stones on the hydraulic conductivity of relatively dry desert soils. *Soil Science Society of America Proceedings*, 39: 37–41.

Mishra, P. K., and S. P. Neuman. 2010. Improved forward and inverse analyses of saturated-unsaturated flow toward a well in a compressible unconfined aquifer. *Water Resources Research*, 46, W07508, doi:10.1029/2009WR008899.

Moench, A. F. 2008. Analytical model and numerical analyses of unconfined aquifer test considering unsaturated zone characteristics. *Water Resources Research*, 44, W06409, doi:10.1029/2006WR005736.

Montagne, C., J. Ruddell, and H. Ferguson. 1992. Water retention of soft siltstone fragments in a Ustic Torriorthent. Central Montana. *Soil Science Society of America Journal*, 56: 555–557.

Mualem, Y. 1976. A new model for predicting the hydraulic conductivity of unsaturated porous media. *Water Resources Research*, 12: 513–522.

Narasimhan, T. N., and M. Zhu. 1993. Transient flow of water to a well in an unconfined aquifer: Applicability of some conceptual models. *Water Resources Research*, 29: 179–191.

National Research Council (NRC). 1996. *Rock Fractures and Fluid Flow: Contemporary Understanding and Applications*. Natl. Acad. Press, Washington, DC.

Neuman, S. P. 1972. Theory of flow in unconfined aquifers considering delayed gravity response of the water table. *Water Resources Research*, 8, no. 4: 1031–1045.

Neuman, S. P. 1973. Supplementary comments on "Theory of flow in unconfined aquifers considering delayed gravity response of the water table." *Water Resources Research*, 9, no. 4: 1102–1103.

Neuman, S. P. 1975. Analysis of pumping test data from anisotropic unconfined aquifers considering delayed gravity response. *Water Resources Research*, 11, no. 2: 329–342.

Neuman, S. P. 1979. Perspective on "delayed yield." *Water Resources Research*, 15, no. 4: 899–908.

Neuman, S. P. 1987. Stochastic continuum representation of fractured rock permeability as an alternative to the REV andfracture network concepts, in I. W. Farmer, J. J. K. Daemen, C. S. Desai, C.E. Glass, and S. P. Neuman (eds.), *Rock Mechanics*. Proceedings of the 28th US Symposium, Tucson, Arizona. AA Balkema, Rotterdam, 1240 pp.

Neuman, S. P., and P. A. Witherspoon. 1969. Theory of flow in a confined two-aquifer system. *Water Resources Research*, 5, no. 4: 803–816.

Ni, C.-F., and T.-C. J. Yeh. 2008. Stochastic inversion of pneumatic cross-hole tests and barometric pressure fluctuations in heterogeneous unsaturated formations. *Advances in Water Resources*, 31, no. 12: 1708–1718, doi:10.1016/j.advwatres.2008.08.007.

Ni, C.-F., T.-C. J. Yeh, and J.-S. Chen. 2009. Cost-effective hydraulic tomography surveys for predicting flow and transport in heterogeneous aquifers. *Environ. Sci. Technol.*, 43 no. 10: 3720–3727, doi:10.1021/es8024098

Nielsen, D. R., J. W. Biggar, and K. T. Erh. 1973. Spatial variability of field-measured soil-water properties. *Hilgardia*, 42: 214–259.

Nwankwor, G. I., J. A. Cherry, and R. W. Gillham. 1984. A comparative study of specific yield determinations for a shallow sand aquifer. *Ground Water*, 22: 764–772.

Nwankwor, G. I., R. W. Gillham, G. van der Kamp, and F. F. Akindunni. 1992. Effect of the capillary fringe during pumping of unconfined aquifers: Field evidence of delayed drainage. *Ground Water*, 30, no. 5: 690–700.

Parker, J. C., and R. J. Lenhard. 1987. Model for hysteretic constitutive relations governing multiphase flow. 1. Saturation-pressure relations. *Water Resources Research*, 23: 2187–2196.

Polubarinova-Kochina, P. Ya., 1952. *Theory of Motion of Ground Water*. Moscow: Gosudarstv. Izdat. Tehn.-Teor. Lit., translated as Polubarinova-Kochina, P. Ya., 1962. *Theory of Ground Water Movement*. Princeton, NJ: Princeton University Press.

Press, W. H., B. P. Flannery, S. A. Teukolsky, and W. T. Vetterling. 1992. *Numerical Recipes in Fortran 77: The Art of Scientific Computing*. Edition 2. Cambridge University Press, 933 p.

Priestley, M. B. 1981. *Spectral analysis and time series: Probability and mathematical statistics, a series of monographs and textbooks*. Academic Press, London/San.

Pruess, K., and T. N. Narasimhan. 1985. A practical method for modeling fluid and heat flow in fractured porous media. *Society of Petroleum Engineering Journal*, 25, no. 1: 14 – 26.

Pyrcz, M. J., and C. V. Deutsch. 2003. The whole story on the hole effect. *Geostatistical Association of Australasia*, Newsletter.

Rawls, W. J., D. L. Brakensiek, and K. E. Saxton. 1982. Estimating soil water properties. *Transactions American Society of Agricultural Engineers*, 25, no. 5: 1316–1320, 1328.

Rehfeld, K. R., P. Hufschmeid., L. W. Gelhar, and M. E. Schaefer. 1989. Measuring Hydraulic Conductivity with the Borehole Flowmeter, Top. Rep. EN-6511, Elec. Power Res. Inst., Palo Alto, Calif.

Robin, M. J. L., E. A. Sudicky, R. W. Gillham, and R. G. Kachanoski. 1991. Spatial variability of strontium distribution coefficients and their correlation with hydraulic conductivity in the Canadian Forces Base Borden Aquifer. *Water Resources Research*, 27, no. 10: 2619–2632.

Rose, C. W., W. R. Stern, and J. E. Drummond. 1965. Determination of hydraulic conductivity as a function of depth and water content for soil in situ. *Australian Journal of Soil Reseach*, 3: 1–9.

Ross, B. 1990. The diversion capacity of capillary barriers. *Water Resources Research*, 26: 2625–2629.

Russo, D. 1997. On the estimation of parameters of log-unsaturated conductivity covariance from solute transport data. *Advances in Water Resources*, 20, no. 4: 191.

Russo, D., and M. Bouton. 1992. Statistical analysis of spatial variability in unsaturated flow parameters. *Water Resources Research*, 28, no. 7: 1911–1925.

Russo, D, I. Russo, and A. Laufer. 1997. On the spatial variability of parameters of the unsaturated hydraulic conductivity. *Water Resources Research*, 33, no. 5: 947.

Scheibe, T. D., E. M. Murphy, C. Xingyuan, A. Rice, K. C. Carroll, B. J. Palmer, A. M. Tartakovsky, I. Battiato, and B. D. Wood. 2015. An Analysis Platform for Multiscale Hydrogeologic Modeling with Emphasis on Hybrid Multiscale Methods. *Ground Water*, 53(1): 38–56. doi: 10.1111/gwat.12179.

Selker, J. 1997. Design of interface shape for protective capillary barriers. *Water Resources Research*, 33: 259–260.

Sharmeen, R., W. A. Illman, S. J. Berg, T.-C. J. Yeh, Y.-J. Park, E. A. Sudicky. 2012. Transient hydraulic tomography in a fractured dolostone: Laboratory rock block experiments. *Water Resources Research* 48. W10532, doi:10.1029/2012wr012216.

Sisson, J. B., and A. H. Lu. 1984. Field Calibration of Computer Models for Application to Buried Liquid Discharges: A Status Report, RHO-ST-46P. Rockwell Hanford Operations, Richland, WA.

Smith, L., and R. A. Freeze. 1979a. Stochastic analysis of steady state groundwater flow in a bounded domain. 1. One-dimensional simulations, *Water Resources Research*, 15, no. 3: 521–528.

Smith, L., and R. A. Freeze. 1979b. Stochastic analysis of steady state groundwater flow in a bounded domain. 2. Two-dimensional simulations. *Water Resources Research*, 15, no. 6: 1543–1559.

Soil Conservation Service. 1975. *Soil Taxonomy*. United States Department of Agriculture Agriculture and Natural Resources Conservation Service, Handbook Number 436.

Stephens, D. B., K. Lambert, and D. Watson. 1987. Regression models for hydraulic conductivity and field test of the borehole permeameter. *Water Resources Research*, 23, no. 12: 2207–2214.

Straface, S., T.-C. J. Yeh, J. Zhu, S. Troisi, and C. H. Lee. 2007. Sequential aquifer tests at a well field, Montalto Uffugo Scalo, Italy. *Water Resources Research*, 43, W07432, doi:10.1029/2006WR005287.

Streltsova, T. D. 1972a. Unconfined aquifer and slow drainage. *Journal of Hydrology*, 16: 117–123.

Streltsova, T. D. 1972b. Unsteady radial flow in an unconfined aquifer. *Water Resources Research*, 8, no. 4: 1059.

Sudicky, E. A. 1986. A natural gradient experiment on solute transport in a sand aquifer: Spatial variability of hydraulic conductivity and its role in the dispersion process. *Water Resources Research*, 22, no. 13: 2069–2082.

Sun, R., T.-C. J. Yeh, D. Mao, M. Jin, W. Lu, and Y. Hao. 2013. A temporal sampling strategy for hydraulic tomography analysis. *Water Resources Research*, 49, doi:10.1002/wrcr.20337.

Tartakovsky, G. D., and S. P. Neuman. 2007. Three-dimensional saturated-unsaturated flow with axial symmetry to a partially penetrating well in a compressible unconfined aquifer. *Water Resources Research*, 43, W01410, doi:10.1029/2006WR005153.

Theis, C. V. 1935. The relation between lowering the piezometric surface and the rate and duration of discharge of a well using ground water storage. *EOS Transactions of the American Geophysical Union*, 16: 519–524.

Tsang, Y. W., C. F. Tsang, F. V. Hale, and B. Dverstorp. 1996. Tracer transport in a stochastic continuum model of fractured media. *Water Resources Research*, 32, no. 10: 3077–3092.

Tokunaga, T. K., J. Wan, and K. R. Olson. 2002. Saturation-matric potential relations in gravel. *Water Resources Research*, 10: 1029–2001.

Tuller, M., and D. Or. 2005. Retention of water in soil and the soil water characteristic curve, in D. Hillel (ed.), *Encyclopedia of Soils in the Environment*. Elsevier, 2005.

Ünlü, K., D. R. Nielson, and J. W. Biggar. 1990. Statistical parameters characterizing the spatial variability of selected soil hydraulic properties. *Soil Science Society of America Journal* 54: 1537–1547.

Van Genuchten, M. Th. 1980. A closed-form solution for predicting the conductivity of unsaturated soils. *Soil Science Society of America Journal*, 44: 892–898.

Van Genuchten, M. Th., and D. R. Nielsen. 1985. On describing and predicting the hydraulic properties of unsaturated soils. *Annales Geophysicae*, 3: 615–628.

Vasco, D. W., and K. Karasaki. 2006. Interpretation and inversion of low frequency head observations, *Water Resources Research*, 42, W05408, doi:10.1029/2005WR004445.

Vasco, D. W., H. Keers, and K. Karasaki. 2000. Estimation of reservoir properties using transient pressure data: An asymptotic approach, *Water Resources Research*, 36, 3447–3465, doi:10.1029/2000WR900179.

Ward, A. L, T. G. Caldwell, and G. W. Gee. 2000. *Vadose Zone Transport Field Study: Soil Water Content Distributions by Neutron Moderation*. PNNL-13795, Pacific Northwest National Laboratory, Richland, WA.

Warrick, A. W. 2003. *Soil Water Dynamics*. Oxford: Oxford University Press.

Watson, K. K. 1966. An instantaneous profile method for determining the hydraulic conductivity of unsaturated porous materials. *Water Resources Research*, 2, no. 4: 709–715.

Wen, J. C., C. M. Wu, T.-C. J. Yeh, and C. M. Tseng. 2010. Estimation of effective aquifer hydraulic properties from an aquifer test with multi-well observations (Taiwan). *Hydrogeology Journal*, 18: 1143–1155, doi:10.1007/s10040-010-0577–1.

White, I., and M. J. Sully. 1992. On the variability and use of the hydraulic conductivity alpha parameter in stochastic treatments of unsaturated flow. *Water Resources Research*, 28, no. 1: 209–213.

Woodbury, A. D., and E. A. Sudicky. 1991. The geostatistical characteristics of the borden aquifer. *Water Resources Research*, 27, no. 4: 533–546.

Wu, C. M., T.-C. J. Yeh, J. Zhu, T. H. Lee, N. S. Hsu, C. H. Chen, and A. F. Sancho. 2005. Traditional analysis of aquifer tests: Comparing apples to oranges. *Water Resources Research*, 41, W09402, doi:10.1029/2004WR003717.

Xiang, J., T.-C. J. Yeh, C.-H. Lee, K.-C. Hsu, and J.-C. Wen. 2009. A simultaneous successive linear estimator and a guide for hydraulic tomography analysis, *Water Resources Research*, 45, W02432, doi:10.1029/2008WR007180.

Yates, S. R., M. Th. van Genuchten, A. R. Warrick, and F. J. Lei. 1992. Analysis of measured, predicted, and estimated hydraulic conductivities using the RETC computer program. *Soil Science Society of America Journal*, 56: 347–354.

Ye, M., R. Khaleel, and T.-C. J. Yeh. 2005. Stochastic analysis of moisture plume dynamics of a field injection experiment, *Water Resources Research*, 41: 3, W03013, doi:10.1029/2004WR003735

Yeh, T.-C. J. 1989. One-dimensional steady-state infiltration in heterogeneous soils. *Water Resources Research*, 25, no. 10: 2149–2158.

Yeh, T.-C. J. 1992. Stochastic modeling of groundwater flow and solute transport in aquifers. *Hydrological Processes*, 6: 369–395.

Yeh, T.-C. J. 1998. Scale issues of heterogeneity in vadose-zone hydrology, in G. Sposito (ed.), *Scale Dependence and Scale Invariance in Hydrology*. Cambridge University Press.

Yeh, T.-C. J., and L. W. Gelhar. 1983. Unsaturated flow in heterogeneous soils, in J. W. Mercer et al. (eds.) Role of the Unsaturated Zones in Radioactive and Hazardous Waste Disposal. *Ann Arbor Science*: 71–79.
Yeh, T.-C. J., and D. J. Harvey.1990. Effective unsaturated hydraulic conductivity of layered sands. *Water Resources Research*, 26, no. 6: 1271–1279.
Yeh, T.-C. J., and C. H. Lee. 2007. Time to change the way we collect and analyze data for aquifer characterization. *Ground Water* 45, no. 2: 116–118.
Yeh, T.-C. J., and S. Liu. 2000. Hydraulic tomography: Development of a new aquifer test method. *Water Resources Research*, 36, no. 8: 2095–2105.
Yeh, T.-C. J., and J. Zhang. 1996. A geostatistical inverse method for variably saturated flow in the vadose zone, *Water Resources Research*, 32, no. 9: 2757–2766.
Yeh, T.-C. J., P. E. Gagnard, A. Guzman, and R. Srivastava. 1993. Numerical simulation of the wicking effect in linear systems. *Ground Water*, 32, no. 1: 2–11.
Yeh, T.-C. J., L. W. Gelhar, and A. L. Gutjahr. 1985a. Stochastic analysis of unsaturated flow in heterogeneous soils. 1. Statistically isotropic media. *Water Resources Research*, 21, no. 4: 447–456.
Yeh, T.-C. J., L. W. Gelhar, and A. L. Gutjahr. 1985b. Stochastic analysis of unsaturated flow in heterogeneous soils. 2. Statistically anisotropic media, *Water Resources Research*, 21, no. 4: 457–464.
Yeh, T.-C. J., L. W. Gelhar, and A. L. Gutjahr. 1985c. Stochastic analysis of unsaturated flow in heterogeneous soils. 3. Observations and applications. *Water Resources Research*, 21, no. 4: 465–471.
Yeh, T.-C. J., A. L. Gutjahr, and M. Jin. 1995. An iterative cokriging-like technique for groundwater flow modeling, *Ground Water*, 33: 33–41.
Yeh, T.-C. J., A. Guzman, R. Srivastava, P. E. Gagnard, and J. Kramer. 1995. Simulation of the wicking effect in a two-layer soil liner system. *Waste Management & Research* 13: 363–378.
Yeh, T.-C. J., S. Liu, R. Glass, K. Baker, J. Brainard, D. Alumbaugh, and D. LaBrecque. 2002. A geostatistically based inverse model for electrical resistivity surveys and its applications to vadose zone hydrology, *Water Resources Research*, 3: 12, Article Number: 1278, doi:10.1029/2001WR001204
Yeh, T.-C. J., C.-K., Lee, K.-C. Hsu, Y.-C. Tan. 2007. Fusion Of Active and Passive Hydrological and Geophysical Surveys: The Future of Subsurface Characterization, "Data Integration in Subsurface Hydrology," D. W. Hyndman, F. D. Day-Lewis, and K. Singha (eds.), AGU Monograph.
Yeh, T.-C. J. et al. 2008. A view toward the future of subsurface characterization: CAT scanning groundwater basins, *Water Resources Research*, 44, W03301, doi:10.1029/2007WR006375.
Yeh T.-C. J., D. Mao, L. Wan, C.-H. Lee, J.-C. Wen, K.-C. Hsu. 2011. *Well Definedness, Scale Consistency, and Resolution Issues in Groundwater Model Parameter Identification*. University of Arizona, Tucson.
Yeh, T.-C. J., Mao, D., Zha, Y., Hsu, K.-C., Lee, C.-H., Wen, J.-C., Lu, W., and Yang, J. 2014. Why hydraulic tomography works? *Ground Water* 52: 168–172, doi:10.1111/gwat.12129.
Yeh, T.-C. J., M. Ye, and R. Khaleel. 2005. Estimation of effective unsaturated hydraulic conductivity tensor using spatial moment of observed moisture plumes. *Water Resources Research*, 41, W03014, doi:10, 1029/2004WR003736.
Yeh, T.-C. J., J. Zhu, A. Englert, A. Guzman, and S. Flaherty. 2006. A Successive Linear Estimator for Electrical Resistivity Tomography, Applied Hydrogeophysics, H. Vereecken, A. Binely, G. Cassiani, A. Revil, and K. Titov (eds.). NATO Science Series, Springer, 383 p.

Zhang, J., and T.-C. J. Yeh. 1997. An iterative geostatistical inverse method for steady flow in the vadose zone, *Water Resources Research*, 33, no. 1: 63–71.

Zhu, J., and T.-C. J. Yeh. 2005. Characterization of aquifer heterogeneity using transient hydraulic tomography. *Water Resources Research*, 41, W07028, doi:10.1029/2004WR003790.

Zhu, J., and T.-C. J. Yeh. 2006. Analysis of hydraulic tomography using temporal moments of drawdown recovery data, *Water Resources Research*, 42, W02403, doi:10.1029/2005WR004309.

Zhu, J., and T.-C. J. Yeh. 2008. Hydraulic tomography to characterization of heterogeneity of unconfined aquifers, paper presented at AGU Fall Meeting, San Francisco, Calif.

Zhu, J., T.-C. J. Yeh, and D. Mao. 2011. Hydraulic tomography to characterize heterogeneity of unconfined aquifers, *J. Nanjing Univ.*, 47, no. 3: 252–264.

Index